Secrets
Of Gem Trade
Second Edition

Secrets
Of Gem Trade
Second Edition

The Connoisseur's
Guide To
Precious Gemstones

With an introduction by Vincent Pardieu
Foreword by Benjamin Zucker

Richard W. Wise

BRUNSWICK
HOUSE PRESS

Lenox, Massachusetts

Library of Congress Cataloging-in-publication Data

Names: Wise, Richard W.

Title: Secrets Of The Gem Trade: Second Edition
The connoisseur's guide to precious gemstones/Richard W. Wise

Description: Second edition Brunswick House Press, Lenox, Ma. Includes bibliography, glossary and index

Identifiers: LCCN 2003103774, ISBN: 9780972822329

Subjects: 1. Gemstones; value & connoisseurship---Handbooks, manuals, etc. 2. Precious stones---Collection and quality---Handbooks, manuals, etc. I. Title

NK5530 W57 2016   736'.2'075

QB103-200285

Printed in South Korea

Cover and book design by Harry Bernard
Copy editing by Rebekah Tressel Wise

Published by Brunswick House Press
P. O. Box 2048
Lenox, Massachusetts 01240
Phone: (413) 443.3280
Email: d.sage@secretsofthegemtrade.com
Website: www.secretsofthegemtrade.com

SAN 255-1365

To my darling wife Rebekah,
partner in my travels
with love and admiration
and to my grandmother Kate.

The author in the shaft of the Consorsio Emerald Mine, La Pita Complex, Boyacá, Colombia. Photo: Ron Ringsrud

CONTENTS

LIFTING THE VEIL

The gemstone trade is an old and venerable one. Evidence of its existence dates back at least seven thousand years. Since medieval times, a few ethnic minorities have dominated the gem business — the Indians on the Asian subcontinent, the ethnic Chinese in Thailand, and the Jews in Europe. The Jewish people, for example, endured discrimination all across Europe. In some countries they were denied membership in the craft guilds which regulated and determined who could work in a specific trade. People denied a livelihood found other means to earn a living. The gem trade was in many respects a natural one for these restless peoples. Secrecy was essential: gemstones are rare and so are good sources. Given the political conditions in Europe at that time, to reveal the inner workings of the business would have been both a tactical and strategic error. Secrecy was not simply a necessity of business; it was a matter of self-preservation. Thus the gem trade became a closed fraternity, its wisdom passed orally from father to son. Secrecy became a habit, one that has persisted until the present day.

I was not born into the trade but came to it in my early thirties, after pursuing other interests for a number of years. Gems, however, have always fascinated me. When I was a child and my friends collected baseball cards, I collected gems and shells. Oh, not real gems, of course, just rhinestones and paste. Some were given to me; others were pried out of discarded pieces of costume jewelry. In those days, Providence, Rhode Island, the city where I grew up, was the center of the costume jewelry industry. A number of adults I knew made a living working at Coro and other factories making costume pieces. They were the source of a number of my treasures.

I began my career in the jewelry business as a craftsman, an apprentice goldsmith. Like many people who matured politically in the 1960s, I graduated from college with a burning desire to right the world's wrongs. I spent several years as a political activist. Also, like many others, I eventually grew disillusioned with social activism. Learning to make jewelry was a sort of therapy, something interesting to do while I sorted out my life and thought about my next step.

In the tiny shop where I worked we used a lot of stones, cabochons mostly: jasper, agate, and turquoise. This was the late seventies, the tail end of the second Arts and Crafts Movement.[1] Precious materials, including platinum and ruby, were considered politically incorrect. "Honest materials" such as sterling silver and turquoise were the materials of choice. What else would you wear with a work shirt and jeans? I enjoyed working with these gemstones, but I can't say that they really moved me. During this time I found myself a partner in a retail craft-jewelry shop.

As the Eighties dawned the business began to receive more and more requests for pieces made of gold and, along with that, a renewed interest in diamonds, sapphires, and other faceted gems. My own appetite stimulated, and, seeing an opportunity for profit, I decided to learn all I could about gemstones.

My years in college and graduate school and the time spent in community organizing had prepared me, I thought, with the necessary research tools to begin my studies. I began searching for books and asking questions of the gem dealers who, with increasing frequency, called at our shop. "Why," I asked, "was one green tourmaline more expensive than another? What are the parameters used to define quality?" I visited museums and attended previews at some of the major New York auction houses, where I saw magnificent jewels. I could not understand why the much smaller gemstones offered me by dealers were not as beautiful. Why couldn't I purchase smaller gems comparable to the large ones I had seen at the major auction houses? The larger stones were, after all, much rarer. The dealers I asked fended off my queries. Their answers were either illogical or just plain evasive. Either they didn't have the answers or they simply were not going to tell me. Later, I found out, it was a little bit of both.

The books I read did little to clarify the issue. The authors had a great deal to say about history, lore, sources, and the physical characteristics of gemstones, but almost nothing to say about quality, connoisseurship, or beauty. I was frustrated, but determined that I would force this trade to yield up its secrets.

I then made three decisions that put me on the right track. First, I enrolled, by correspondence, in the Graduate Gemology program offered by

1 It is interesting to note that the rediscovery of handcrafts reached its peak in the 1 970s, almost exactly one hundred years after the heyday of the Arts and Crafts Movement of the late nineteenth century. The two movements shared much of the same folk ideology -- the celebration of handwork and the use of nontraditional materials.

the Gemological Institute of America. Second, I took a subscription to Gems & Gemology, the journal of the institute. Third, and most importantly, I decided to take my questions to the source. I had always had a yen to travel, and the romance of the exotic entrepôts of Asia had taken hold of my imagination.

My first trip took me to Thailand, the ancient center of the ruby and sapphire trade. A friend and colleague with a number of years in the gem business provided me with introductions. I visited the mining centers of Chanthaburi and Trat, and the great trading city of Bangkok. That first trip marked the beginning of my real education. Other trips followed: I visited the pearl farms of the Tuamotu Islands north of Tahiti, the opal mines of the Queensland outback, the mining districts of Brazil, Tanzania, and Burma and Colombia — observing, always observing, and asking questions.

Still, the quest was not an easy one, and it is far from over. It may surprise the reader to know that much of the information contained in this book has never been available before. A trade that has kept its secrets for thousands of years does not yield up its wisdom simply for the asking. Sources must be protected and contacts safeguarded. However, this is the age of information, and excessive secrecy does the trade a disservice. The public distrusts what it does not understand. It is no coincidence that the expanded interest in gemstones, specifically colored gemstones, coincides with the opening up of the trade by an intrepid band of young gem cutters and entrepreneurs who were part of the crafts movement of the 1970s. Also, many of the stones included here are fairly new discoveries, and I am among the first to consider them seriously from a purely aesthetic point of view.

My gemology studies provided some of the connoisseur's tools, but by no means all. Much of what I have learned has been gleaned by asking questions and by comparing one gem to another and interpreting what I saw.

I began writing early in my travels. It provided a legitimate reason for asking questions, and helped me to organize and reinforce the information I was able to glean from my sources. During a visit to the black pearl farms of the Tuamotu Archipelago, on a pier overlooking the lagoon on Manihi Atoll, I wrote the first draft of my very first article. This effort was followed by many more.

In the course of my travels I have been tutored by some of the world's great experts. To those who have patiently answered my questions, I am profoundly grateful.

One of the most important things I have learned is to trust my own eye. Some of my best acquisitions have been newly discovered stones or unusual examples of well-known gems that have been disparaged by veteran dealers. The gem trade is deeply conservative and will often dismiss newly discovered stones (or new sources of traditional gems) whose characteristics differ from those of traditional sources. The passage of time between a gem's discovery and its grudging acceptance by the trade is ripe with opportunity for one who views the gem with an educated but unprejudiced eye.

Another great lesson learned about acquiring gems — if you see it and can afford it, grab it! Gems are often found in large concentrations or pockets, so a large number of stones will appear on the market at a given time. This gives the false impression that the particular gemstone is in plentiful supply. It is, but the supply is fleeting. Once the initial find has stimulated the market's appetite and supply slows, prices will rise. In many cases the source pipeline quickly dries up; the best stones are snapped up, the price rises, and an opportunity to acquire a beautiful stone at a reasonable price has passed, perhaps forever.

PREFACE TO THE REVISED, ENLARGED 2ND EDITION

This second edition has been enlarged and largely re-written. Five new introductory essays and ten new chapters have been added together with numerous photographs. Though the principles of connoisseurship are in some sense immutable, much has changed and a great deal has been learned since I took a deep breath and put the 1st edition to bed in 2002. First the changes:

The ruby mines at Mong Hsu have played out and new deposits with unique characteristics have been discovered in Tanzania, Madagascar and Mozambique. The discovery of hot pink spinel near Mahenge, Tanzania, together with the discovery in North Vietnam of quantities of a little known variety of spinel colored blue by Cobalt have profoundly altered both the market and the connoisseurship equation in spinel. The venerable natural pearl, more or less disregarded since the rise of the pearl culturing, has returned to prominence. Cuprian tourmaline, with characteristics that differ markedly from Brazilian Paraíba, have been discovered in Mozambique.

The ruby, sapphire and spinel chapters have been completely rewritten and after much study, additional chapters on jade, natural nacreous pearls, conch pearl, demantoid garnet, peridot, moonstone and sunstone have been added, together with introductory chapters on all except demantoid garnet. A new introduction to blue white diamonds explaining the historical development and true meaning of the term along with a chapter on Golconda (type IIa) diamonds has also been added.

My travels have continued. I have spent time buying emerald in Bogotá and visited the mines of Boyacá, broadening and deepening my understanding of emerald. Several excursions along Jade Row at the Hong Kong Gem Show, together with a visit to the jade market in Guangzhou, China, has sharpened my understanding, I hope, of that inscrutable gem. I have returned several times to Bangkok and been privileged to speak with many of the great experts that reside in that city and have visited the spinel mines and markets of Luc Yen, North Vietnam.

As always the reader will judge the quality of my efforts.

Happy reading,

Richard W. Wise

INTRODUCTION TO THE SECOND EDITION

BY VINCENT PARDIEU, G.G.

I read the first edition of *Secrets of the Gem Trade* in 2003 and fell in love with it. It was the perfect book. It provided just the information I needed to round out my gemological studies:

Secrets of the Gem Trade approaches gemology from an angle very different from most classic gemological texts. Secrets of the gem Trade is unique. It, will not teach you how to separate a ruby from a garnet using a refractometer. It will teach you the criteria an expert jeweler/gemologist employs to judge quality. The second edition also adds chapters on eleven gems not covered in the first edition and updates the reader on the latest finds along with adding lots of additional images of exceptional gemstones.

As a young gemologist with a degree in science as well as two diplomas in gemology, I was comfortable identifying gems, but had seen a very limited number of fine stones. As most of the natural stones a gemology student will observe during his studies are small, low quality stones, I, along with other students wondered if we were studying gemology or junk-ology. For me, as a non-native English speaker, it was very difficult to analyze the beauty in gems and find the words to describe what I saw. Reading *Secrets of the Gem Trade* helped solved my problems. Nothing can replace the experience of holding a gem and studying it with your own eyes; however, Secrets of the Gem Trade gives you the tools to both understand and communicate what you are seeing.

Secrets of The Gem Trade was written by Richard W. Wise, an American jeweler and Graduate Gemologist with over thirty years experience traveling the world and sourcing gems to set in his original jewelry creations. As most gemstones will end up set in jewelry, being able to get a successful jeweler to explain why he appreciates this or that gem is invaluable. As one of my teachers used to say: "It is easier to convince people using their arguments rather than yours. So if you want to succeed in any business associated with people, start studying them, then adapt to them and finally success will come".

In an age where people spend more and more time on social media and depend on the Internet for all the answers to their questions about gemstones, one of the most difficult things to find is a mentor, someone who will take the time to share a lifetime of experience with you, someone who will guide you towards an understanding of quality and connoisseurship and help you communicate that understanding to others. In *Secrets of the Gem Trade* Richard Wise becomes that mentor. The book is a gold mine. I highly recommend it as one of the most useful books I have ever read about gemstones. Actually, I would have loved to have written it myself.

Enjoy,

Vincent Pardieu

Mr. Pardieu is the Senior Manager, Field Gemology, at The Gemological Institute of America (GIA) Laboratory in Bangkok.

Secrets of the Gem Trade, the second edition of Richard W. Wise's *Connoisseur's Guide to Precious Gemstones* builds substantially on the first edition, both are based upon the author's personal observations over thirty-five years of buying gems around the world, together with a meticulous examination of more than three centuries of gemological literature. The second edition breaks new ground in his discussion of blue-white diamonds and includes perhaps the most comprehensive and sensitive overview of the aesthetics of jade to be found in the English language. His chapters on spinel explain in detail how and why new sources have reinvigorated the market and stimulated demand for this venerable gem.

Very much the historian, Wise quotes with ease Egyptian and Medieval European as well as ancient Indian texts demonstrating how consistent the concepts of connoisseurship have remained over the millennia. He reveals many secrets of the trade, carefully putting the reader in the mind of the cutter and gem dealer, explains selling strategies rarely discussed outside the fraternity. Wise's Four Cs of Connoisseurship reintroduces *crystal*, that important and evocative combination of color and transparency that pre-modern writers referred to as *water*, a quality unfortunately neglected in modern grading systems.

One can tell that Wise is a good listener. He has gleaned much wisdom from his conversations with dealers, gem cutters and connoisseurs in Rangoon, Bogota, Nairobi, Bangkok and other exotic locales. Wise was lucky to get his start in this industry as a goldsmith, which gives him a special sensitivity to how gemstones are fashioned into jewelry creations.

Secrets of the Gem Trade is an important book which will remain indispensable for many years to come. It is a marvelous guide to understanding the exhilarating, changing pace in the world of gemstones.

Benjamin Zucker

Author, Gems & Jewels, A Connoisseur's Guide

Acknowledgments

Many people have aided me in my quest for knowledge; some willingly, some not so. I would like to thank the following: C.R. "Cap" Beasley, whose "short" courses in gemstones, together with numerous conversations, provided many insights. Barry Hodgin, my first dealer in Thailand, showed much patience early on. Ronald Sage of Papeete introduced me to the world of the black pearl. Sidney Soriano provided me with introductions on my first trip to Asia. The late Don Thompson was a true adventurer and good friend. Nittin and Jaswin Pattni took me under their wing on my first trip to East Africa. The late great Campbell Bridges, who took us to the mines, explained the geology of tsavorite and hosted my wife and me on our second trip to East Africa. Thanks to David Stanley Epstein author of *The Gem Merchant* for help in translation. My good friend Joseph Belmont, who showed great patience, introduced me the best in ruby, sapphire and spinel and was my companion and mentor on journeys to Burma and Vietnam. The late Vince Evert was my guide to the Queensland outback. Thanks to Ron Ringsrud who guided me to the emerald mines of Colombia. Thanks to Cát Viêt Thái and Vuong Kho Pha, our gracious hosts in Vietnam. I enjoyed numerous conversations with Vincent Pardieu on spinel, ruby and sapphire. and with Michael Cowing on diamonds, and thanks to Richard Hughes for some frank discussions and good advice on publishing. Thanks to photographer Robert Verspui and to Mikola Kukharuk of Nomads for some excellent photographs. I would also like to thank Charlie MacGovern, my first dealer, and Dr. N.R. Barot, Joe Crescenji, Stephen Hofer and Nick Hale, who taught me much about color. Thanks also to Mongol Perdicci, Si and Ann Fraser, Lou Wackler, Bernd Munsteiner, Michael Dyber, Wimon Manorotkul, the late Dana Schorr, Fuji Voll, the Elawar family, and Paulo Zonari.

I would like to thank the following for their help in preparing the manuscript: Jeff Scovil who supplied a majority of the photographs and Rebekah, my wife and companion in many travels, who read, reread, then read it over again Also, the late Henry Maseil, Allen Kleiman and Damien Cody and Benjamin Zucker for their suggestions and editorial comments. Special thanks to Catherine Allen and Eunice Kwok of Sotheby's. Thanks to J. J. Rousseau, Berkshire House Press, for some sage advice.

PART I

BECOMING A CONNOISSEUR:
ESSENTIALS

Natural alluvial ruby rough from a Tanzanian streambed.
Photo: Richard W. Hughes, courtesy Lotus Gemology.

Imagine a time before the existence of the world we know, long before the invention of artificial colors, aniline dyes or LED lighting. Picture a group of our remote ancestors, a party of Neolithic hunters, clothed in animal hides and armed with spears, picking their way single file across a shallow mountain stream. One man stoops to drink. Out of the corner of his eye, he catches a tiny flash of color against the stream's sandy bottom. He pauses, looks around warily, then scoops up the object and holds it up to the sun. The oddly shaped pebble glows with the rich hues of blood and fire. The hunter's curiosity is sparked, but he looks up, sees his comrades disappearing over the stream's high bank. He stuffs the stone into a leather pouch tied to his waist, hefts his spear and scrambles to catch up.

That night a chill wind blows across the rugged escarpment. The hunters huddle in the shelter of a rocky cave, roasting the day's kill over a warming campfire. The spitting and crackling of the meat and the flow of its rich red juices jogs the hunter's memory. He rummages in his pouch, extracts the curious pebble, squints and examines it in the flickering firelight. The man is amazed. The pebble glows like a hot coal yet it remains cool to the touch. What is this magic? The hunter's heart throbs with excitement. The man's curiosity turns to awe and perhaps fear.

Other men crowd around, comment and

reach for this new wonder, but seeing the greed reflected in their eyes, he snatches it back and stuffs it into his pouch. Later, alone, he examines the thing noting its straight edges

Sunny yellow primordial eye candy! A close up of natural topaz rough plucked from a stream just outside Ouro Preto, Minas Gerais, Brazil. Photo: R. W. Wise

and curious octagonal shape. Then, seeking to reach the gem's internal fire, he places it on a large flat rock and strikes it with his hammer stone. No matter how many times he strikes it, the pebble is unaffected. Next, he saws at it with his knapped flint skinning tool, the hardest object he knows, but the flint makes no mark on the pebble's surface. All he accomplishes is to chip and dull his blade.

What happened next is pure speculation. Perhaps, fearful of the tiny object's magical properties, he brings it to the shaman, the wisest man in his village, who tells him that the stone is a manifestation of the sprits and appropriates it for himself. Perhaps he trades it to another hunter or, his woman sees and covets it. He wraps it with a thin strip of wet leather, dries the leather in the sun and ties it, the first pendant, triumphantly around her neck.

Curiosity..., Wonder..., Desire...! How many similar scenes play out daily in jewelry stores across the world? How often I have observed it working with clients? Like a moth transfixed by the flame, our interest in these curious and beautiful natural creations seems instinctive. The first gem may have been the transparent pebble described above or a sunny yellow crystal caught in the firelight and pried from a cavern wall. Perhaps in the search for flint to knap into tools, a particularly beautiful piece of translucent agate caught a young girl's fancy. Humanity's desire to possess and adorn seems instinctive. The first known jewelry, shells perforated to make a necklace, were discovered in a Moroccan cave dating back one hundred thousand years.

The first gems were curiosities appreciated for their beauty and their unusual form. There were no preconceived notions of preciousness. Perhaps one of a handful of crystals had more pronounced color, was more transparent or possessed greater perfection of form. The standards were instinctive, gut level. Even today, an untrained person shown a box containing several exceptional stones of the same variety will, in the vast majority of cases, instantly select the finest stone. The affinity is immediate. After years of contemplating the beauty of gems, I have determined that the basic principles of connoisseurship can be deduced from a thoughtful contemplation of a single fine gemstone.

For the budding connoisseur, the problem is that in a jewelry store one can see hundreds-- at a gem show it is possible to see thousands – of stones. Under such circumstances, the eye is dazzled and the mind goes numb. Without a thorough understanding of principles, the budding but inexperienced aficionado is like the proverbial fatted lamb, ripe and primed for the slaughter.

Chapter 1

Precious Gems:
The History of a Concept

*"The clumsy modern category of "precious" stones
has little relevance when applied to the ancient world."*

–Jack Ogden, 1982

Preciousness: Ancient Concept or Modern Prejudice?

In the West the distinction between precious and semi-precious appears to be a relatively recent one. The idea that one material was precious while another was merely semi-precious simply did not exist in ancient times.[2] The word *semi-precious* itself entered the English lexicon only in the 19th century.

In ancient Egypt, for example, the color, not the type of material, appears to have been the primary criterion of value. Egyptian taste in jewelry favored solid bars of vivid color, particularly blue and orange. Opaque and semi-translucent gems such as lapis lazuli, coral, turquoise, carnelian, and sard were highly valued. The masterpieces of ancient jewelry, such as those unearthed in the tomb of the boy king Tutankhamen, were beautifully worked in gold by skilled craftsmen. These treasures included gems such as turquoise and

Ancient fake! Egyptian engraved faience (glass) bead from the reign of Amenhotep II (1391-1353 B.C.) dyed to resemble lapis lazuli. Faience beads have been found in many fine pieces of ancient jewelry, often side by side with natural gemstones such as turquoise and coral. Photo: ©1999 Christie's Images

carnelian alternated with stones of faience[3] (a ceramic glass of melted feldspar) dyed to resemble a specific gemstone; in short, a fake! Was this due to a rarity of materials? It was obviously not a question of price. Were the Egyptian craftsmen misled by clever forgeries? Doubtful! The Egyptians simply placed a higher value on visual beauty than on the pedigree of the materials themselves.

2 Jack Ogden, *Jewellery of the Ancient World* (New York: Rizzoli International, 1982), p. 90.

3 Lois S. Dubin, *The History of Beads: From 30,000 BC to the Present* (New York: Harry N. Abrams, 1987), p. 42.

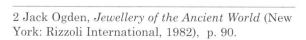 

Today with our rigid notions of what is precious and what is not, this seems odd. Would Cartier or Tiffany consider offering gold jewelry set with glass, plastic, or synthetic gems? Yet the glassmakers of ancient Egypt enjoyed royal patronage.[4] The point is that the idea of preciousness is fluid. The popularity of gem materials has waxed and waned over the millennia. The truth of this becomes clear when we consider that much of the gem wealth found buried with the pharaohs of Egypt at Babylon, and in the royal tombs of ancient Sumer, is what many today would label as semi-precious.[5]

Roman carnelian cameo circa 2nd century A.D. One of the most coveted gems of antiquity, today carnelian is consigned to the semi-precious backwaters. Photo: © Christie's Images.

Descriptions in the Bible also clearly demonstrate that the ideas of the ancients concerning the hierarchy of precious materials differed markedly from our modern view. In Revelations (21: 9-21) an angel describes the heavenly city of Jerusalem as "having the glory of God; and his light was like a stone most precious, even like a jasper stone clear as crystal. . . And the foundations of the wall of the city were garnished with all manner of precious stones. The first foundation was jasper; the second, sapphire; the third, chalcedony; the fourth, emerald; the fifth, sardonyx; the sixth, sardius; the seventh, crysolite (topaz); the eighth, beryl. " Of the twelve gems named, only emerald and sapphire are commonly referred to as precious today, and although emerald was known to the ancient world, we know that sapphire was almost certainly the ancient name for lapis lazuli.[6]

4 Dubin, *History of Beads*, p. 43.

5 Ibid.

6 Although the ancient Egyptians, from whom the Hebrews no doubt derived their notions of gemstones, knew emerald, some distinguished scholars believe that "emerald" (bareketh) was the name given to light green serpentine. George Frederick Kunz, *The Curious Lore of Precious Stones* (1913; reprint ed., New York: Dover Editions, 1971) pp. 292-301.

The idea of a hierarchy of preciousness may have originated in the ancient East and worked its way slowly westward. In the ancient texts of India, the oldest gemstone market, gems were broadly classified as *Maharatna* and *Uparatna*, more and less valuble. Of the nine gems of particular importance in those early times; diamond, pearl, ruby, sapphire and emerald were classified as precious. Topaz, jacinth (red zircon), coral and lapis lazuli were of lesser value.[7]

It is important to keep in mind that beauty was not the sole reason gems were valued in the ancient world. From earliest times gems have been esteemed for their magical qualities, as religious symbols, as talismans, as symbols of rank and status, and for their purported medicinal value.

In the Egypt of the pharaohs, carnelian symbolized blood. In ancient Sumer lapis lazuli represented the heavens. In classical Greece a man supposedly could drink his fill and remain sober if he drank his wine from a cup made of amethyst. To avoid eyestrain, the Roman emperor Nero reputedly viewed gladiatorial contests through an emerald lens.

In ancient China, badges made of gem materials were used to denote rank. Mandarins of the first rank wore red stones such as ruby and red or pink tourmaline; coral and garnet were reserved for bureaucrats of the second rank. Blue stones such as lapis lazuli and aquamarine symbolized the third rank. Mandarins of the fourth rank wore rock crystal. Other white stones indicated the fifth rank. Here again, color, not gemstone type, seems to have been the defining criterion.[8]

Gems were also valued as much for their talismanic or medicinal value as for their beauty. These arcane beliefs and associations persist today, but they no longer have any effect on value or the idea of preciousness, particularly as judged in the marketplace.

Seal Stone Engraving: Value Added

The carving of gems became an important art in ancient times with the introduction of seal stone engraving about 3500 BC by the Babylonians. Gemstones were engraved intaglio with mythical scenes which appeared in relief when the stone was impressed in clay. These engraved gems became the official signatures of kings, nobles, and high-ranking officials of the court. In ancient Mycenae, seal engraving reached a high degree of sophistication by the late Bronze Age. A group of seals recovered from Mycenaean shaft graves at Dendra (on the Greek mainland) shows a mastery of technique as well as a lyric sensibility equaled only by the Greek masters of the classical period and never since.[9]

The earliest seals were engraved in

7 Radha Krishnamurthy, *Gemmology in Ancient India,* Indian Journal of History of Science, (27)3 Should follow this order: Title of Journal Volume, no. Issue Number (Year): Page Number(s) 1992, p. 251. Other texts included as many as thirty-two gemstones.

8 Kunz, Curious Lore, p. 256.

9 K. Demakopoulou, ed., *The Aidonia Treasure* (Athens: National Archaeological Museum, 1996), p. 51.

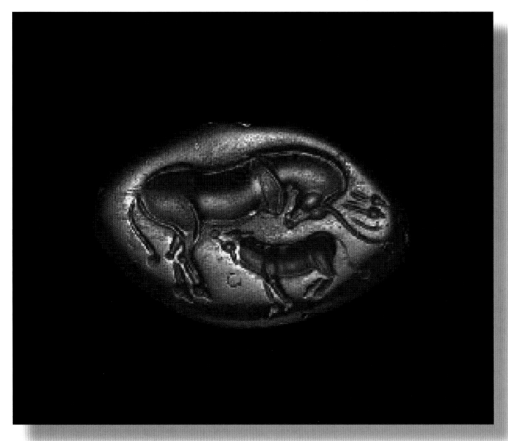

Minoan/Mycenaean carnelian seal stone (1450-1300 B.C.) This masterwork of the engraver's art was often applied to mediocre quality gem material such as this dark opaque carnelian (sardius). Photo: © 2001 Christie's Images

Greek seals and more than ninety percent of Roman intaglios were carved of carnelian and the darker orange agate called sardius. The 13th century Arab gemologist Al Tifaschi, who did admit of a hierarchy of gems, classed carnelian amongst the "royal" or finest gems (*al ahdjar al Mulukiyya*)[10]. Today the stone barely makes the semi-precious list, but carnelian was unquestionably one of the precious stones of antiquity.

relatively soft stones such as serpentine and steatite; these stones could be carved using bronze tools. However, by the 12th century BC, hard stones such as agate, amethyst and garnet became the materials of choice. Engraving these stones (over six on the Moh's scale of hardness) required a more sophisticated technique: even iron, the hardest metal then known, was too soft to engrave carnelian agate.

Carnelian, the eighth stone of the of the high priest's breastplate described in the book of Exodus, was the gem of choice for engravers from the Bronze Age until late Roman times. Fully fifty percent of

By classical times seals were in use throughout the lands bordering the Inland Sea. Experts in this craft enjoyed high status. Some of the best quality gem material is found in Mycenaean gems unearthed at Aidonia. These are carved in the finest translucent layered carnelian. They are the exception. By Roman times,

---

10 Huda, S.M.A., *Arab Roots of Gemology, Ahmad ibn Yusuf Al Tifaschi's Best Thought on the Best of Stones*, (London: Scarecrow Press, 1998), p.24. Al Tifaschi distinguished between *al ahdjar al Karima* gemstones that were rare and precious and those that were "royal" *al Mulukiyya*.

some of the finest masterworks of the engraver's art were executed in opaque, relatively mundane pieces of dark orange carnelian and sard, demonstrating that the beauty of the material itself was of secondary importance. The real preciousness of the gem lay in the artistry and the quality of execution.[11]

The Middle Ages: Shifting Values

In medieval Europe, superstitions centered around the religious, talismanic, and medicinal properties of gemstones were accepted without question. Many of these beliefs had been passed down from ancient times in the writings of the Roman scholar Pliny and repeated in the works of the 7th century bishop Isidore of Seville. The medieval mind, obsessed as it was with questions of sin, death and the torments of Hell, proved fertile ground for the growth, dissemination and acceptance of such beliefs.

In those times, each gem was valued for its ability to protect its wearer from evils both physical and spiritual. "Coral, which for twenty centuries or more was classed among the precious stones," cured madness

and assured wisdom.[12] Emerald was considered to protect the wearer against all manner of enchantments. Carnelian drove out evil and protected the wearer from envy. Lapis lazuli was a sure cure for quartan fever. Sapphire also offered protection from envy and was thought to attract divine favor. Chrysoprase protected the thief from hanging.

So universal was the belief in the magical and medicinal qualities of gem materials in the Middle Ages that it is impossible to discuss the value of gemstones in those times without reference to these arcane beliefs. Was the emerald valued for its beauty, or for its supposed value as a treatment for diseases of the eye?

Diamond: the Invincible

Diamond's fluctuating popularity on the gemstone hit parade further illustrates the point. Diamond was unquestionably the preeminent gemstone in India from as early as the 5th century BC. India in those far-off times was the only source of diamond and had a flourishing gem-trading industry. The Romans, too, placed diamond at the very pinnacle of preciousness. By early medieval times in the West, however, diamond had fallen to number seventeen on the bestseller list. As late as the 16th century, the celebrated Italian goldsmith Benvenuto Cellini placed diamond third after ruby and emerald, with a price of only one eighth of what a ruby would bring. Writing in 1565, Garcia ab Horto, an early European traveler who described his trip to the gem fields of India, placed diamond at number three, but considered emerald, not ruby, to be the

---

11 Greek philosopher Theophrastus, in his treatise on gemstones, written toward the end of the fourth century BC, uses the word *perittotera,* which Calley and Richards translate as "precious." See E.R. Calley and J.C. Richards, *Theophrastus on Stones* (Columbus, Ohio: Ohio State University Press, 1956), p. 45. Other translators, notably Eichholz, translate *perittotera* as meaning "unusual." Professor C.J. Fuqua of Williams College states that Theophrastus uses the term in the sense of *more unusual,* not *more precious.* Theophrastus does not use the superlative degree, "most unusual," in this passage. There is no hierarchy involved with *perittotera.* (C.J. Fuqua, personal communication, 1999.)

12 Kunz, *Curious Lore,* p. 69.

Natural bipyramidal diamond crystals. Prior to the 16th century the technology did not exist to either cut or polish a diamond. These natural six-sided crystals were highly valued for their transparency and perfection of form.

most precious gem of all.

One prominent scholar, Godeherd Lenzen, maintains that diamond's early popularity in the western world was based not on its beauty, but on its durability and hardness. The characteristics that make diamond so desirable today — brilliance, dispersion, and transparency — are qualities that occur naturally only in well-formed, transparent diamond crystals. In Roman times, the technology did not exist to fashion or polish diamonds. Transparent well-formed crystals either were retained and sold in India (where they were highly valued) or bought up along the long overland trade route before they reached Rome. Thus, due to the rarity and desirability of fine crystals and the length of the overland trade route between India and Rome, the uncut rough stones that made their way to the ancient Mediterranean were of inferior quality; the attributes of beauty which make the diamond so avidly sought after today were necessarily

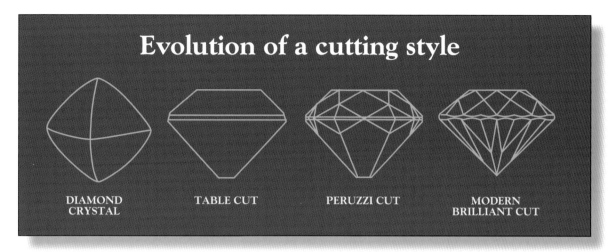

Evolution of a cutting style

| DIAMOND CRYSTAL | TABLE CUT | PERUZZI CUT | MODERN BRILLIANT CUT |

Moving from left to right: The bipyramidal natural diamond crystal suggests the basic outline of the table cut, one of the earliest cutting styles. The Peruzzi cut in the next logical step, facets added in an attempt to manipulate light. The modern brilliant cut shows a very subtle change of proportions and fifty-five precisely placed facets.

unknown to the ancient Romans.[13] Therefore, Lenzen argues, diamonds could not have been valued for beauty at all, but must have had some other attraction. The Greeks named diamond *adamas*, a word

that means invincible. This obviously relates to the gem's legendary hardness, a virtue much admired in imperial times. Was it diamond's "invincibility" that made it so attractive and valuable to the Romans?

To be fair, by the 17th century diamond achieved its current preeminent position in the gem world. The Portuguese subjugation of Goa in west-central India in the 16th century opened up more direct trade routes, increasing the flow of finer diamond rough to the West. The necessary technology for revealing the diamond's unique beauty – polishing, cutting, and cleaving – was in place in Europe by the middle part of the century[14]. Diamond's preeminence is also a direct result of the development of the brilliant cut in the late 17th century.

13 Godeherd Lenzen, *The History of Diamond Production and the Diamond Trade* (London: Barrie & Jenkins, 1970), pp. 18-19. *The Arthashastra of Kautilya,* written sometime between the 5th and 6th centuries BC, characterizes a good diamond as one that is "regular in shape, capable of . . . reflecting light brilliantly in all directions." Some scholars have held that this is a description of a cut diamond, and conclude that diamonds were cut in India as early as the 5th century BC. Lenzen maintains that the diamond described here is a perfectly formed natural crystal, and that such rare crystals were greatly desired in India and never found their way as far as Europe. According to Lenzen, the diamond crystals familiar to the Romans would have been gray, misshapen, barely translucent, and not at all beautiful. See also Kautilya, *The Arthashastra*, ed. and trans. L.N. Rangarajan (India: Penguin Books, 1992), p. 775. The 2nd century BC writer Damigeron states "the best diamonds are found in India, second best in Arabia and the rest in Cyprus." Lenzen, a scholar, not a trader, would perhaps not be aware that a stone may be purchased in Arabia but not necessarily remain there if a better price could be obtained in Rome. Damigeron, *The Virtues of Stones*, trans. Patricia Tahil (Seattle: Ars Obscura, 1989), p. 10.

14 Lenzen, History of Diamond Production, p. 105. Although the crude process of faceting — by rubbing one diamond against another to wear each stone down — was known as early as the 14th Century, the technology for cleaving and sawing was not developed until the 17th Century.

The 116 carat blue precursor of the Hope
Diamond, brought to Europe by the French
adventurer Jean Baptiste Tavernier, was
recut in 1683, on the orders of Louis XIV,
into the sixty-eight carat stellate brilliant
that became known as The French Blue.
This important technological advance
in the lapidary arts unleashed, for the
first time, diamond's full potential — the
astonishing brilliance and fire for which the
gem is justly revered.[15]

15 Another famous 17th Century gem is the
35.56-carat Wittelsbach Blue. First mentioned in
1667 as part of the collection of Empress Margarita
Teresa of Austria was also cut as a stellate
brilliant, probably either in Venice or Lisbon. vid.
Dröschel, J. E., et al., *The Wittelsbach Blue, Gems
& Gemology*, The Gemological Institute of America,
Winter 2008, pps.348-352

worldwide.[20] A rigid price structure is associated with each of these grades.

Precious Versus Semi-Precious: a Distinction Without a Difference

The question "Is it precious or semi-precious?" is an expression of pure market snobbery. As has been shown, the term precious has had different meanings at different periods in different cultures. In earliest times, it had no meaning at all. The term semi-precious is today as meaningless as the term semi-pregnant or semi-deceased. That this term is still in general usage points only to the fact that many of the gemstones described as such still lack a degree of market acceptance or are out of fashion and may be purchased at relatively low prices.

Several gem species and varieties discussed in Part II of this book are fairly recent discoveries: tsavorite garnet, tanzanite and malaya garnet were completely unknown just seventy years ago. In many cases examples of these new precious gemstones are rarer and

more beautiful than those gems that have traditionally been labeled precious. In just the past twenty years, recently discovered gem varieties such as Paraíba tourmaline as well as the long neglected gem species spinel (known since the 10th century) have become the fashion of the day and as a result have increased dramatically in demand and price, often eclipsing prices asked for stones traditionally labeled precious.[21]

To the astute aficionado "semi-precious" should translate as "buying opportunity." The true lover of gemstones looks at the object without regard for the verbal baggage it may carry with it. If the foregoing discussion has demonstrated anything, it is that the whole idea of preciousness is fluid and changeable. In the world of gemstones, if it is rare and beautiful, and if demand is strong, it is precious.

Beauty Versus Pedigree: the Question of Origin

Each time a new pocket of gemstones is unearthed, stones from the new location are compared with those produced by traditional sources ~ usually to the detriment of the newer source. This is a crutch and yet another manifestation of the innate conservatism of the gem market, one controlled by professional dealers. From the connoisseur's perspective this misses the point entirely. The point is to look at the stone. The most conservative dealer will always fall back on a stone's pedigree. These traders are often those who have prospered,

20 In fact, GIA codified and adopted the traditional grading system that can be traced at least as far back as the 4th century BC. *The Arthashastra of Kautilya* describes good diamonds as "regular in shape and reflecting light brilliantly in all directions . . . which have the whiteness of a shell or of rock crystal . . . unblemished, smooth, heavy, lustrous, transparent . . . ." Rangarajan, *The Arthashastra*, pp. 775-778. The Arab scholar Ahmad ibn Yusuf al Tifaschi writing around AD 1250 divided diamond qualities into two categories: *zayti*, those with a slight yellow body color, and *billawri*, those that are colorless like rock crystal. Tifaschi held that the former were of the highest value. See Samar Najm Abul Huda, *Arab Roots of Gemology: Ahmad ibn Yusuf al Tifaschi's Best Thoughts on the Best of Stones* (London: The Scarecrow Press, 1998), p. 118.

21 Prices for pink, red and blue spinel have increased over 1,000% since the first edition of this book (2003) was published.

have well-heeled clients, and can pay the highest prices.

In the gem marketplace, a fine stone with a famous pedigree will command an extraordinary premium. A fine natural sapphire from Burma will often sell for twice the price of a comparable gem from Sri Lanka (Ceylon). A comparable natural Kashmir, a sapphire discovered in a small deposit on one side of a stony hillock in colonial India, will bring at least twice that! These premium prices are based solely on the stone's geographic origin. Astronomical prices are regularly paid for stones that carry this sort of geographic pedigree. Depending on the local geology, a given gemstone will have slightly different visual characteristics than examples from another location. This is usually the result of the mix of minerals specific to a particular geographic setting. For example, the iron-rich environment of the central Thai provinces of Chanthaburi and Trat lends ruby from this area a distinctly brownish cast, whereas ruby from the iron-poor soil of northern Burma generally lacks a brownish hue. Burmese ruby is better, generally speaking, because it looks better than ruby from Thailand, and it looks better because the physical environment is more favorable.

As a result of these localized differences, gemstones from different areas develop reputations based on the general look of the stones from that specific source. This

Traditional mining in the ancient "Valley of the Serpents," Mogok, Burma. This age-old mining method consists of a twinlôn, a hand-dug vertical shaft drops forty feet to the gem-bearing layer, or byon, where a narrow horizontal shaft is dug into the gem gravel. The gravel is raised by bucket and sorted. Photo: R. W. Wise

is simple branding and has led to marked price differentials just discussed.   In the marketplace, Burma ruby, Kashmir sapphire, and Paraíba tourmaline will command premium prices simply because they hail from the specific areas named. This is not a scam but it is a snare! The true connoisseur wishes to acquire a beautiful gem. Each individual gemstone has a specific personality and must be evaluated on its face. A given stone from Burma may very well be inferior to a given gem from Thailand, and not worthy of the premium asked. The Thai stone, by contrast, may be a particularly fine example and worth collecting despite the general reputation of Thai ruby.[22]

Here lies the snare! The connoisseur must ask, does the beauty of this stone from this source justify paying a premium price? The answer generally may be yes, if the best stones from the area in question are truly the best of their kind. However, it is important to stay focused. The aficionado collects gems, not generalizations. The quality of the specific gem under consideration should be evaluated on its merits, its individual beauty without reference to its geographic origin.

The Rarity Factor

The relationship between beauty and price is, at best, problematic. People have preferences. In the marketplace preference creates demand, which is a primary determinant of price. Are yellow stones intrinsically more or less desirable, more or less beautiful, than blue? Obviously not! Such partiality is clearly subjective. All colors are created equal. Yet a fine blue sapphire commands a much higher price than a fine yellow sapphire. This is purely a function of subjective preference, which manifests itself as market demand.

In the gem world, beauty drives demand and rarity drives price.[23] This is a catchy little phrase, but what does it mean? More to the point, what weight should the connoisseur give to the rarity factor when deciding on an acquisition?

There are two categories of rarity, *actual* and *apparent*. Some gem varieties are found in very small deposits and can be classed as actually rare. Other varieties, colorless diamond for example, are in such high demand that, though relatively numerous, may be difficult to find in the marketplace. Gems which fit into this second category are apparently rare. Another appropriate term would be market rarity. From a price perspective, apparent (market rarity) is the more important. Unless the item is in demand, then its actual rarity doesn't matter very much. Fine amethyst is actually quite rare, yet due to relatively lackluster demand for finer qualities, its price remains relatively low. Collecting can really get interesting when a gem is both apparently and actually rare. These are stones that are in short supply and also in high demand. Alexandrite, Paraíba tourmaline and blue diamond are good examples of gems that are both apparently and actually rare. Gems that fall into this category will command the very highest prices.

It is fair to say that, with the exception of colorless diamonds of less than five carats, the finest examples of all gem species

---

22 *Generally speaking diamonds have little geographic dissimilarity because they are formed not within the crust, but deep in the earth's mantle. However, as we shall see, certain types of diamonds, specifically the low nitrogen type IIa "Golconda" colorless diamond, originally sourced in India, will command a premium.*

23 The sole exception to this rule is fancy color diamonds. See the Introduction to Colored Diamonds.

and varieties are, at least, apparently rare and difficult to find in the marketplace. From the connoisseur's viewpoint the very finest examples of any gemstone are rare and difficult to obtain. Amethyst is a type of quartz, one of the earth's most abundant minerals. Even so, the deep Siberian quality described in Part II, though relatively inexpensive, is extraordinarily difficult to find. The author has sorted through thousands of parcels of cut and rough amethyst at the source in Brazil and Africa and come away without a single example of the finest quality of this relatively common gem.

In almost all cases rarity increases with size. (Tanzanite is somewhat the exception).[24]

Before the 18th century, diamonds came mainly from India, and were extremely rare, especially in Europe. The Indian sources, chiefly the Golconda mines in the Indian province of Hyderabad, were essentially already mined out when diamonds were discovered in Brazil in 1725.[25] Diamond exports from Brazil from 1730 to 1787 increased total world diamond supplies as much as twenty-fold. Due to this abundance, between 1730 and 1735 the diamond market went into free-fall and rough diamond prices dropped seventy-five

percent.[26]

With the discovery in the late 19th century of vast diamond reserves in southern Africa, huge supplies of diamonds began to enter the market. Newer discoveries in Russia, Australia, and, most recently, Canada, have kept supply strong. These discoveries, coupled with improvements in prospecting and recovery methods, have created a virtual glut of colorless diamonds under five carats.

In 1992 it was estimated that if all diamonds produced by Indian and Brazilian sources from antiquity to that date were totaled, that number would be equal to just twenty-two percent of the total world production of the previous five years (1987-1991). In fact, the annual production from Australia's Argyle Mine in the early 1990s was approximately equal to the total amount of diamonds produced in India and Brazil from antiquity to 1869.[27] In the past two decades, cut diamond production has increased from fifty million carats (gem quality) annually to a peak of one hundred

---

24 Smaller tanzanite (one-four carats) in the finest quality is rarer than gems over five carats..

25 This is the traditional date given by most sources. The first date mentioned in the literature is 1714. J.P. Cassedenne, "Diamonds in Brazil," *Mineralogical Record*, vol. 20 (1986), pp. 325-335. Diamonds from India came from several sources;

26 Lenzen, *History of Diamond Production*, pp. 50, 126. DeBeers was not the first syndicate to control the market. The diamond market was saved by the simple expedient of monopolistic practices on the part of the Antwerp Diamond Cutters Guild, which controlled prices in the 18th century to such a degree that lower prices for rough were not passed on to the jeweler and consumer as lower prices for cut stones. Cut stone prices remained stable.

27 A.A. Levinson et al., "Diamond Sources and Production: Past, Present and Future," *Gems & Gemology*, (Winter 1992), p. 236.

sixty-eight million carats.[28]

In the case of diamond, an apparent rarity, maintaining the price structure is created by high demand coupled with a carefully controlled distribution system. The monopolizing organization, variously called the cartel, the syndicate, or simply DeBeers, took control of the diamond market in 1889.[29] Colorless diamonds are not actually rare, but the syndicate (through selective distribution and a careful hoarding of reserves) insured that supply did not exceed demand.[30] The price of diamonds, as with all other gems, is based on beauty — plus supply and demand. The difference is that demand for diamonds was mightily stimulated by advertising, and supply was ruthlessly controlled by the DeBeers cartel.

As of this writing, the iron control formerly exercised by DeBeers is no more. By 2012 De Beers' share of the diamond production had been reduced to twenty-nine percent.[31] A half dozen major producers now share control of the diamond market.[32] What does that bode for the future? One would assume that DeBeers loss of market control would lead to greater competition and lower prices. That was my prediction in 2002. However prices remained fairly steady even during the Great Recession of 2008. Does this reflect the simple law of supply and demand? No, De Beers may be a pale shadow of its former self, but its specter still walks the earth. Major producers responded to the turndown by cutting mining output by as much as fifty percent and some producers stockpiled rough stones. Total production fell from one hundred sixty-five million in 2008 to one hundred twenty-four million carats in 2009. In 2014 production jumped to just over one hundred thirty million carats

32 Russell Shor & Robert Weldon. "An Era of Sweeping Change in the Diamond and Colored Stone Production and Markets," *Gems & Gemology*, The Gemological Institute of America, Vol. 46, No. 3, (Fall 2010), 166–187.

28 Russell Shor & Robert Weldon. "An Era of Sweeping Change in the Diamond and Colored Stone Production and Market, *"Gems & Gemology"* ,The Gemological Institute of America, Vol. 46, No. 3,(Fall 2010), 166–187.

29 Stefan Kanfer, *The Last Empire: De Beers, Diamonds, and the World* (New York: Noonday Press, 1993), p. 106.

30 Kanfer, *The Last Empire*, p. 339.

31 Russell Shor, "Rough Diamond Auctions: Sweeping Changes in Pricing and Distribution." *Gems & Gemology,* (Winter 2014), 236.

Working the polishing wheel. Emerald polishing, Argotty workshop, Bogata, Columbia. Photo: R. W. Wise.

Chapter 3

Connoisseurship:
Rethinking the Four Cs

"In selecting precious stones you must mentally ask yourself the following questions: Is their transparency conspicuous? Are they like a dew-drop hanging from a damask rose leaf; that is, are they of pure water and do they possess the power of refraction to a high degree? Or, are they transparent and colored; and, if the latter, have they a play of color? Lastly, have they notable imperfections?"

–E.W. Streeter, 1879

Over the centuries experts have developed a series of criteria for judging the objective aspects of beauty in gemstones. Lately these criteria have been given a catchy title--the Four **Cs**. Although something of an oversimplification, these categories — *color, cut, clarity,* and *carat* weight — are useful, if incomplete. The traditional four **Cs,** or at least the first three **Cs** (weight has nothing to do with quality), can be used as criteria for evaluation of the beauty and quality of all gemstones. I propose to omit carat weight and substitute, or rather reintroduce, a neglected, almost forgotten, but necessary criterion, crystal, as the fourth **C.** *Crystal,* also known as diaphaneity or transparency, is, as will be demonstrated in the following chapter, the true fourth **C** of gemstone connoisseurship.

Beauty is a balancing act between a number of factors. Color, clarity, cut, and crystal are the four factors—abstractions, really —which can be used to analyze and discuss the beauty of a gemstone. None of these criteria by themselves is sufficient to define a beautiful gem. All four are necessary conditions, without which the

Gem cutting the old fashioned way. This Thai gem cutter estimates facet angles by eye and cuts freehand. Bo Rai, Trat Province, Central Thailand. Photo: R. W. Wise.

stone will simply not make the grade. Of the four factors, cut is most easily quantifiable, and the only one which depends completely on the ingenuity of man.

The idea of the *four* **C**s is useful. The concept would be more useful still if each **C** were of equal relative importance in the connoisseurship of all species and varieties of gemstones. Unfortunately, they are not. **C**olor is of primary importance in the appreciation of colored gemstones. **C**ut is the primary criterion in the evaluation of colorless diamonds. **C**rystal, as we shall see, is a key to connoisseurship in all gemstones but central to the appreciation of translucent cabochon-cut and phenomenal gemstones.

Quality and connoisseurship in colored stones, by far the broadest class, will be discussed first, followed by phenomenal stones (stars and cat's-eyes), diamonds, and the pearl. Opal exists in a world of its own and will be dealt with in detail in Part II.

Color: First Among Equals

By convention there is color, but in reality there are only atoms and space.

–Democritus, 460 BC

Common Perceptions

We are all familiar with the basic color wheel. Sir Isaac Newton is said to have discovered that light is composed of seven colors (red, orange, yellow, green, blue, violet and indigo) when he introduced a glass prism between a beam of sunlight and a white wall and noted the rainbow projected thereon. If you repeat Newton's experiment you may be surprised to find that the actual spectrum is composed of five colors: red, yellow, green, blue and

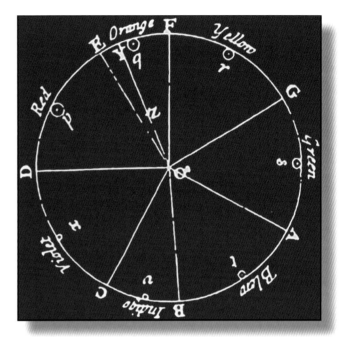

Newton's original drawing of his color wheel containing seven hues.

violet. Newton himself was colorblind. He recruited a friend to break up the spectrum for him and insisted on adding orange and indigo, a dark-toned violet, to make seven because, as Newton said, "seven corresponded to the seven intervals of our octave."[33]

In the 1960s a group of cultural anthropologists conducted a worldwide survey to determine how the world's cultures defined color. Representatives of each society surveyed were shown a specially prepared color chart containing all colors — some three hundred twenty-nine separate hues.

The scientists studied the color terms native to twenty languages. The most "primitive" cultures surveyed had only two concepts, or two words they used to

33 Edwin Ardener, *Social Anthropology and Language*. See David Lewis-Williams, See also *The Mind in the Cave, Consciousness and the Origins of Art*. Thames a Hudson, London, 2002, Kindle Location. 2100.

describe all the colors on the chart: *black* and *white* or *light* and *dark*, what we would call tones. Interestingly enough, all the cultures with two concepts had the same two concepts. Other slightly more "advanced" societies had three words: *light*, *dark*, and *red*. Still more advanced cultures added a word which meant *green or yellow*. Those cultures that had a fifth word had a term for *yellow* and one for *green*. Languages that contained a sixth term added *blue*; those with seven added *brown*. Cultures whose language included eight or more terms contained words for *purple*, *pink*, *orange*, *grey*, or some combination of these colors.[34]

By way of conclusion, the anthropologists determined that all human societies share (at most) just eleven basic color concepts: *red*, *orange*, *yellow*, *green*, *blue*, *violet*, *purple*, *white*, *gray*, *black* and *brown*.[35] Although we may amplify these basic color concepts in different ways, adding sophisticated variations such as *umber* and *chartreuse* to describe exotic mixed hues, the fact remains that all the gemstones discussed in this book can and will be described using these same eleven terms.

All human beings share the same perceptual apparatus. We receive raw data through our senses, and our perceptual

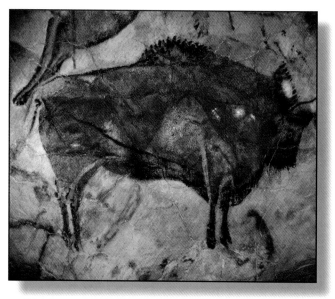

Black, white and red, the primitive palate! The artist who painted this 18,000-year-old rendering of a bison in Spain's Altamira Cave used all three. Hematite or red ochre, charcoal or manganese for black and kaolin or ground calcite mixed with animal fat for white. Photo: Remessos.

apparatus refines and orders this data for us. The German philosopher Immanuel Kant called this apparatus "the faculty of apperception." The philosopher was seeking to explain how it is that human beings perceive and understand what they see.

All that we see, feel, and taste is processed through the funnel of our sensory apparatus. Sense data, then, takes on the contours of that funnel. According to Kant, the human mind impresses certain categories or forms upon the data the senses deliver to the brain. He called these mental forms *a priori* categories. [36] *Space* and *time*, for example are *a priori*

34. Brent Berlin and Paul Kay, *Basic Color Terms, Their Universality and Evolution* (Los Angeles: University of California Press, 1969), pp. 2-3. Berlin & Kay claimed that the words used, in each case exhausted the entire spectrum as represented by the 239 color samples. However, later research suggests that some cultures simply have no term for certain hues. See Asifa Majid & Stephen Levinson, "The Senses in Language and Culture". *The Senses and Society Journal*, Volume 6, Number 1, Bloomsbury Journals, March 2011 Special Issue, pp.12-13.

35 Ibid.

36 Immanuel Kant, *The Critique of Pure Reason*, trans. Norman Kemp Smith (London: St. Martins Press, 1964), pp. 65-74. *A priori* refers to categories that exist before sense data, and through which the data of the senses passes and is understood. Color itself, or at least our manner of perceiving it, is an *a priori* category impressed by the mind upon the raw data of the senses.

categories which do not exist in the world at all, but only in the mind. Odd though this may seem at first, ever see an animal, aside from Alice's white rabbit with his watch, that seemed to have any concept of time? How is it that all societies share just eleven basic color concepts? It is because we share an identical perceptual apparatus and, therefore, process sensory data in the same way. The idea of color, however, is not a sensation, it is rather a concept and, in some sense, a cultural *a priori* category that our mind uses to filter the visual sensations that we do experience.

Blessed with color vision, humans have a somewhat unique view of the world, a view conditioned by our unique sensory apparatus. The family dog cannot perceive colors; his world exists solely in black and white. By contrast, this same dog, with his arguably superior hearing, lives in a world with a much richer soundscape. We are different species, and the machinery of our perception is different. Luckily, all humans, as members of the same species, possess a more or less identical sensory apparatus.

Some skeptics would argue that none of this can be proved. Isolated as I am within the prison of my individual consciousness, I may be seeing a color I call blue. You, on the other hand, may be receiving the identical sensation and call it red. Eventually, of course, one of us will incur so much disagreement that we will redefine the sensation. Inversely, my visual impression may be the opposite of yours even though I call it by the same name.

The fact is, none of this matters very much so long as we agree on price.

Human beings differ from the family pet in other ways. We have the faculty of judgment. We make qualitative decisions about the raw data our sense organs deliver to us. The eye is a part of the brain. We see colors but are free to prefer the color blue to yellow, pink to red, etc. Because we have similar faculties, we often have similar tastes. Opinions, however, are personal and subject to change. In the gemstone marketplace, price differentials represent the current majority opinion expressed as market demand.

The Perception of Color

Color has many interesting properties. Colors can be pure, intense, warm, dark or cool. Orange, red and yellow are warm colors; blue, green and light purple are cool. Colored objects can appear large, small, close, or distant. The brighter the color the larger the object appears. Yellow is the "biggest," most highly saturated color, followed by orange, red, green, and blue. In the fine arts, specifically in painting, color alone can be used to create the illusion of depth. Warm colors come forward; cool colors recede. In modern times, cubist and constructivist painters fomented an artistic revolution by using the visual laws of color to create, on their canvases, a sense of depth without the use of perspective.

Color has direct physiological and psychological effects on the viewer. Red causes the lens of the eye to thicken, blue flattens the lens.[37] Wassily Kandinsky, the first painter to create totally abstract pictures, believed that the use of certain colors alone could induce specific emotional states in the viewer.

The fact that certain colors evoke connotations which bridge cultures and have remained constant over the centuries proves Kandinsky's thesis. Red is the hot color of blood and the color of war. To become angry is to see red. Blue is the color of the heavens. When a person is depressed he has the blues.

In art class we learned that the color wheel divides into primary and secondary colors called hues. Red, blue, and yellow are primary in the sense that they are irreducible, unmixed pure hues. Secondary colors are those created by mixing two or more primaries. For example, mixing yellow and blue makes green. However, the rules of the color wheel have to do with paint and pigments. The rules change when color, transparency and light are considered.

When two colored slides are projected on a screen, green and red light will mix and create yellow. A similar mixture of paints would yield a dirty olive brown.[38] Colored light contains eight chromatic hues: red, orange, yellow, green, blue, violet, pink and purple. The first six hues are called *primary spectral hues*: the colors of the rainbow or the colors seen when white light is refracted through a prism. The last two, purple and pink, are *modified spectral*

hues: purple is the hue that lies halfway between red and blue on the color wheel, and pink is a lighter-toned, less saturated hue of red. Any of the eight hues can play the part of the primary or secondary hue in a gemstone.

Color as it is applied to connoisseurship in gemstones requires further definition. Color is divided into three components: *hue, saturation,* and *tone*. These categories are relatively common in color science where they are variously labeled: *hue, value, and chroma; or hue, intensity, and tone.*

Hue

Hue is the technical equivalent of "color" as that term is used in normal speech: What color is it? In gem grading parlance, color is a general term; red and blue are *hues*. Nature exhibits relatively few pure hues. The colors we see in objects are a mixture of hues. Thus, for technical precision in describing color it is necessary to divide hues into primary, secondary, and occasionally tertiary categories. Primary is used in the sense of dominant, the majority hue. A sapphire that is greenish blue has a primary hue of blue and a secondary hue of green. Just as in language, the term greenish is an adjective modifying the noun blue; green is a secondary hue modifying the primary or dominant hue, in this case blue.

A royal blue sapphire has a primary blue hue. The term royal is simply a more poetic way of describing the combination of the primary blue with the secondary hue, which in this case is purple or purplish blue. Put more precisely, the hue may be described in percentages. In the above example,

37 L. Moholy-Nagy, *Vision in Motion* (Chicago: Paul Theobold, 1947), p. 155.

38 Moholy Nagy *Ibid.*, p. 159.

royal blue sapphire is mostly blue, perhaps eighty-five percent, with an admixture of up to fifteen percent purple. The use of percentages is more useful because it is more precise.

Faceted gemstones exhibit two types of color, refracted color and transmitted color. When a beam of light is directed toward the top, or crown, of a faceted gem, some of that light will enter the gem, reflect internally, and be refracted back to the eye. The color of that refracted light is called the *key color*. This is the color of the sparkle or brilliance. The color of a gem is evaluated by observing the key color. This is a vital point. *Body color* results from light that is transmitted through the gem. The quality of these phenomena, body color and key color, almost always differ at least in darkness or lightness—the component of color that is called tone.

The key color is the color used to evaluate faceted stones, except colorless diamond, which is judged by its own unique set of rules (see Chapter 5). This distinction may be confusing at first. To differentiate between key and body color, the gem is turned table or face down under or above a light source. A gemstone is designed to generate brilliance in the face-up position. Placed face down the gem will not sparkle, it will glow. The color seen is the body color. The gem can then be turned over and viewed face up. In this position the color of the sparkle may be observed and the differences in hue (key versus body color) will become apparent. Key color is often lighter in tone and paler in saturation than the gem's body color. Body color can

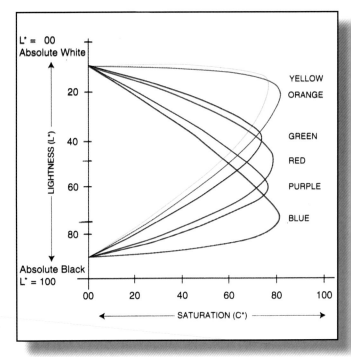

Color gamut limits: The graph illustrates the point at which saturation and tone combine to produce the most vivid hue. Note that yellow and orange achieve maximum brightness at relatively light tones (twenty to thirty percent). Above these tonal values, saturation (brightness) of the hue rapidly diminishes. Blue and purple require fairly dark tones (sixty to eighty percent) to achieve maximum saturation; one hundred percent would appear as absolute black and totally opaque. Tonal values below these gamut limits begin to look washed out. The most desirable combinations of saturation and tone in gems of these hues follow these curves fairly closely; e.g., the finest yellow diamond is fairly light and bright; the finest blue sapphire, deep and rich. This graph objectively demonstrates the very simple dictum that, in almost every case, the brighter the hue of a gemstone, the more beautiful it is and the more desirable it will be in the marketplace. Stephen Hofer, Collecting and Classifying Coloured Diamonds © 1998 W.N. Hale, Hale Color Consultants

be particularly seductive in lighter-toned highly crystalline gems such as aquamarine where the body color may be richer, more highly saturated and more prominent than the refracted color. Take care! If the eye becomes lazy, the result can be costly.

Saturation

Saturation, the second of the three components of color, refers to the brightness of the hue or the quantity of color. The greater the quantity of color, the brighter it will be. Saturation may be described as vivid or dull or somewhere in between. Although some hues are brighter than others, pure hues are always vivid! International orange, the color used in buoys and life jackets, and the red of a stop sign are examples of particularly vivid hues. "Day-Glo" colors are vivid hues. In this system of evaluation, the neutral, non-spectral colors gray and brown are not classified as hues; they are considered to be saturation modifiers or masks.

Like a splash of mud on a Hawaiian shirt, the addition of gray and brown dull the hue. Some gem varieties have a tendency toward brown, some toward gray, rarely both. Blue sapphire may be grayish, but rarely brownish, whereas red tourmaline will often have a brown modifier and these modifiers may vary in intensity depending upon the type of light the gem is viewed under. A grayish greenish blue sapphire is a dull greenish blue sapphire. A brownish red tourmaline is a muddy-hued red tourmaline.

Gray or brown masks are often like a light film that is itself of low saturation and very light tone and is difficult to see. The effect — dullness or muddiness — is visible and, since pure hues are always bright, we infer from this that a gray or brown mask is present. A trained eye and careful observation are often necessary to see the mask. Another clue which may help is that brown gives the impression of warmth, whereas gray is cool. If the hue appears dull and cool, the mask present is probably gray. If the hue seems to be dull and warm, the mask is likely brown.

Saturation Modifiers: from Minus to Plus

What good are rules if there are no exceptions? Life would be so much simpler, consistent, and boring. When gray and brown are themselves highly saturated and dark enough to be dominant, they too begin to act the part of a hue. This is particularly true of brown. Thus dark-toned gray and brown, if they are dominant, become hues, whereas grays and browns that modify, or dull a spectral hue (such as blue) are saturation modifiers or masks. Both yellow and orange, when they are dark toned, appear brown.

To sum up: brown and gray are not generally considered to be hues, but are classed as saturation modifiers or masks, unless the stone is primarily brown or gray; then either may be considered a hue, and may even have spectral hues as modifiers — for example, reddish brown. Examples of beautiful brown-hued stones include fancy color "cognac" diamond and brown tourmaline.   Gray may be dominant in diamond, moonstone and spinel.

Tone

Tone — that is, lightness and darkness — is the third component of color. It can best be described as the addition of black or white to a hue. A dollop of black paint added to a bucket of robin's egg blue yields a darker blue. The more black is mixed in,

"Tonal range of purple hue. From right to left, tonal percentages graduate in approximately ten percent increments from ten to eighty percent tone. Darker tone yields a richer and more vivid hue." Photo: Tino Hammid.

the darker the tone: first sky blue, then royal blue, then midnight blue, and finally enough black is added to overcome the hue and the paint turns completely black. White added to a color has the opposite effect.

For the sake of analysis, tone is described as a percentage. A transparent quartz crystal or a windowpane is zero tone. No tone = no color. A lump of coal or a crow's wing is one hundred percent tone. One hundred percent tone is opaque black. Too much tone snuffs out hue, saturation and transparency. When this occurs the stone is overcolor. Tone refers to the key color, not the body color.

In gemstones the beauty of the color is a certain balance of hue, saturation, and tone. For each color the optimum percentages differ. If the tonal levels are too high the stone is described as being overcolor, or too dark. If the key color is too light in tone the hue appears pale and washed out.

The two attributes of saturation and tone are abstractions which, in gemstones function together to define the beauty of the hue. Color scientists have long recognized that there is an optimum combination of saturation and tone for each hue. This is the point at which saturation and tone produce the most vivid hue. These points are called gamut limits.[39] For example, the most vivid tone for yellow is twenty percent, while the most vivid tone in blue is about eighty-five percent, red eighty percent, and green seventy-five percent.[40] Beyond these limits, as the hue darkens, it loses saturation and moves from vivid towards dull.

Not surprisingly, market desirability closely parallels these gamut limits for gems occuring in the basic spectral hues. The optimum tone for ruby is eighty percent, sapphire and tanzanite eighty to eighty-five percent, and emerald and tsavorite seventy-five percent. Reduced to simplest terms, *the brighter, the more saturated the hue, the better the hue*. The rule must be applied somewhat gingerly because a gemstone rarely exhibits an absolutely pure hue. Changing taste also plays a role. Purple gems deviate from this norm. Amethyst reaches the peak of beauty and market desirability at seventy five to eighty percent tone. The gamut limit graph demonstrates that the hue purple achieves its gamut limit at a much lighter sixty percent tone. In the case of amethyst this means that a richer hue is generally preferred over a brighter hue. This is also true of red, yellow, orange and green gems.

A particularly well-formed hexagonal emerald crystal. Conosco Mine, Boyacá, Colombia. Photo: Jeff Scovil, courtesy R. W. Wise, Goldsmiths.

39 Nick Hale, personal communication, 2002.

40 Stephen C. Hofer, *Collecting and Classifying Coloured Diamonds* (New York: Ashland Press, 1998), p. 172, fig. 13-3. The chart pictured page six shows saturation gamuts for opaque color pigments; however, gamut limits for transparent media are higher, although the relative saturation between different hues remains the same. Zero has no color, absolute black would be one hundred percent tone. Nick Hale, personal communication, 2002.

The color of a gemstone can be accurately described using this terminology. For example, a fine bright, dark, purplish red ruby is better described as a ruby with a ninety percent vivid primary red hue, and a ten percent purple secondary hue of eighty percent tone (with little or no gray mask present).

Cut: The Geometry Of Beauty

The Faceted Gem

Traditionally, the relative importance given each of the four Cs has differed depending on whether it was applied to colored gemstones or colorless diamonds. In colored gemstones the most important criteria is, of course, color. In colorless diamond, which by definition has no color, it is brilliance, or life, a function of cut, that is of first importance.

The refinement of the lapidary art and the invention of faceting in 15th century Europe ushered in the modern period of gem appreciation. Proper cutting and polishing are required to achieve a gem's full potential. By the end of the 16th century, beauty became the primary criteria in the evaluation of gemstones.[41] Before then, as we have seen, religious, talismanic, and medicinal uses of gemstones were at least as important, and often of greater importance, than beauty in the value equation.

Although the shaping of gems, mostly as cabochons, had reached a high degree of sophistication in the Middle East by the third millennium BC, faceting was not developed until the Renaissance, and it was a European invention. Even today, cutting as a grading criterion is emphasized a great deal more in the West than in the East. In Asia, color is by far the predominant factor in grading gemstones; the perfection of cut,

what we call "make," is at best a secondary consideration.

The outward form of the gem crystals themselves suggested the first cutting styles. Diamond, for example, normally occurs as a bipyramidal crystal; i.e., its normal form, what crystallographers call its "habit," looks like two pyramids joined together at their base. Such perfect crystals are quite rare. Before the invention of faceting in the 15th century, uncut diamond crystals were often set into jewelry. Diamonds could not be cut but they could be cleaved. It is probable that the first attempts at faceting were an attempt to restore the bipyramidal symmetry of the natural diamond crystal.

Once the technology was developed, the next step followed closely. Knock the point off one end of a well-formed diamond crystal, polish the sloping sides, and presto—a fine example of one of the earliest faceting styles, the table cut! Viewed in profile the table cut retains the same basic outline of the natural crystal. A few more modifications and the table cut evolves into the star, the stellate brilliant, the old mine cut and, finally, into the modern round brilliant, the cutting style which graces the engagement finger of millions of young women. [42] Other cutting styles have a similar history: the hexagonal profile of the emerald crystal suggests the basic outline of the emerald cut. The cabochon, perhaps the oldest cut of all, is little more than a natural stream pebble which has been flattened on one side and polished.

41 Lenzen, *History of Diamond Production,* p. 110.

42 Herbert Tillander, *Diamond Cuts in Historic Jewellery 1381-1910,* (London: Art Books International, 1995).

As technology developed, cutting gradually achieved greater importance. Today fine cutting must be considered a necessary factor in the connoisseurship equation. That is, a well-cut stone may not necessarily be fine, but a fine stone must be well cut.

Natural uncut gem material, commonly called rough, is found in all sorts of shapes and sizes. The crystals may have grown together or in distorted shapes. Other minerals, sometimes other crystals, may have formed and been trapped within them. In the millions of years since its formation, the rough may have been cracked in a landslide, crushed by earthquakes, or pulverized by glacial action. Rough gem material may have been displaced from the place of its formation by erosion, tumbled down mountain streams, or ground into pebbles by the action of ocean tides.

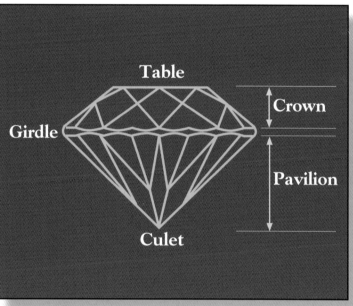

This drawing illustrates the basic parts of a faceted gemstone. Although the gem pictured is a round brilliant cut, almost all cutting styles: oval, pear shaped or square have a table, crown, girdle, pavilion, and culet.

The cutting of gemstones is a craft, but the ability of a lapidary to coax and maximize the beauty inherent in the rough is an art. The shape and size of the cut gemstone is dictated in part by the size, shape, and clarity of the piece of raw material placed before the lapidary. Rough is expensive and is sold by weight. The objective of the lapidary is, thus, twofold: first, to create a beautiful gem; second, to maximize the yield in carat weight from the rough material and thereby maximize the price. Bigger and heavier translate into greater profit, particularly when fudging the proportions makes the difference between "holding the carat" or cutting a sub-carat gem. This is particularly true in Asia. In Bangkok the asking price for a one-carat (1.00+) blue sapphire may be four times that asked for a similar 0.85-carat stone.[43] Little surprise that the choice between maximizing beauty or maximizing profit often tilts in the direction of commerce. In many cases the loss of weight necessary to create a beautiful stone is unacceptable to the dealer.

Beauty versus profit! These two objectives are often at odds. If the dealer's

---

43 The modern metric carat is one-fifth of a gram. The carat is divided into one hundred points. A half-carat is written 0.50. The carat was originally based on the weight of the carob bean.

wares are criticized for poor cut, he will often shrug and use the term *native cut* implying that the gems were fashioned by technologically challenged primitives. This is both an ethnic slur and a very convenient bit of disinformation. Cutters in Asia, Africa, and South America are very familiar with their indigenous materials and are exceptionally skilled. For example, I have seen lapidaries in upper Burma produce precisely cut sapphire using primitive foot-operated cutting wheels with buffalo horn fittings. Poor cutting is almost always the result of a well thought out weight-retention strategy. From the connoisseur's perspective, there is no excuse for poor cutting. The aficionado should remember that a poorly cut gem is the result of a conscious decision made before the rough is put on to the cutter's wheel.

Beginning in the early 17th century, the major preoccupation of the lapidary arts has been the development of proportions and facet patterns which would maximize the brilliance, or life, of a gemstone. Cutting is all about brilliance. Brilliance is defined as the total amount of reflected light reaching the eye from the internal and external surfaces of the facets. The color of the refracted light, the color coming from inside the gem, is what is called the key color of the gem.

Viewed face up, a faceted gemstone is far from uniform in appearance. It is a complex mosaic, a chromatic jigsaw puzzle, a shifting crazy quilt of color. On close examination, some facets appear bright, others dull; some one color, some another. Faceted gemstones are designed so that light enters through the crown of the stone, bends, or refracts into, and reflects within the gemstone. It then returns through the crown to the eye. Jewelers sometimes talk about gems gathering light from the back and sides. They will even suggest that prong settings, particularly high ones, will improve the stone's brilliance. This is a misunderstanding of the geometry of light and the objective of the lapidary. In a properly cut gem, light is not gathered from the sides or back of the stone; it enters through the top or crown of the stone, is

Aquamarine crystal with faceted gems. Photo: Harold & Erica Van Pelt, courtesy Kalil Elawar.

Beauty trumps size. The oval 3.30 carat pink topaz (top) recut by John Dyer to a 2.30 carat radiant eliminating the window. Note the lack of brilliance toward the center (table) of the oval. The radiant pictured (bottom) exhibits the strong multi-color effect typical of topaz. Photo: John Dyer.

and the perception of color in a faceted gemstone. For example, a "blue" sapphire may exhibit blue and greenish blue along with tonal variations of these hues depending upon the facet being viewed, further complicating judgment. It would seem to follow that singly refractive gems, those that do not cause a light ray to split, would exhibit colors that are uniform in hue and differ only in saturation/tone and that is true... in part.[44]

Green, blue and purple singly refractive gems will exhibit variations in saturation and tone only in the face-up mosaic. Singly refractive gems of all other spectral hues may exhibit different hues, usually hues adjacent on the color wheel viewed face up. This is due to their relative positions in three dimensional color space. Some hues overlap. For example, brown underlies both orange and yellow. A de-saturated, dark toned orange becomes brownish as does yellow. Pink covers the full range of red plus some portions of orange and into the purple. Thus, pale, light toned red becomes pink or orangy pink. This means that certain faceted garnets, spinels and colored diamonds will, like their doubly refractive cousins, exhibit varying hues viewed face-up.

reflected internally, and is refracted back out of the crown. As we will see, only a poorly cut stone can derive any real benefit from an open setting.

Optically speaking there are two types of gems; those that divide light as it refracts into the gem in two colored rays and those that do not. The former are called doubly refractive, the latter are singly refractive. Most gems other than diamond, spinel and garnet are doubly refractive. Each light ray that enters a doubly refractive gem is divided in two with each sub-ray containing a portion of the visible spectrum. This further complicates the face-up mosaic

Brilliance is partly a result of faceting but more directly the result of the gem being cut to the proper proportions.

44 Hofer, Collecting and Classifying Coloured Diamonds, pp.152-153.

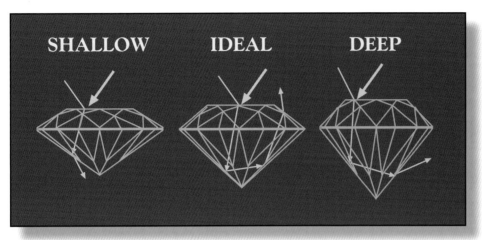

SHALLOW   IDEAL   DEEP

*"Tonal range of purple hue. From right to left, tonal percentages gradu-
ate in approximately ten percent increments from ten to eighty percent
tone. Darker tone yields a richer and more vivid hue." Photo: Tino Hammid.*

Proportions refer to the ratio between the length/width of the stone against the depth. Light bends when entering a substance; this is called refraction. The angle of the bend depends on the density of the substance and is different for every gemstone species. The angle of refraction can be measured and is called the refractive index. If the gem is properly proportioned (length-width to depth) most of the light entering the crown will be bent so that it reflects several times within the interior of the stone and exits through the crown, causing the gem to dance in the light. In improperly proportioned gems light will leak out the pavilion of the stone reducing brilliance.

Window, Fish-Eye, or Belly

Three basic cutting strategies are used to maximize the size and/or weight of a gemstone. They are commonly referred to descriptively as cutting a window, a fish-eye or a belly. The first strategy, which maximizes the weight as well as optimizes the size of the gem, is called cutting a "window." In this case, the lapidary cuts a

stone that is both wider and shallower than is required for maximum light return (brilliance). This gives the illusion that the stone is actually bigger because it has a greater diameter. Unfortunately, the depth-to-width ratio is too small and the pavilion is too shallow. As a result, instead of refracting back through the crown to the eye, light passes right through the center of the stone and causes a lens effect or window beneath the table of the stone. If a windowed stone is placed over a sheet of newprint, the words printed on the page will be visible. In a properly cut gemstone, the words on the page will not be visible because the light entering the gem through the crown is bouncing off the internal pavilion facets, returning the way it came, and exiting through the crown in the form of brilliance. A lens effect, on the other hand, allows light to move in both directions, resulting in a loss of brilliance at the center of the gem; the larger the window, the greater the loss of brilliance.

If the stone has been cut in such a way that the face or diameter is fifty percent larger than normal but is so shallow that it loses fifty percent of its brilliance, where is the net gain? When the stone is viewed at a distance, it is the light return, or brilliance (not the diameter of the stone) which is most apparent. Sometimes a window will be cut in a gem that is overcolor or has poor crystal; i.e., the stone is only semi-translucent. By cutting a lens, the craftsman can assure that some color, primarily transmitted or body color, will be observed under the table, through the window, in the face-up gem.

In our culture, bigger often translates as better and more expensive. A dealer or jeweler can often entice a buyer to purchase a stone with a larger diameter simply because it is a higher carat weight, particularly if all the stones the buyer sees are poorly cut. A small amount of windowing is difficult to avoid. The connoisseur will be offered many windowed gems. As with any grading criterion, the question is how much does the effect disturb the eye and detract from the beauty of the gem? A bit of a window may be acceptable in an otherwise beautiful but dark toned gem. But by any standard, badly windowed gems should be judged as de facto cabochons; unless the collector is looking for a cabochon, badly windowed gems are poor bargains and should be avoided.[45]

Cutters will sometimes cut a "well". A well is not quite a window. Looking at the stone in the face-up position, one can see right down to the pavilion facets, but not quite all the way through the gem. I have seen older diamonds and some emeralds cut in this way. One of Bogota's most respected lapidaries, Adolpho Argotty, once

Diamond exhibiting darkness beneath the table, known as the fisheye effect. Note the bright spot at the center of the table, that is a window caused by a flat faceted culet.

told me that he preferred cutting a well because it showed off the gem's crystal.[46]

Another strategy used by the cutter to preserve weight is to cut the stone overly deep, or too deep for its width. In this case, some of the light entering the gem will bounce around the inside of the stone and exit through the side of the pavilion rather than exiting through the crown, causing the gem to appear dark beneath the table. In trade parlance, this is referred to as a "fish-

---

45 Cabochon, from the French meaning *little head*, is a facet-less round-topped gem which looks like a gumdrop. It is little more than a natural stream pebble that has been polished. Cabochons, or "cabs," are normally cut from badly flawed rough and although they can be quite beautiful sell at discounts of as much as seventy-five percent less than faceted gems of similar color. Opaque gems such as turquoise and lapis lazuli are normally cut *en cabochon*.

---

46 Adolpho Argotty, personal communication, 2006

eye," meaning that the center of the stone is dark like the iris of a fish's eye, with some brilliance from the crown surrounding it like a halo. Again, a gem to be avoided.

Stones cut with a rounded or bulbous profile below the girdle are referred to as "bellied." Cutting bellied stones is another common practice used to retain weight. A bellied stone may also exhibit a fish-eye or window. However bellied stones may show no defect in brilliance but, like an overweight man with a spare tire around his middle, they are simply out of proportion. Since the largest percentage of a stone's weight is centered at the girdle, bellying a stone can increase its weight by a surprisingly large degree.

As a rule of thumb, the crown in a well-made faceted stone is approximately forty percent of the total weight, whereas the pavilion is sixty percent. The face is measured at the girdle. A stone with a little fat around the belly will show a bulbous, rounded profile between the girdle and the bottom tip of the stone (culet). Stones cut in this manner have a smaller face for its weight class than a properly cut stone.

Most ruby and sapphire are bellied. Two one-carat size sapphires may have markedly different faces. One may measure 6mm at the girdle; another only 5.5mm. The additional weight will usually be found encircling the pavilion just below the girdle. Any additional weight above what is necessary should not be considered in the evaluation and pricing of the gem; e.g., a 5.5mm round sapphire should weigh approximately 0.87 carats and this should be considered as part of the desirability

equation. The optimum dimensions of a perfectly cut stone will differ for each gem species.[47] In the author's view, up to twelve percent extra weight is acceptable in a rare stone.

Occasionally a little fat around the girdle can have a positive effect. When the pavilion of a gem is cut at a sharp angle, the stone may face-up beautifully when viewed perpendicular to the table. It may either wink out, going totally black or window as the gem is tilted five to ten degrees from the perpendicular. A bellied gem will sometimes retain a bit of sparkle when the table is tilted five, ten or even fifteen degrees away from the eye. This panorama of brilliance is a definite plus and may completely compensate for a bit of a bulge.

Performing the Test

The *quantity* or percentage of brilliance a gemstone exhibits can be measured by a straightforward test. Hold the stone face up directly under an overhead light. Diffused fluorescent light is best. For this test, the kelvin temperature of the light is not important. Stones that are not perfectly round or square should be held along the long axis with a pair of gem tweezers. Then tilt the stone toward you to a forty-five

---

47 Each gem species— tourmaline, spinel, tanzanite, etc.— has a different density or specific gravity (s.g.). Carat is a unit of weight equal to one fifth of a gram. The higher the specific gravity, the denser the substance and the smaller it will be per carat. Some gems weigh more than others. A perfectly cut round one-carat ruby has a specific gravity of 4.00 and an optimum dimension of 6mm. A 6mm round tourmaline, by contrast, has a specific gravity of 3.06 and perfectly cut, will weigh only 0.87 carats, thirteen percent less.

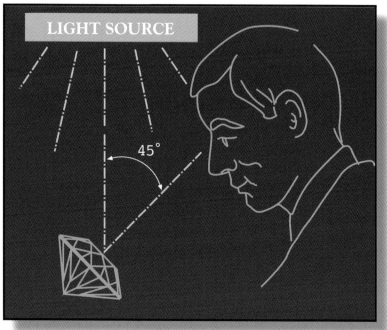

LIGHT SOURCE

45°

The correct viewing angle for evaluating the brilliance of a gem. At this angle, the lower half of the gem's face is evaluated, then the stone is rotated 180 degrees and the percentage of brilliance exhibited by the other half is measured. The sum of the two numbers equals the total brilliance of the gem.
© The Gemological Institute of America

degree angle toward the perpendicular.[48] This will allow viewing of the gemstone without your shadow intruding between the eye and the light source.

This procedure will allow you to observe the brilliance of the bottom half of the stone. At this viewing angle, the upper half of the stone will usually remain dark. Observe the bottom half of the stone: if half of the bottom half of the stone is reflecting light, it is twenty-five percent brilliant. To view the brilliance of the other half, rotate the stone one hundred eighty degrees with the bottom half again tilted toward you. Again mentally draw a horizontal line dividing the face of the gem

in half. Divide the total brilliance of the other half, in half and add the two figures together. The sum derived is the total brilliance of the gemstone. If the second half reflects light from one half of the surface, if added to the initial figure (25% + 25% percent = 50%) which yields the total brilliance of the gem. The remaining fifty percent of the gem may exhibit a window or show extinction; that is, the portion appears gray to black and exhibits no brilliance. This part of the face-up gem has no life; it is, in short, dead.

As a supplementary test of the stone's liveliness, rock the gem back and forth directly under the light source. Hold it just as you would if it was set in a ring. Gems are dynamic; they are not designed to sit still. When mounted in jewelry and worn, gems move relative to the light. Next, place the stone between the first and index fingers. Placing the stone between the fingers simulates the effect of prong-setting the gem. These procedures will give you an idea of how the gem will perform when mounted. If the stone is windowed, it will likely darken toward the center.

As previously mentioned, the lapidary sometimes purposely cuts a slight window in the center of a dark-toned gem, usually ruby or sapphire. A stone cut this way will look livelier in a jeweler's parcel containing a number of gems; viewing the stone between the fingers simulates the effect of setting the stone and closing the window. The gem should be wiped thoroughly with

---

48 Cabochon cut stones, including opal and phenomenal stones (stars, cat's-eyes, moonstones), must be viewed face up directly beneath the light source. As discussed in Chapter 5, the phenomena (star, cat's-eye) will not be completely visible if the stone is tilted toward the viewer.

a gem cloth before, during, and after this procedure. Slight amounts of dirt or grease from the skin on the stone's pavilion will materially diminish the liveliness of the gem.

When viewing a stone, avoid direct sunlight. Very few gemstones hold up well in direct sun and the glare caused by light reflecting from the surface of the facets makes it difficult to evaluate either the cut or the color. If natural light is used, turn your back to allow the light to diffuse around your body and hold the stone with the table of the gem at an angle directly perpendicular to the sun's angle and far enough away from the body so that its shadow will not intrude. This alternative method allows one hundred percent of the face of the stone to be viewed at one time. For the mathematically challenged, it also eliminates the need to use arithmetic to calculate the gem's total brilliance.

To summarize, the total area above the crown viewed in overhead lighting constitutes one hundred percent of the gem for the purpose of measuring brilliance. The percentage of the gem which is dark, or shows a window or lens effect, is deducted from the equation to arrive at the total brilliance of the stone. Barring truly disturbing faults in symmetry and proportion, a colored gem that exhibits brilliance over eighty percent of the crown is a one hundred percent success. As a standard of judgment, fifty percent brilliance is acceptable and over eighty is exceptional. Total brilliance under fifty percent classifies the gem as undesirable.

Proportion and Symmetry

In some gem species, color is unevenly distributed through the gem. This is particularly true of ruby and sapphire and amethyst. Bands of color alternating with zones of colorlessness are common in these gems. Often the cutter must sacrifice something in proportion and symmetry to mix the color so that it will appear uniform when the gem is viewed in the face-up position.

This can lead to overly deep or bellied stones, off-center culets, odd facet patterns, and unusual or asymmetrical shapes. These are all faults, but they are more acceptable in gem varieties which exhibit color banding, such as ruby and sapphire. Banding can usually be seen by viewing the stone through the side or by placing it face down on a white surface. The relative importance of these faults depends very much on how they affect the overall beauty of the stone when viewed face up. A stone with a striated or blotchy color appearance face up is far less desirable than one with odd proportions. In such cases, symmetry should be viewed as a minor fault.

In a perfectly cut stone, over ninety percent of the light entering through the crown bounces off the pavilion facets and is returned to the eye. In practical terms, the greater the percentage of a gem's crown that refracts and reflects light, the more successful the cut and the more beautiful the stone.

Two other cut factors not directly related to brilliance affect the overall desirability of a gem. These factors are *proportion* and *symmetry,* and they refer to the eye appeal and craftsmanship of the cut stone. The outline of the stone should be symmetrical. The facets should be symmetrical and evenly distributed. The girdle should not be noticeably thick. The other issue is proportion.

In the market, certain length-width ratios are preferred, affecting the value in some gemstones. In diamond, for example, the ideal length-width ratio for an emerald cut is 1.5-1.75. Stones with length-to-width ratios over 2.00 are considered to be "lean" looking, and those with of 1.25 to 1.10 are called "squarish." Deduction due to these strict ratios are mostly limited to colorless diamonds.

Due to rarity, colored stone proportions can be all over the map. Tourmaline and topaz are normally cut with length-to-width ratios of five to one and more. This is because the rough crystals are themselves long and narrow. Such stones are strikingly beautiful in part because of their long narrow outline. In these cases the extended shape has no effect on value. On the other hand, if the stone is supposed to be round and it looks more like a flat tire, this is not pleasing to the eye and will have a definite and dramatic effect on value. Symmetry faults are for the most part more detrimental than odd proportions.

Certain shapes will command a premium within specific gemstone species. For example, ruby and sapphire are normally cut as ovals, cushions, and pear shapes simply because the outline of the rough lends itself to these shapes. Round shapes are rare, potentially more brilliant,

18k yellow and white gold brooch by R. W. Wise, Goldsmiths featuring a quartz holographic cut gem by Tom Munsteiner. The bubble image is cut into the back of the stone. Photo: Jeff Scovil, courtesy R. W. Wise, Goldsmiths.

and command a premium in most gems. Emeralds, a notable exception, are usually shaped in the octagonal step cut, also called the classic emerald cut, not only to retain maximum yield from the rough but also because many connoisseurs rightly believe that this shape maximizes the satiny beauty of emerald. Round emeralds actually sell at a discount.

Gems cut to standard or calibrated sizes often are priced at a premium for reasons unrelated to the beauty of the gem. Such stones are referred to as calibrated.

Calibrated gems fit in standard-size jewelry mountings and are, therefore, in great demand by commercial jewelers. A collector can often find an uncalibrated stone at a much more attractive price simply because it will not fit into a standard commercial setting. Non-calibrated stones present a buying opportunity for the aficionado. Non-standard stones can be set by a skilled goldsmith at a price somewhat higher than that of setting calibrated stones. Moreover, handmade settings can be a good investment. In many cases the skilled craftsman will make a setting which enhances the beauty of the stone. This is not any sort of fraud. Improving the look of a set gemstone is part of the goldsmith's art. By experimenting with the angles of a handmade setting it is possible to correct light leakage and make the stone both more brilliant and improve the apparent color. This also suggests that the aficionado should be very cautious about purchasing a mounted gemstone. Beware the jeweler who is unwilling to dismount a pricey prong-set gem for inspection.

Finely cut stones do command a premium. Though poorly cut gems are of little or no interest to the connoisseur, the price of a colored gem will normally be discounted ten percent for a gem which exhibits less than sixty percent brilliance. Stones with brilliance less than forty percent are discounted fifteen percent and more. Colored gems that show more than eighty percent brilliance will command

a ten to fifteen percent premium.[49] Price differentials for cut in colorless diamonds can vary as much as seventy percent between fine and poorly cut stones. A detailed discussion of diamond cutting can be found in Chapter 5.

New cutters, New Rules

For the last four hundred years, the major preoccupation of the lapidary arts has been the pursuit of faceting designs, which optimize brilliance. This has led to a sort of tunnel vision, continually narrowing, so that by the beginning of our own century, the lapidary was concerned almost exclusively with the cutting of the four basic symmetrical shapes: round, oval, step cut, and pear.

In the early 1980s, the German master lapidary Bernd Munsteiner introduced a new style of gem cutting to the United States. Immediately dubbed "Munsteiners," or "fantasy cuts," these gemstones were fashioned with asymmetrical outlines and faceting patterns more reminiscent of optical sculpture than traditional gemstones. Though ridiculed by conservatives, innovative jewelry designers and consumers embraced Munsteiner's fantasy cuts. After four centuries, the market was hungry for something new. Though little noted at the time, a revolution had begun, a lapidary

---

49 Ted Themelis, *Mogok: Valley of Rubies and Sapphires* (Los Angeles: A&T Publishing, 2000), p. 213. Themelis refers specifically to premiums charged for ruby and sapphire by the gem dealers of Mogok, Burma; that is, in the oldest gem market on earth. The range described in this book is more general and based on average premiums charged around the world.

Vision in Motion! Margaret De Patta, pendant w/rose quartz "opticut" by Francis J. Sperisen. The white gold shape is distorted and moves and changes shape as either the wearer or observer moves. Courtesy Oakland Museum

renaissance, and the only major change in gem-cutting philosophy since the cabochon gave way to the point cut.

Encouraged by Munsteiner's success, a whole generation of new cutters has emerged. The term "new cutters" is used simply for lack of a better one. These are lapidary artists, mostly Germans and Americans, whose chief interest is not just cutting a well-made brilliant, but creating a tiny work of art.

Given recent history it is natural to conclude that the creative cutting movement has its roots in Germany. This is incorrect! The father of the new cutting movement, who prefigured Munsteiner by more than fifty years, was not a German, but an unassuming American pioneer by the name of Francis J. Sperisen.

Beginning in 1941, Sperisen, a self-taught lapidary working out of a studio in San Francisco, began collaboration with Margaret De Patta, a metalsmith, who is today considered the doyenne of American art jewelers. Sperisen worked with De Patta, cutting unusually shaped gemstones to complement her jewelry.

De Patta was herself a student of the constructivist artist and founder of Chicago's Design Institute, Laszlo Moholy-Nagy. Moholy-Nagy, an important Hungarian-born artist trained at the Bauhaus, established a training curriculum for young artists which emphasized experimentation and the use of new materials such as thermoplastics. He also had an abiding interest in motion and light he called "vision in motion." It is apparent that the fruit of the Sperisen/De Patta collaboration was firmly based on Moholy-Nagy's aesthetic. He believed that light may be considered an artistic medium equal to color in painting and sound in music. He suggested that De Patta, "catch your stones in the air. Make them float in space. "Don't," he advised her, "enclose them."[50]

De Patta's most famous piece, currently in the Oakland Museum of Art, she made a pendant in ebony and white gold in the

50 Ursula Ilse-Neuman and Julie M. Muniz, *Space Light Structure, The Jewelry of Margaret De Patta*, Museum of Arts and Design, New York, (2012), pp. 31-40

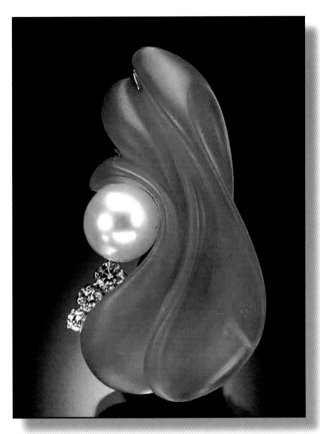

A semi-translucent Mojave Blue Agate gem sculpture by Steve Walters set w/pearl and diamond by R. W. Wise, Goldsmiths. Photo: Jeff Scovil, courtesy R. W. Wise, Goldsmiths.

shape of the inverted letter "Y", mounted behind — and meant to be seen through — a Sperisen double lens cut quartz (pictured). This gemstone, the precursor of the modern opposed bar cut, was deliberately cut as a lens (windowed) and exhibits no brilliance. The effect was a conceptual *tour de force*, a realization of what art critic Norbert Lynton calls "the Constructivist ideal of massless sculpture inserted into space." As the viewer or the pendant moves, light, distorted as it passes through the stone, causes the metal mounted behind the stone

to appear to move and change shape.[51]

The connoisseurship of gem sculpture requires that the collector look beyond the traditional standards of cut. Color, clarity, and crystal are as important in gem sculpture as they are in any other gem, but the intention of the lapidary artist differs from the usual objective of simply maximizing brilliance. In some cases the artist is looking to create internal mirrors or other holographic effects within the gemstone that are, in a sense, a direct contradiction of the traditional objective; brilliance is not the primary intention. Munsteiner's gems, along with those fashioned by a number of American lapidaries, are designed to be placed in settings while retaining versions of the traditional girdle, crown and pavilion. These gems should be evaluated face up. Others are true sculptures and should be evaluated in all three dimensions.

When evaluating gem sculpture it is important to remember that they are works of art; as in the fine arts, balance, proportion, and the intention of the artist must also be considered. The stone may be beautiful but the effect cannot be judged simply by measuring the total amount of brilliance. A window, for example, may be part of the artist's intention. Body color

51 Richard W. Wise, "A Cut Above: The New Cutters and The American Lapidary Renaissance", *National Jeweler Magazine*, (July, 1996) c.f. p.40.

may be as important as key color, and crystal, as we shall see, is of real import. The connoisseur must retain an open mind. The question is, has the artist been successful? Does it work? Beauty, however, remains the defining criterion.

The Cabochon: Judging the Cut

The word cabochon is derived from the French meaning "head." It is the oldest cut, being little more than a stream pebble flattened on one side and polished. Judging a cabochon requires a slightly different set of rules. Cabochons or "cabs" have no facets and are designed to transmit rather than reflect light. A fine cabochon does not sparkle, it glows. In modern times the cabochon has fallen into disfavor; for the most part, only opaque gem material, star and cat's-eye stones, and stones judged to have too many visible inclusions to be faceted, are cut en cabochon. In a faceted stone, light is reflected internally. Multiple internal reflections will often cause visible inclusions to appear doubled or tripled like an image in a fun-house mirror. Because a cabochon transmits and does not reflect light, this sort of doubling does not occur. A cabochon ordinarily will be priced significantly lower than a faceted stone of the same gem variety.

A well-made cabochon which is not overly included can be a very beautiful gemstone. The fact that cabochons may be purchased at significantly lower prices than those asked for a faceted stone of the same gem variety should not be seen as a deterrent, but rather viewed as a buying opportunity. The true aficionado will approach each gemstone as through the eyes of a child as if each were a thing entirely new.[52]

Crystal: a Defining Criterion

Cabochons may be cut in a variety of ways: high domed or low domed with flat, concave, or convex backs. None of that really matters very much — what matters is the result! A well-cut cab will glow with characteristics similar to a stone that has been faceted. The strength of the glow is substituted in the evaluation equation for the percentage of brilliance in a faceted stone. It is the degree of transparency, or what I have chosen to call crystal (see Chapter 5) that is the key issue here. If the stone has good crystal and is well cut, it will glow. Cabochons, particularly phenomenal stones such as stars and cat's-eyes, must be examined with the light perpendicular to the girdle of the gem; that is, directly

---

52 I once employed a European-trained master goldsmith; he was an older man who, over the years, had worked for some of the finest jewelers in the world, including Harry Winston. He knew a good stone when he saw one. He showed me some of his jewelry, which included stones that had numerous visible inclusions. "Why," I asked him, "do you use such flawed stones in your work?" "Ja, ja," he replied, "I know they are not so good but I like these stones because the inclusions are like internal facets, they make the stone more interesting and, I think, more beautiful. And, besides, they are much cheaper." Jürgen Sierau, personal communication, 1988.

overhead. The two remaining Cs, clarity (visible inclusions) and color, are judged in the same way as faceted stones. In the case of the cabochon, however, the color evaluation is based on the body color: the hue, saturation, and tone of the light transmitted by the stone. There are no facets, therefore no key color to be evaluated.

Clarity: Blemish or Beauty Mark?

Clarity is a concept that is easily understood. No complex analysis is necessary, just the use of the eyes.

Inclusions and Flaws

"Rather than regarding inclusions in colored stones as harmful, in small sizes and numbers that do not in any way detract from their beauty they should be regarded as adding to desirability, for they provide identifying characteristics."

–GIA Colored Stone
Course,1980

Gem formation is a dynamic series of processes which continue as the crystals form. Cracks, foreign bodies, minerals, even other crystals are normally *included* in this process. Hence the word *inclusion*, the technical term gemologists use to describe these tiny intruders. The world inside a gemstone is a complex and

fascinating one, a miniature self-contained universe. Inclusions tell the story of a gemstone's genesis and often provide clues as to where and how it was formed. It is, in fact, inclusion study which allows gemologists to identify and separate rubies formed in Burma, for example, from rubies formed elsewhere.

Note that the more common term, *flaws*, has not been used in this discussion. This is because inclusions become flaws only when their presence materially affects the beauty and/or the durability of a gemstone. This bears repeating: inclusions become flaws if their presence is substantially detrimental to the beauty and/or the durability of the gem.

The aficionado must understand that there is a real distinction to be made between inclusions and flaws. In fact, the

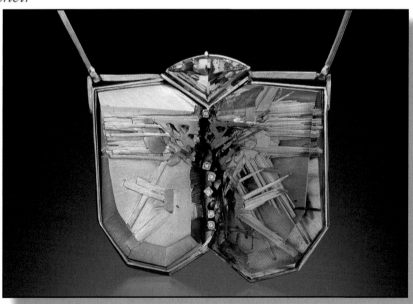

Inclusions play a central role in this contemporary pendant. The right half of the piece is a gem sculpture in quartz. A combination of hair-like rutile and grainy black graphite inclusions are the centerpiece of this flat asymmetrical gem. The goldsmith's work echoes the pattern in the plaque on the left. Quartz sculpture by Glenn Lehrer. Photo: Jeff Scovil; courtesy of R.W. Wise, Goldsmiths, Inc.

use of the term flaw should be avoided. *Inclusion* describes the situation; "flaw" is pregnant with negative meaning. As has been said throughout this volume, a gem is all about beauty; Beauty is its sole raison d'être. A small inclusion, or a series of them, invisible to the naked eye and does not create any sort of defect possibly leading to breakage may normally may be ignored in the equation.

In some cases, inclusions make a major contribution to the beauty of a gemstone. For example, the "star" in a star sapphire (see Chapter 5) is produced by inclusions of rutile crystals, long thin hair or straw like structures without which there could be no star. In other cases, rows of microscopic breaks in the gem produce the very desirable billowing moon-glow effect in moonstone whercas myriads of tiny dust-like inclusions create the signature "sleepy" glow of Kashmir sapphire.

Inclusions are also very helpful in determining how, where, and when the gem material was formed. Important questions such as country of origin and natural versus synthetic formation may be answered by studying inclusions.

There are two standards for judging clarity: the visual, using the naked eye; and the microscopic, using ten power magnification to examine the interior world of a gem. The microscopic standard is used for colorless diamonds; the visual standard for all other gems, including,

to some degree, fancy color diamonds.[53] Diamonds that are flawless under ten power are termed *loupe clean* or *flawless*. This leads to a question: if a diamond has a tiny inclusion invisible to the eye and barely visible under 10X, is the stone flawed?[54] The answer is technically yes, even though the inclusion *does not materially affect the beauty or durability of the given gem.* (For a full discussion of diamond clarity grading, see Chapter 5.) With colored stones the issue is straightforward. Colored stones with no visible inclusions when viewed by the naked eye are termed eye-clean or eye-flawless.[55] Visual inclusions will result in most cases in a discount from the price of an eye-clean gem. The amount of the discount depends on how much an inclusion disturbs the eye and affects the beauty of the stone. This judgment involves an individual decision

---

53 The rarer the color, the less influence clarity has on price. It is relatively more important in browns and yellows and of little importance in reds and pinks. See John King, et al., "Exceptional Pink to Red Diamonds: A Celebration of the 30th Argyle Diamond Tender", *Gems & Gemology*, The Gemological Institute of America, (Winter 2014), p. 278.

54 Prior to the invention of the microscope by Robert Hooke in 1665, the eye standard was used for all gemstones.

55 Current Federal Trade Commission guidelines make no distinction between grading clarity in colored stones and diamonds. Using the term *flawless* to describe an eye-clean sapphire that has visible inclusions under 10X magnification is technically a violation of federal law (FTC 23.26). In order to remain on the good side of the FTC, I will use the term *flawless* to describe gems that have no visible inclusions under 10X magnification and *eye-flawless* will be substituted for the term *flawless* to describe the clarity of colored gemstones.

which the aficionado must make when considering a purchase of a visibly included gem. From a connoisseur's perspective, no gem, with the exception of emerald, can be judged fine if it has eye-visible inclusions.

It is a good idea for the budding aficionado to become proficient with the jeweler's loupe. This basic instrument is available in many shapes and sizes and strengths. The ten-power or 10X loupe is the most practical because it has enough depth of field to be able to keep the whole gemstone in focus while providing sufficient magnification to see inside it.

Chapter 4
Connoisseurship: Secrets of the Trade

"Nor trust o'ermuch the treacherous candleshine, Your eye for beauty's warped by night and wine. When Paris judged the three and gave the prize to Venus, there were clear and cloudless skies. Night hides each fault, each blemish will condone, The hour can make a beauty of a crone. To daylight pearls and purple gowns refer; Of face and limb let day be arbiter."

–Publius Ovidius Naso (Ovid), 43 BC -18 AD

Color: Primary Hues

Dividing color into its primary and secondary hues is very useful and good practice for the aficionado. In many, though not all, cases, colored gemstones are evaluated by how closely the color of the stone approaches one of the eight spectral hues. This is particularly true, (as will be discussed in the appropriate chapters), with ruby, emerald, tsavorite and spessartite garnet, tanzanite, and sapphire. The very finest ruby would be, in theory, one with a primary hue of one hundred percent red with no secondary modifying hue. In nature few stones even begin to approach this ideal. A gem with a primary hue of eighty-five percent or better will be a high scorer, provided that it has vivid saturation and proper tone for its type. A visually pure orange sapphire will be more desirable than a slightly yellowish orange sapphire. In short, a gem which exhibits a visually pure single hue is usually more desired that one which shows both a primary and secondary hue. This is a generalization with a number of exceptions, as will be discussed in the sections on

padparadscha sapphire, malaya garnet, alexandrite, and topaz.

The Visualization of Beauty: Light

Gemstones are creatures of the light. Not one of the qualities discussed thus far can even be seen, let alone appreciated, without light. Jewelry store shoppers can often be heard to say that such and such stone looks great in the store display, implying that the use of proper lighting is somehow misleading. No one would try to read this book in the dark, yet who would suggest that the book's content, the ideas contained within these pages, cease to exist when the lights are turned off?

In preindustrial times the world was much simpler. There were only a few kinds of light: sunlight, moonlight, and firelight. Thanks to modern technology, we now can choose from many different types of lighting. Incandescent light is light produced by a flame (lamplight, candlelight, firelight). Daylight is light from the sun or from a fluorescent lamp expressly designed to reproduce it. Specialized types of incandescent and fluorescent lamps have been developed which emphasize specific

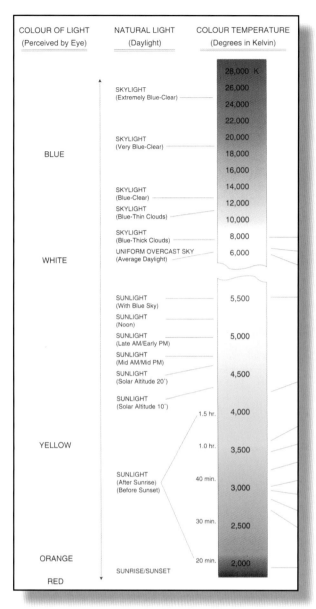

COLOUR OF LIGHT (Perceived by Eye)	NATURAL LIGHT (Daylight)	COLOUR TEMPERATURE (Degrees in Kelvin)
		28,000 K
	SKYLIGHT (Extremely Blue-Clear)	26,000
		24,000
		22,000
	SKYLIGHT (Very Blue-Clear)	20,000
BLUE		18,000
		16,000
	SKYLIGHT (Blue-Clear)	14,000
		12,000
	SKYLIGHT (Blue-Thin Clouds)	10,000
	SKYLIGHT (Blue-Thick Clouds)	8,000
WHITE	UNIFORM OVERCAST SKY (Average Daylight)	6,000
	SUNLIGHT (With Blue Sky)	5,500
	SUNLIGHT (Noon)	
	SUNLIGHT (Late AM/Early PM)	5,000
	SUNLIGHT (Mid AM/Mid PM)	
	SUNLIGHT (Solar Altitude 20°)	4,500
	SUNLIGHT (Solar Altitude 10°)	
	1.5 hr.	4,000
YELLOW	1.0 hr.	3,500
	SUNLIGHT (After Sunrise) (Before Sunset) 40 min.	3,000
	30 min.	2,500
ORANGE	20 min.	2,000
RED	SUNRISE/SUNSET	

Image courtesy: Stephen C. Hofer

portions of the visual spectrum, and each projects light of a different color temperature. Light no longer simply illuminates, it also creates. Lamps that are slightly yellowish are used in the home-furnishing industry to add a warm and fuzzy feeling to displays. Pinkish lighting in beauty parlors adds a healthy glow to the client's skin. With the introduction of the LED, it is possible to alter the lighting temperature and thus the color of the light.

Historically, daylight — specifically north daylight or light coming through a north-facing window at noon — has been the standard for evaluating color in gemstones. This is because noon daylight is relatively white and therefore balanced. However, gem dealers have long recognized that the quality of daylight changes at different times of the day and in different parts of the world. North daylight in the Sonoran desert is qualitatively different from north daylight in Bangkok.

In fact, the quality of natural daylight is affected by several variables, including geographic latitude and air quality. In cities with high levels of air pollution, such as the gem capitals of Bangkok and New York, daylight is noticeably more yellowish. In addition, the relative strength and color composition of daylight changes as the day progresses. "Don't buy blue sapphire after two in the afternoon" is a bit of advice I heard on my first buying trip to Bangkok. This dealer's truism is based on the fact that as the day progresses the color composition of daylight also progresses, from red through yellow and into the blue range. The hue of a blue sapphire viewed in natural light late in the afternoon will appear more highly saturated than the same stone viewed in the yellowish light of midmorning.[56]

The color temperature of light is measured in kelvin units. Daylight ranges from 2,000 to 6,000 kelvin; that is, from red to orange to yellow then white at noon

56 Viewed in daylight, red, orange, blue, and violet gemstones look best early in the morning and late afternoon. Yellow, green, and purple, and colorless gems look their best during the midmorning and midafternoon.

and following the reverse order back to red as the day comes to a close at sunset. Noon daylight is the most balanced. North daylight at noon falls between 5,500-6,500 units kelvin.[57]

A number of grading laboratories now use fluorescent lamps emitting light within the above parameters. Several companies produce daylight-equivalent fluorescent and LED lamps used by gem professionals for grading and display purposes.[58] This type of lighting is available at relatively low cost and fits standard fluorescent fixtures. The availability of daylight-equivalent lighting creates the opportunity to control and stabilize the viewing environment. If gems are consistently viewed in a single lighting environment, one variable, at least, is removed.

Although the old fashioned incandescent light bulb has been outlawed in Europe and being phased out in the United States, jewelers and gem dealers prefer and continue to use an incandescent

spotlight because spots, like the sun, throw a concentrated beam of light which maximizes the brilliance of a gem. LEDs have begun to replace these incandescent spots. Fluorescent lighting is diffused and produces a brilliance that is relatively subdued by comparison. Fluorescent light of sufficient power to show off a gem will light up a retail store like the inside of a fish market and the result is not generally flattering to precious metals or to a woman's skin. Standard incandescent lighting, light from an old fashioned light bulb, is rated at around 3,200 kelvin; this means that incandescent light is distinctly yellowish[59]. Fluorescent lighting, by contrast, can be adjusted to any kelvin temperature from a balanced white (5,500-6,500 kelvin) to distinctly bluish (7,000+ kelvin).

Even though daylight — specifically north daylight between 5,500 and 6,500 kelvin — is the standard, colored gems should be viewed in incandescent light as well as in daylight or daylight-equivalent fluorescent lighting. When evaluating a gemstone, the aficionado should ask to see the gem in both types of lighting. This is a good idea even if, as is often the case, dealers have daylight-equivalent lighting in their offices or, at gem shows, inside their booths.

Ideally a gemstone should be beautiful in all types of lighting. It is not necessary to remember all the technical aspects of lighting. However, it is important for the

---

57 Facing north toward blue sky adds blue to the color of daylight.

58 Duratest Corporation markets a daylight equivalent called "vitalite" (5,500 kelvin) which is used by the American Gemological Lab. The MacBeth division of the Kollmorgen Corporation produces several daylight-equivalent fluorescent lamps rated between 5,500-6,500 kelvin. The Gemological Institute of America uses 6,500-kelvin lighting manufactured by MacBeth in all its grading environments. The American Gemological Lab uses Duratest vitalite for grading colored gemstones. Recently, System Eickhorst, a German firm, introduced the lighting system Dialite Pro X, based on my recommendations. This system uses Solux 4,700-kelvin halogen lights coupled with 6,000-kelvin fluorescent tubes, making it possible to view a gem in a good average daylight environment. See Richard W. Wise, "Light Up Your Life," *Asia Precious Magazine*, Guide to Industry Services, (1998), pp. III-V.

59 New types of low-volt incandescent bulbs have recently been introduced with kelvin temperatures as high as 5,000 bringing them close to the daylight standard.

aficionado to understand that lighting quality may vary due to a number of factors, to be aware of these variables, and to observe the gem in all types of lighting as part of the evaluation process.

Diaphaneity or Crystal: The Fourth C

"So long as the depth of its colour does not interfere with the transparency of a sapphire, the darker it is in colour the more highly is it prized."

–Max Bauer, 1904

Gemstones are divided into three classifications: transparent, translucent and opaque. With a few notable exceptions, the gem species and varieties discussed here fall into the transparent category. It seems self-evident that degree of transparency would be a consideration in grading gemstones in this category, particularly when it comes to questions of quality and beauty.

"Crystal" is probably the most misused word in the gemstone lexicon. Technically, a crystal is a substance with a definite atomic structure. Most gemstones are cut from crystals. Scientifically speaking, glass is the opposite, a substance without a definite crystal structure. Still, when the words "crystal clear" are used to describe certain types of glass, which are by definition noncrystalline, the meaning is, well...clear! In gemstone connoisseurship, crystal, the common term for diaphaneity, refers to a limpid quality, an ultra-transparency which very fine gemstones will exhibit. Crystal

becomes increasingly important as we ascend toward the pinnacle of gemstone quality.

A Gem of the Finest Water

"It is interesting to note that the terms FINEST WATER and PURIST WATER...mean not only a diamond free from any color, and freedom from flaws, but also one of exceptional transparency...like a drop of pure spring water."

–Robert M. Shipley, 1936

The idea that diaphaneity, transparency, or crystal is an important criterion in the evaluation of gemstones is not new. It is one, however, that has been largely ignored in the development of modern, so-called scientific gem grading systems. In earlier times, the combination of color and transparency was referred to as the gem's water. There was a hierarchy: a gem was classified as first water or second or third water.[60] Gems that had both poor color and little transparency were referred to as byewater. The most exemplary stones were

---

60 Lewis Feuchtwanger, *Treatise On Gems,* (Hanford, N. Y., 1838, pp. 60-61. Feuchtwanger makes the point. "The value of polished Diamonds depends on the following conditions...2nd The Purity, Faultlessness and Transparency."

said to be "gems of the finest water."[61]

As early as the 4th century BC, Kautilya, the ancient Indian Machiavelli, lists among the qualities of a "good gem . . . [that it be] transparent and reflecting light from inside."[62] Sometime around 1433 the Chinese admiral Ying-yai Sheng-lan uses the term water to describe a particularly "clean, clear" quality of amber.[63] The famous Jean Baptiste Tavernier, 17th century gem merchant, used the phrase *gem of the finest Water* to describe exceptionally transparent diamonds, gems, and pearls he encountered on his six voyages to India.[64]

Crystal: That Warm Fuzzy Feeling; The Exception That Proves The Rule

Perhaps due to the fact that this criterion is so obviously self-evident, the gemological community has been hesitant to embrace crystal or transparency as an important grading criterion. Since the book's first publication, several reviewers have either misunder-

"A Gem of the finest water." Blue sapphire is often deep and dark. This 2.05-carat Ceylon sapphire exhibits fine color and exceptional transparency, diaphaneity or crystal. Photo: Jeff Scovil, courtesy R. W. Wise, Goldsmiths.

stood or criticized this concept, often both at once. One reviewer suggested that crystal is really a subset of clarity;[65] the problem with this argument is that clarity is, by definition, all about inclusions. If a diamond has no inclusions under 10x magnification, it is, by definition, flawless. However, it is possible for a gemstone that is technically flawless to appear murky or sleepy. In such cases, the gem simply is lacking in transparency. Another reviewer suggested that crystal was totally subjective, forgetting that the terms *transparent, semi-transparent, and translucent,* etc., have very specific meanings in the science of gemology. Sometimes, as with certain types of sapphire and spinel, this is a result of a large number of microscopic inclusions barely visible under extremely high

61 In the 6th edition of the *Dictionary of Gems and Gemology*, published by GIA, Robert M. Shipley defines "water" as a term "occasionally used . . . as a comparative quality designation for color and transparency of diamonds, rubies and other stones. . . ." p. 243. Color and transparency together equal "water." Shipley goes on to mention the same hierarchy as in Feuchtwanger — first water, second water, etc. Clearly even from earliest times, the quality of diaphaneity was recognized as an important and indispensable criterion in gemstone connoisseurship. See also, Jeffries, David, *Diamonds and Pearls*, 1750, 4th edition, p.73

62 Rangarajan, *The Arthashastra*, pp. 777-778.

63 Ying-yai Sheng-lan Ma Huan, *The Overall Survey of the Ocean's Shores*, 1433 (Bangkok: White Lotus Press, 1970 reprint), p. 111.

64 Jean Baptiste Tavernier, *Travels In India*, 1676, trans. V. Ball (New Delhi: Munshiram Manoharlal Publishers Pvt. Ltd., 1977), Vol.II, p.98 .

65 Boehm, Edward. "Book Review: Secrets Of The Gem Trade." *Gems & Gemology*, (Summer, 2004).

magnification.

Perhaps the most obvious objection is to be found in the legendary sapphires of Kashmir, the finest of which exhibit a sleepy or fuzzy scintillation and transparency resulting from light scattering as it is refracting through, and reflecting from myriad (almost sub-microscopic) inclusions within the gem. One expert has suggested that Kashmir sapphire is the exception which proves that crystal lacks the necessary universality to be considered the fourth C[66]. If so, it is the exception that proves the rule. For Kashmir sapphire to exhibit that wonderful glow, a certain density of the characteristic tiny inclusions must be present together with *a strong underlying transparency.* If the stone has reduced transparency, no glow! Too few inclusions, no glow; too many inclusions and a reduced transparency again, no glow. Transparency is an absolutely necessary condition for the phenomenon to exist.

When we say a stone is limpid, clear, or crystalline, we are talking about a quality quite distinct from clarity. Connoisseurs have likened

this quality to a "dew-drop" (Streeter), "a clear mountain stream" (Shipley, Zucker, Bauer) or "light passing through a vacuum" (Hammid). Often you will hear dealers refer to such stones as simply "having good crystal" (Belmont).

From a strict grading perspective,

A crystal with Crystal. A highly diaphanous 9-carat natural emerald crystal from the La Pita Complex, Boyacá, Colombia. Photo: R. W. Wise, courtesy R. W. Wise, Goldsmiths.

66 Richard W. Hughes, personal communication, 2011. Mr. Hughes is the author of the seminal work: *Ruby and Sapphire.* He is particularly fond of the "light scattering inclusions," (*ibid,* p.213) which are responsible for the warm and fuzzy glow in Kashmir sapphire and which, when present, dramatically enhance the gem's price in the marketplace. Hughes maintains that this phenomenon reduces transparency, and given that single exception, disqualifies transparency, crystal or diaphaneity as a universal criterion in the judgment of gemstones. He ignores the fact that "misty" simply describes a quality of crystal relevant specifically to just one type of sapphire.

the inclusions which create the fuzzy appearance in some gemstones are not visible or at least not resolvable under 10X magnification and have no effect on the clarity grade given a particular stone. In the case of Golconda or Type

IIa diamonds, the presence or absence of the ultra-transparency so characteristic of these gems would have absolutely no impact on the clarity grade listed on a laboratory grading report, but is a quality readily recognized by diamond connoisseurs.

Tiny inclusions are one, but only one, of the possible causes of poor crystal. Poor diaphaneity has several causes. Consider a strongly blue fluorescent diamond. Such stones are often described as visibly "oily", due to a loss of transparency when the diamond fluoresces in natural daylight. Fluorescence will be noted in a grading report, but its presence or absence cannot be said to affect the stone's clarity grade.

Many varieties of gemstones tend to lose something when viewed in certain lighting. Incandescent light is the usual culprit. Most varieties of tourmaline and garnet and some varieties of corundum "close up," "muddy," or "bleed color" when exposed to the light of an ordinary light bulb. The cause of these phenomena is not completely understood though none appear to be caused by the presence of inclusions in the gemstone. The point is, the gem loses some transparency and some of its beauty. Therefore, the additional criterion, crystal, is needed to describe this quality.

Crystal must be judged in various lighting environments. Different types of light have distinct color temperatures. As described above, north daylight at noon, the traditional gemstone-grading standard, is balanced between yellow and blue at 5,500 degrees kelvin. As kelvin temperature decreases, light becomes yellower; as the temperature increases, the light becomes

bluer. Incandescent light at 3,200 kelvin is distinctly yellowish and most LED lighting is distinctly blue. The lighting temperature determines the color of the light, and that in turn impacts the visual appearance of the gem being viewed in that light.

The tendency of gems to change appearance, or to lose color, between natural daylight and incandescent light has traditionally been called "bleeding." In blue sapphire, for example, one of the qualities that makes a Kashmir stone so desirable is that it doesn't bleed color. Due to an absence of chromium, the color of a fine Kashmir sapphire will remain unaltered as the lighting environment is changed. The color of blue sapphire from other sources may wash out; that is, lose color (hue, saturation, and tone) in bright sunlight.

The appearance of some gemstone varieties, including some ruby and sapphire and most varieties of garnet and tourmaline, will also change as the lighting environment changes. However, it is not quite accurate to use the term bleeding to describe the result. I doubt if any jewelry professional who regularly works with colored gemstones has failed to notice these changes. And although language lacks precision there is little choice but to use it to describe the visual effect. Tsavorite garnet seems to close up (lose transparency) in incandescent light while rhodolite garnet turns muddy and brownish. Green and blue tourmaline pick up a gray mask and appear dull and sooty like the chimney of an oil lamp. Pink to red tourmaline acquires a muddy brownish mask which reduces transparency. Not all the effects are

Three green tourmalines photographed in incandescent light. The two stones at either end maintain their transparency, the center stone closes up and turns sooty. Note the elongated bow-tie shaped black bands of extinction in the stones to the left and right. Photo: Jeff Scovil, courtesy R. W. Wise, Goldsmiths.

negative, aside from its loss of transparency: Thai ruby turns a purer red, losing its purplish secondary hue, when viewed in incandescent light.

These changes affect not only color (hue, saturation, and tone) but crystal as well. Such effects are general, but not universal. For example, ninety-eight percent of all rhodolite garnet will muddy, or turn brownish, losing both transparency and color saturation in incandescent light. This leaves only about two percent retaining both color and crystal under the light bulb. All other Cs being equal, if the stone is of high color, clean and well made, this two percent constitutes the crème de la crème of rhodolite garnet. The same may be said for pink tourmalines which do not muddy, green tourmaline that does not gray out, and tsavorite which retains its open color in incandescent light.

Crystal becomes relatively more important in higher quality gemstones; however, it is fair to say that the diaphaneity of diamond with a clarity grade of I3 is not particularly significant. Blue sapphire is a good example. Fine sapphire, particularly heat treated blue sapphire, is tonally quite dark, so dark that flashes of color seem to rise to the surface like a gas bubble erupting from a tar pit; under the table there is often no life at all.

One collector with a very good eye describes sapphire as having a "heart of darkness."[67,68] Lapidaries will sometimes intentionally cut a window under the table in an attempt to lighten such a stone. An exceptional sapphire which has deep color together with good crystal is entirely different. Visually the gem is like a deep well, blue and bottomless. Such a stone will glow even when viewed from the side and its price will be dramatically higher.

"No pearl is completely transparent, but among different specimens there are many degrees of translucency. Different qualities of pearls are described as in the case also of diamonds, as being of different "waters," and the differences between them depends upon the amount of light transmitted in each particular case".

–Max Bauer, 1904

Historically crystal has also played a part in the discrimination of the finest

67 Cora N. Miller,: Personal Communication, 2009

68 Joseph Belmont, Personal Communication, 1998

pearls. Before the introduction of cultured pearls, (most of which are seeded with an opaque sphere ground from the shell of a freshwater mollusk), transparency, or at least translucency, was a characteristic very much valued in the finest pearls.[69] In his 17th century *Travels In India*, gem merchant Tavernier describes the world's paramount pearl (circa 1670), a gem at that time in possession of a minor prince of Muscat. "This prince possesses the most beautiful pearl in the world, not by reason of its size, for it only weighs 12 1/16 carats, nor on account of its perfect roundness; but because it is so clear that you can almost see the light through it."[70] Tavernier repeatedly uses the term water to describe the quality of pearls.[71]

As we have shown, diaphaneity, transparency, or crystal is a necessary

---

69 George Frederick Kunz and Charles Hugh Stevenson, *The Book of The Pearl: The History, Art, Science and Industry of the Queen of Gems*, 1908, (reprint ed., New York: Dover Editions, 2003), pp. 370-371. Kunz suggests that the method to be used to "know good pearls" is to let a ray of sunlight fall on the pearl at an angle while the gem is held in a black velvet cloth. The light will penetrate into the skin and reveal if the pearl is speckled or has any defects.

70 Tavernier, *Travels*, vol. 2, p. 87. A pearl of this weight would probably be no more than 10mm in diameter. Most cultured pearls contain an opaque bead implant made from the shell of a Mississippi River clam. A relative lack of transparency (crystal) is probably the only visual qualitative difference that can be found between natural and cultured pearls.

71 Transparency in pearls can often be observed in antique jewelry. Pearls in jewelry made before 1930 will, in all probability, be natural. Often such pearls, particularly small ones, will show a distinct translucency; the Japanese use the term "wet" to describe this look. This is a phenomenon unknown in their cultured cousins except, as we shall see in Chapter 24, in tissue-nucleated Chinese freshwater pearls.

grading criterion that deserves more than just a footnote in the discussion of quality in gemstones. Several factors including microscopic inclusions, ultraviolet fluorescence, and the color of the lighting environment may affect crystal.

The distinctions made here are real in that they reflect observable phenomena which affect the beauty and desirability of gemstones. They are real also because they reflect demonstrable price differentials in the marketplace. From a grading perspective, crystal is a distinct and vitally important criterion; without this standard, it is impossible to describe the finest gemstones adequately. In short, crystal is the true fourth C of gemstone quality evaluation and connoisseurship.

The three Cs of gem grading; (color, cut, and clarity) are a good shorthand method to categorize the basic criteria used in the evaluation of beauty in gemstones. These three Cs are like the "three Rs" — they are the bedrock basics. The addition of the fourth C, crystal, completes the basic curriculum. All four are necessary conditions and together they are almost sufficient. To be considered fine, a given gemstone must achieve high marks in all four, a low grade in any one means that the gem you are looking at is not a fine example. In next section the discussion turns to the more subtle nuances of connoisseurship.

The Face-up Mosaic: Multicolor Effect

Variously known as pleochroism, dichroism and dichroic effect, these terms refer to the fact that the face-up mosaic (key color) of a faceted gemstone is far from uniform.[72] It is a complex mosaic, a shifting crazy quilt of color. Some facets are bright, others dull; some one color, some another. In the connoisseurship of gemstones, multicolor effect is simply the tendency of a gemstone to show multiple hues or tonal variations of the same hue when viewed face up. [73]

As every gemologist knows, light behavior – refraction, reflection, and total internal reflection – is a function of proportion and cut. One light ray may enter the gem through the crown, bend and exit through the pavilion. Other rays will totally reflect, remain inside the gem, bouncing like a pinball around the interior facets and eventually exiting through the crown. The behavior of each light ray, the length of its path, and its eventual exit point determines ,,what we see in the *face-up mosaic of the gem*.[74]

The length of the path which light follows determines the saturation and the tone of the color we see. The longer the light path, the more color the ray picks up in its rapid passage through the interior of the gemstone. In addition, the ray also loses some color through selective absorption.

Multi-color effect in spessartite garnet. Spessartite is singly refractive and here exhibits tonal variations of the hues orange to yellow. Photo: Jeff Scovil, courtesy of R. W. Wise, Goldsmiths.

This explains why, in a singly refractive gem, some facets may exhibit differences in darkness or lightness (tone) or appear duller or brighter (saturation) than others. In a doubly refractive gem, the scene is made more complex by the splitting of each ray into two components, each containing a portion of the visible spectrum. In dichroic gems, additional hues may be added to tonal variations, further complicating the face-up mosaic.

Traditionally, gemstones such as ruby

---

72 Pleochroism, refers to the property of doubly refractive gems to exhibiting two or more colors when viewed in different crystal directions in incandescent light. Most gems are dichroic, meaning that they show two colors. I prefer multi-color effect because pleochroism is limited, by definition, to incandescent lighting and singly refractive gems which also may exhibit multi-color effect in either type of lighting.

73 Multicolor effect is a new term chosen to replace the more common *dichroic effect* because that term is confusingly similar to the term *dichroism*. Dichroism is a physical property of light. When light enters a dichroic substance it divides into two distinct rays, each containing a portion of the visible spectrum. Most gems other than diamond, spinel, and garnet are dichroic and one, tanzanite, is trichroic. Multicolor effect is a term that describes the tendency of a gem to show two or more hues, or variations in saturation and tone of a single hue when viewed face up. Multicolor effect is a property which may or may not be caused by dichroism, but it can also be found in colored diamond, garnet and spinel, gem varieties which are not, by definition, dichroic.

74 Thanks to Stephen Hofer for coining this evocative term.

Multi-color effect in tanzanite. Tanzanite is one of the few trichroic gems meaning that it may exhibit three colors.; red, blue and brown. The 2.50 carat unheated gem pictured shows two colors, purple and blue, in the face-up mosaic. Photo: Gene Flanagan, Precision Gem

the sun. This phenomenon is prevalent in blue sapphire. Multicolor effect is classified as weak, medium, and strong.

Sometimes multicolor effect is a fault, but sometimes it is not. In some gemstone species it materially contributes to the stone's appeal. In primary color gems, such as ruby, sapphire, emerald, and tsavorite, (where the ideal is a visually pure single hue – pure red or green, for example) – the addition of any other hue or a change in saturation and tone is less than desirable. In such cases, the multicolor effect is a defect. In gems such as tourmaline and topaz, a strong multicolor effect which does not materially detract from the stone's beauty, can be and is a definite plus. The key issue is how it adds to or subtracts from the overall beauty of a given stone. In the case of topaz and tourmaline, the stone does not truly bleed color.[75] A little purple adds to rather than subtracts from the beauty of a red tourmaline. A bit of red at either end of the gem adds to the attractiveness of a peach-colored topaz. Multicolor effect and its impact on the beauty of specific gem varieties will be addressed in Part II.

Daystones, Nightstones

and sapphire have been judged in part by their reaction to two specific types of lighting—natural light and incandescent lighting, (the modern equivalent of light from a candle). Stones which lose or bleed color when moved from natural to incandescent light are considered less desirable than those that hold their color in both types of light. In ruby and sapphire, multicolor effect takes the form of tonal variations of red and blue. In a visibly dichroic ruby, for example, some of the reflections will remain the same as when the stone was viewed in daylight while others will become a lighter red or pinkish. The stone appears to bleed or lose color as lighting environments are shifted. The lighter-toned reflections, the pinkish flashes, give the color a bleached-out appearance, like a curtain partly faded in

75 The term "bleeding" describes a situation where the stone loses color or becomes paler or lighter in tone and less saturated as the lighting environment shifts from daylight (or daylight-equivalent) to incandescent lighting.

Some gem varieties look their best in direct sun or skylight.[76] Others look good in fluorescent light but fall apart when observed under the direct sun. Some individual gems will hold up well and may even appear more beautiful when viewed in subdued incandescent light. Opal is famous for this characteristic. The phenomenon is a product of a number of factors. The negative side, as discussed above, may manifest itself in a several ways, including color bleeding, reduced transparency, and dulled saturation resulting from the addition or intensification of a gray or brown mask.

Some gemstone varieties are said to prefer or look their best in the light of a certain kelvin temperature. Stones that look their best in daylight I term *daystones*; those that put their best foot forward in incandescent lighting I have dubbed *nightstones*. Though the old one hundred watt incandescent light bulb has now been outlawed in Europe and the United States (and it looks as if incandescence is likely to be replaced with LED lighting), the concept is still useful because light from a flame is still relevant and because many jewelers and gem dealers continue to use incandescent spotlights to bring out the sparkle of gems.

Most gemstones are daystones, although

topaz, aquamarine, and alexandrite are true ladies of the evening. Some dealers have suggested that all singly refractive gems look the same in both types of light, but this is not true. Garnets are most definitely daystones. It is important to remember that individual members of a gem variety will act differently. The terms daystone and nightstone provide a convenient aid for the aficionado to mentally classify gems and to think about their behavior in different kinds of light. A good rule of thumb: check a daystone in incandescent. If you like the look of the gem under the light bulb, you will love it in daylight.

Buying Gemstones: Some Strategic Advice

It is difficult at best to advise anyone on the proper strategy for buying a gemstone. Each transaction invariably involves a relationship between two people, the buyer and the seller. However, some advice on what not to do may be helpful.

Reinforcement: Developing Color Memory

The judgment of color in gemstones takes practice. Some experts maintain that the mind has no memory for color. Certainly it is true that this facility can be acquired only with training. Even experienced dealers will often carry comparison stones, gems of known quality, to compare with gems being offered for sale. Connoisseurship can be obtained and maintained only by constantly viewing and comparing gemstones. Comparison is the key. Experts achieve their mastery by constantly viewing and comparing one

76 To view a gem in skylight, turn your back on the sun to see the gemstone with the light diffused around your body. This viewing environment is superior to looking in direct sun as it eliminates glare. Skylight is bluer than direct sunlight, particularly when you face north. Many gemstones bleed or lose some of their saturation and tone in direct sun. This is not a fault, but rather is to be expected. For comparison purposes a gemstone should be observed in all possible environments: sunlight, skylight, shaded natural light, fluorescent light, and incandescent light.

stone to another.

Dealers have a strategy of their own for taking advantage of the buyer's lack of color memory. The technique is simple. The dealer will begin by showing his lowest grade parcels, one after another. The objective is to "wash the eyes," to break the buyer down, frustrating him, by showing him so many poor quality stones that after a while even mediocre gems begin to look good.[77]

One method for remembering color was for centuries a closely held secret among the ethnic Chinese gem dealers of Thailand.[78] We will call this memory aid reinforcement. As stated above, the mind has no innate color memory. How then does the aficionado develop this sense? The clue is to be found in the Thai word which denotes the finest ruby color: *kim-kim*, which translates as "salty." Saltiness is, as we know, an attribute of taste. In this situation, the dealer attempts to reinforce the sense of sight with that of taste. This reinforcement takes place on the most basic level. Primary hue is identified with a primary taste. Thus, the dealer both sees and tastes the salty red color of the ruby.

Other well-known attempts at reinforcement try to back up sight with sight. This approach has caused much confusion through the centuries. In the East, the first three colors of fine ruby are called pigeon blood, chicken blood, and beef blood. Such terms are not easy to define nor are they of much practical use —try squeezing the first two drops of blood from a pigeon's nose, as the ancient Hindus advise, some afternoon while browsing among the rubies at Tiffany's![79] The problem, of course, is that a sense cannot reinforce itself. Other examples of this method, such as comparing the color of a gem to that of a specific flower, may have a strong poetic appeal but on a practical level is fraught with difficulties. Flowers have been used as a comparison for centuries; little wonder, for like a beautiful gem they are a visual burst of pure color, a balm to the eye. Unfortunately each blossom may be of a slightly different hue. Furthermore, the beauty of a flower is fleeting while the perfection of a gem is eternal.

The Pick of the Litter

Dealers love to sell parcels of gemstones. This is often because the dealer himself purchased the stones in a parcel, and he would rather sell several stones than just one. If the aficionado develops his passion beyond the level of novice, he will often find himself looking at a number of stones contained in a paper envelope or parcel. Just as inevitably he or she will be offered entire parcels by dealers at what are said to be "bargain" prices. Such offers can be very

---

77 Richard W. Hughes, *Ruby & Sapphire* (Boulder, Colorado: RWH Publishing, 1998), p. 225. I had firsthand experience with this selling strategy. On my first trip to Bangkok I spent most of a week looking at one low-grade parcel after another. Luckily I split my time between two brokers, spending about three hours a day with each. (The exercise was not completely in vain; the broker who showed the low-quality stones also provided a free lunch.) Toward the end of the week I did manage to purchase a two-carat ruby at a bargain price.

78 Barry Hodgin, personal communication, 1986. Anthropologists refer to such analogies as cross-modal.

79 Souindro Mohun Tagore, *Mani Mala: A Treatise on Gems* (1879; reprint ed., Nairobi, Kenya: N.R. Barot, 1996), vol. 1, pp. 241-242.

tempting. The collector should resist such temptation with every fiber of his being.

The construction of parcels is a minor art form and is practiced with care by gem dealers. It is good to remember that gems always look better grouped together in a parcel. Every parcel can be divided into at least three portions. First there are the eyes, the finest stones in the parcel, so called because they flash their lovely lashes at us. They call to us because they are beautiful,

Far from arbitrary, the construction of parcels is a minor art form.

but they are also like the sirens of Odysseus: they mean to tempt and shipwreck us, and they will play us false. The second portion consists of the mediocre stones which are usually the largest number in the parcel. They look pretty good when viewed together but, for the connoisseur, pretty good is not good at all. The savvy aficionado is aware of this and always isolates and views each stone singly. The third section of the parcel

contains the dogs. This is a non-technical term and refers to the real bow-wows of the parcel. Usually there are at least as many dogs as there are eyes.

Dealers will normally quote two prices, the parcel price and the pick price. The difference between the two prices should give the buyer some idea just how large the disparity is between the eyes and the dogs of the parcel. If there is no disparity there is no reason the dealer should not allow selection at the parcel price. Selection from a parcel is called the pick: always pick! The dealer can always sell the inferior stones at a reduced price; the collector's objective is to pick the eyes out of the parcel, preferably at the parcel price. Sometimes the dealer will ask the buyer to make a selection before he establishes the price. This puts the buyer at a distinct disadvantage. If selection is allowed, the price should be stated before the buyer does the work of picking.

Pairs and Suites

A matched pair of stones is worth approximately fifteen percent more than two single stones. Three or more matched stones are called a suite. Add ten percent per carat for each stone added to the suite. Parcels sometimes give the aficionado the opportunity to pick a pair or a suite. Dealers, on the other hand, will usually charge less for a larger pick from a parcel. Tactically, the aficionado should first establish the pick price before making a selection, then select from the parcel. By long-established custom in the trade: once a price has been established, it cannot be withdrawn. The buyer will often have to remind the dealer of that fact, sometimes emphatically, once a particularly astute selection is made. Picking a pair or a suite may justify the higher selection price.

Investing in Gemstones

Gemstones are hard assets. Historically they have been seen as a hedge against inflation and the breakdown of more abstract forms of investment such as stocks and bonds. Gems are a small portable concentration of wealth. Their size has made them an excellent choice for those who wish to hide assets and for refugees who have been forced, for political or religious reasons, to flee their homes. One somewhat tongue-in-cheek theory has it that gemstones are a permanent store of value. According to this theory, gems never increase or decrease in value but have remained stable for thousands of years while currencies have fluctuated wildly.

Investments such as stocks and bonds trade in an orderly market. The New York Stock Exchange, for example, guarantees that its members' stocks can be traded at any time the exchange is open. Investments of this kind are liquid. Stocks and bonds have an established value and exchanges operate on volume so the commission on any given transaction is very small.

Gemstones do not trade in an orderly market. This is something of an understatement. There are no exchanges to facilitate trading, though in recent years auction houses have fulfilled the role to some extent. Each gemstone is unique and so, therefore, is its price. Gems normally pass through many hands before they reach those of the collector. When the investor wishes to sell he must often resort to the wholesale market. The gap between retail and wholesale must be bridged if the gem investor is to make a profit. These are real problems and the collector should be aware of them before deciding to invest.

Auctions provide a potential option for the collector-investor. Since the 1970s major auction houses have provided a venue for buying and selling particularly fine and rare gemstones. Fancy color diamonds have benefited most from the publicity and excitement generated by the auction process. As of the end of the 1980s fancy color diamonds held nineteen of the twenty record per-carat prices paid for gemstones.

Since the publication of the first edition of this book, other gem varieties have benefited as well. The auction market has changed dramatically. Once the more or less exclusive preserve of wholesale dealers, fully fifty percent of bidders are retail buyers.[80] Kashmir sapphires, Burmese rubies, and Paraíba tourmalines regularly achieve "retail" prices at auction. In 2011 Christies Hong Kong achieved a record price for Kashmir sapphire, and in 2012 a 6.04 carat Burmese ruby sold at the same venue for $4,620,000. In 2015 both of these records were broken. It is fair to

---

80 Russell Shor. "Auction Houses, A Powerful Market Influence on Major Diamonds and Colored Gemstones," *Gem & Gemology*, The Gemological Association of America, (Spring 2013), p.14

say that the auction market now tends to set the prices, and prices within the trade often bump upward after a record setting auction sale. The reader should, however, take to heart the words "particularly fine" and "rare" in this context. The collector wishing to build a portfolio should follow a very simple strategy — buy only the best. Gemstones of mediocre quality usually sell below wholesale at auction.

Gemstones do increase, sometimes dramatically, in value. Witness, for example, price increases in Paraíba tourmaline, Kashmir sapphire, and fancy color diamonds in the past decade. Gemstones are a long-term investment. Gemstone prices seesaw, like most other investments, but over the long term values have risen significantly.

Recently, online retailers have begun to post bar graphs reputed to show increases in historic auction prices of high-end gems. It is important to keep in mind that gems are not stock certificates. As we shall see in the chapters on fancy colored diamonds, even certificated gems with identical grades, such as fancy vivid blue, are not identical. There is a range of variation in saturation/tone in any given grade. Graphs of this sort can be very misleading. As of this writing (December, 2015) the market catering to those capable of spending many millions on a fine colored diamond is very hot, but the market for the other ninety-nine percent is as cold as ice.

The Certificate Game

The past two decades have seen a proliferation of independent gemological

laboratories. Gem labs issue grading reports which are often referred to as "certificates" in the gem trade. These reports may cover a number of grading issues including country of origin, the presence or absence of color and clarity treatments, and actual quality grading of a given gem. Grading reports are an important tool for the collector. They are not, however, the be-all-and-end-all. Yes, they look impressive, but it is also too easy for the timid or inexperienced aficionado to become overly impressed or too dependent on them. This tendency is what experts call "buying the cert." Laboratories certify gemstones but unfortunately no one certifies laboratories.

There are a number of highly qualified gemological laboratories: the Gemological Institute of America (GIA-GTL), American Gemological Laboratories (AGL), the European Gemological Laboratory (EGL), and the American Gem Society (AGS), to name just the more important American labs. In Europe, there is Gubelin and SSEF and in Bangkok, GRS and AIGS and Lotus Gemology. However, each of these laboratories has it's own methodology, and often these methods do not overlap. There are no uniform requirements, no universally accepted methods. Grading expertise is another variable. Depending on what sort of service is required — country of origin determination, presence of treatments, gem grading — each lab approaches these issues in its own way. Thus, it is possible for one lab to issue a report concluding that a gem is untreated while another determines that the stone has been heat enhanced. One lab may call a sapphire pink while another calls it "padparadscha" simply because of a disagreement over definitions. Dealers are very aware of these differences and will try to use a lab which will provide the most useful certificate.

The serious collector will find it advisable to become conversant with the reports issued by the major gemological laboratories as well as with their reputed strengths and weaknesses. Better still, look at the gem; look at a lot of gems. Again, there is no real shortcut to connoisseurship.

Caveat; Buying Online; A Picture May Tell a Thousand Lies

Since the publication of my first edition (2003), the Internet has become an important gemstone marketplace. Dealers from all over the world post images of their wares online and more and more business is conducted on the Internet. Buyers seeking bargains believe that they have found a shortcut which enables them to compare prices with quality by comparing images on the web. Online gurus who populate gem forums claim to be able to advise newbies on making a purchase by simply comparing images posted by sellers. These forums are very useful for education and the exchange of ideas, but they do buyers a great disservice in making such claims.

Small differences in quality don't make a great deal of difference in the appraisal of commercial gemstones. However, as quality approaches theoretical perfection minor differences in hue, saturation, tone, cut, clarity and crystal assume much greater importance. Is it a superior stone or a superior photograph? Some gems, like

the late Marilyn Monroe, are simply more photogenic than others. In the stratosphere of gemstone quality, tiny nuances, in color, levels of clarity, dimensions of cut and degrees of transparency make for very large price differentials. Online images rarely capture these subtle nuances, making image comparisons at best problematic and at worst, misleading.

Online buyers compare images; but rarely compare actual gemstones. Buyers order only the stone that is most photogenic online or gets the best review in the forums without ever seeing the ones they reject and, therefore, never know what they might have missed. I personally know of a number of instances where buyers have purchased inferior gems at higher prices just because they liked the image. Since these forums encourage anonymity and treat genuine credentials as spam, the buyer rarely if ever knows who these "experts" are or what, if any, qualifications they may have.[81] Caveat emptor! Sellers court these online gurus and due to the anonymous nature of these forums, some participants may be shills for online sellers.[82]

Let's take an example that everyone knows, the D flawless, the crème de la

---

81 Separating the wheat from the chaff: I was informed by the moderator of one online diamond forum that adding my Graduate Gemologist (GG) credential as part of my signature was "borderline spam." Another forum moderator refused to allow the posting of "author, Secrets Of The Gem Trade, The Connoisseur's Guide." as part of my signature.

82 Online forums function best as idea exchanges and can serve a real educational function. One of the best is Gemologyonline.com. This forum is a community of gemologists, welcomes questions and offers an online encyclopedia called The Gemology Project and online gemology courses as well.

A quick tweak in Photoshop and a ruby with low saturation (at bottom) becomes a gem. Photoshop can also easily eliminate inclusions. Photo: Jeff Scovil.

crème of colorless diamonds. I choose this example because diamonds are very precisely graded. According to *The Guide*, a well-respected industry publication, the current wholesale price of a 1.00-carat D-IF is 31% higher than the very next color grade, E-IF. Compared by clarity grade, (D-Fl-D-VVS1) the spread is slightly less, about 29%. A similar comparison between diamonds with a color grade of L and O shows only a few hundred dollars separating

the two grades. These subtle differences will not be visible in an online image.

When considering colored gemstones, similar price differentials apply and the grading equation becomes much more complex. Colorless diamonds (Chapter 5) are graded based on minute tonal variations of yellow. Color itself, as we have seen, breaks down into two additional components, hue and saturation, which must be added to tone in the quality equation. A gemstone's color is composed of primary, secondary and sometimes tertiary components plus masks; a top color ruby for example, may be seventy-five percent red, fifteen percent purple, five percent orange and five percent gray. A blue sapphire with ten percent (or less) greenish secondary hue, invisible in the online image, will sell at a dramatically lower price than a stone with a pure or slightly purplish blue hue. Similarly, an emerald that is seventy-five percent green with a ten-fifteen percent secondary blue hue can sell for double the price of a slightly yellowish stone. These slight differences in hue do not show up even in professional images. This makes online comparison between two images deceptive and of little or no value.

There are a host of variables, each of which will fundamentally alter the color you think you see. The image itself was taken by a specific camera in a specific lighting environment. Each make and model of camera has plusses and minuses when it comes to accurately rendering color. Some are good with greens, some with blues, others with reds—most digital

cameras have great difficulty with both green and red. The color temperature of the specific lighting will make a big difference is the color you see. Artificial light can be controlled to emit almost any hue. All these variables affect the apparent color of gemstones.

A picture may tell a thousand lies. Apparent color may also be easily altered with editing software. *Photoshop*, the world's most popular image editing software employs sophisticated tools which can easily alter the hue/saturation/tone of an image. With five minutes of training, the most unsophisticated *Photoshop* user can be taught to turn a ruby into an amethyst or turn it as green as an emerald.

In some cases the image will distort what the eye sees. Award winning gem photographer, Robert Weldon, makes the point that due to the limits of depth of field, the camera's lens will compress inclusions into a single plane, increasing their prominence. This compression can lead to particular difficulties when trying to accurately render images of expensive gemstones such as emerald, where the difference in value between an eye-clean gem and one with eye visible inclusions will be dramatic. Online images are normally in JPG format. This format is created, Weldon points out, by subtracting information from the original high-resolution image.[83]

With colored gemstones, the eye standard replaces the loupe standard: what the eye sees is what is important. In most varieties of colored gemstones, the difference between eye-clean and visibly

included is dramatic. It is similar to the difference between a diamond graded flawless and another graded SI2. Assuming that this gem was offered by two sellers on different websites, a buyer viewing Image A on one website and Image B on another might wrongly conclude that the second seller was offering gemstones at a better price.

A well-lighted gemstone may appear to perform better, have a finer cut than one that is less well lit. The lighting environment is not visible in the image and multiple light sources of the type normally used by professional photographers can mask real deficiencies in cut.

Each brand and type of monitor renders an image differently. 24 bit monitors differ markedly from 16 bits and from manufacturer to manufacturer and even day to day. Jennifer Robbins author of Learning Web Design tells her readers: "*Let go of precise color control. Yes, once again, the best practice is to acknowledge that the colors you pick won't look the same to everyone, and live with it. Precise color is not a priority in this medium where the colors you see can change based on the platform, monitor bit-depth, or even the angle of the laptop screen.*"[84]

---

84 Jennifer Niederst Robbins, *Learning Web Design*, 4th Edition, O'Reilly Media Books, 2012.

---

83 Robert Weldon personal communication, 2011.

Grading cultured South Sea pearls. Photo: Robert Vespui.

Chapter 5

Connoisseurship: Grading Special Cases

"Good gems are hexagonal, rectangular or circular in shape, pure in color, easily settable in jewelry, unblemished, smooth, heavy, lustrous, transparent and reflecting light from inside. Any gem of faint color, lacking lustre, grainy, blemished with holes in it, cut badly or scratched is bad."

–Kautilya, 400 BC

Connoisseurship in Phenomenal Gemstones: Stars and Cat's-Eyes

A cat's-eye is an optical effect that can be produced in a number of different gemstones by the way the inclusions are oriented

A 14.40 carat blue star sapphire exhibiting a very well formed six-rayed star. The hexagonal arrangement of rutile inclusions responsible for the phenomenon is visible in the background. Photo: Jeff Scovil, courtesy Mine Design.

in the cutting of the finished stone. The inclusion responsible is normally rutile, a mineral that crystallizes in long hair- or needlelike structures. Rutile hairs (crystals of titanium oxide) often form in dense

Exceptional alexandrite showing a bright, straight, well-formed cat's-eye running directly down the center of the cabochon. Photo: Harold and Erica Van Pelt; courtesy of Kalil Elawar.

parallel masses like broom straw along a specific crystal direction while the gem crystal, still in molten form, is solidifying. The cutter fashions the stone en cabochon

so that it includes rutile needles going in only one direction. When light shines directly on the stone, a band of light is produced which runs perpendicular to the direction of the needles. The phenomenon can also be observed in a spool of thread. In a gem, its translucency and the dome of the cabochon, cut convex to concentrate the reflection, can produce a sharp line which resembles the iris in a cat's eye. This same phenomenon is the cause of the star in star sapphire.

Sapphire and ruby crystals are hexagonal. As sapphire crystals grow, the rutile inclusions growing inside the crystal are forced into alignment along the six directions of crystal growth. If the cabochon is oriented and cut properly, it produces three crossing bands of light, each ray at a sixty-degree angle and perpendicular to each of the six concentrations of rutile needles, forming a six-sided star.

Phenomenal stones such as stars and cat's-eyes have special grading criteria in addition to the four Cs. Color, clarity, cut and crystal are still part of the equation, but of overriding importance is the phenomenon itself. Ideally, in star stones the star should be distinct, perfectly centered, and have all six rays or legs straight and of equal length and strength. The star is tested by positioning the stone directly beneath a single ordinary incandescent light source or by placing the stone in direct sunlight. Ordinary light and a single light source are key. Two light bulbs will produce twin stars. Stones that require the concentrated beam of a flashlight or Maglite to bring out a distinct star are much

less desirable. The stone must be viewed directly under the light source to check the centering of the star or eye.

The more distinct, centered, and perfectly formed the star or eye, the finer the stone. A strong punchy phenomenal effect is a necessary condition for a stone to

Interior view of a blue star sapphire showing the hexagonal arrangement of rutile crystal inclusions which create the star. Similar inclusions aligned in a single direction in a transparent chrysoberyl cabochon will create a cat's-eye. Photo: John Koivula

be considered fine, but it is not, by itself, sufficient to make a determination.

Hue/Saturation/Tone

The judgment of color of star and cat's-eye gems follows the same rules as faceted stones of the same variety. Thus the finest color (hue, saturation, and tone) of a star sapphire mirrors that of faceted blue sapphire; the same applies to ruby.

However, the color seen in phenomenal stones is the body or transmitted color, not the key (reflected) color used in defining quality in faceted gemstones. The finest color in star ruby and sapphire mirrors the finest color seen in faceted stones. A fine star ruby should be a medium dark tone, have at least an eighty-five percent pure vivid red primary hue, with no more than fifteen percent secondary hue of purple, orange or pink. Star sapphire should be a medium dark-toned vivid primary blue hue, with no element of green and no more than fifteen percent purple secondary hue. In practice, most stones of these hues are heated to dissolve the inclusions, then faceted, so any gem with a pure red or blue hue, regardless of tones above fifty percent, is worthy of consideration. Stars can be found in other sapphire colors as well, including pink and purple.

Crystal

The fourth C plays a defining role in the evaluation of cat's-eyes and star stones. Phenomenal stones are often dark toned and only semi-translucent. Stones of this description will often exhibit a distinct star or eye. Once this is established, the more translucent and limpid the stone, the more visually pleasing it will be. Crystal is at least co-equal with color in phenomenal stones. A high degree of transparency is a hallmark of the very finest stars and cat's-eyes. The finest star or cat's eye stone is one that has both a clean, limpid body color together with a bright well-formed star or eye.

Cat's-eye stones can be thought of as one-legged stars, except that in the evaluation of a cat's-eye, the stone is rotated three hundred sixty degrees under the light source. In the finest cat's-eyes, the eye will appear to expand and contract, winking like the iris of a cat's eye. As with star stones, the eye should be distinct, centered, and straight as an arrow.

Cut

Star and cat's-eye stones must be cut en cabochon and are normally cut either oval or round. If the shape is oval, the eye and the star follow the long axis of the stone.

In the marketplace, ruby and sapphire star stones are less valuable than faceted stones of similar color. Heating technology makes it possible to "burn out" the rutile inclusions responsible for the star phenomenon, clarifying the crystal and improving the color so that a more valuable faceted gem can be cut from the resulting rough. For this reason, there are very few high quality star sapphires and rubies, and most of what is available will be somewhat deficient in color.

Connoisseurship in Natural Fancy Color Diamonds

Unlike colorless diamonds, which are evaluated face down by analyzing the stone's body color, fancy color diamonds, like all gems of color, are evaluated face up by analyzing the stone's key color. To rieterate, key color is the color of refracted light, the color of the brilliance, the sparkle as distinguished from the body color, the light seen facedown, transmitted through the body of the gem.

As a class, fancy color diamonds are so rare, and the prices they command are so

Which is more desirable, Fancy Intense or Fancy Deep? The language on the certificate isn't much help. The question should be, which is more beautiful? — but it's a difficult call. The 2.18-carat emerald cut on the left was graded Fancy Intense Blue by GIA-GTL. The 2.47-carat gem at right carries a grad of Fancy Deep Blue (the same grade as the Hope Diamond). © The Gemological Institute of America.

high, that rarity, not beauty is the engine which drives this market. The fancy color diamond market can also be said to be certificate driven. Fancy color diamonds are usually sold with a written color description and guarantee of natural origin issued by a gemological laboratory. These reports are known as "certificates" in the gem trade. By far the most influential certificates are those issued by the Gemological Institute of America's Gem Trade Laboratory (GIA-GTL). [85] It is fair to say that virtually all-important fancy color diamonds are accompanied by a grading report issued by GIA-GTL.

In 1994 GIA-GTL completely revamped its grading criteria for fancy color diamonds. The new grading standards raised the bar for certain rare colors of diamond to receive one of GIA-GTL's coveted fancy color labels. Fancy color is a name traditionally applied to all hues of colored diamonds. In the new standards the GIA linked the color grade to the rarity of the hue! This will require a bit of explanation.

GIA-GTL uses twenty-seven hue names but only nine grades for classifying fancy color diamonds. The nine grades are based on the saturation and tone of the hue. Moving from pale to vivid and light to dark the nine designations are *faint, very light, light, fancy light, fancy, fancy intense, fancy vivid, fancy deep*, and *fancy dark*. Fancy deep and fancy vivid were designations added in 1994.[86]

These categories have become the industry standard for describing color in diamonds. The nine grades are not applicable to all colors---The first three grading levels, faint, very light, and light--- do not carry the prefix "fancy" and are used on GIA-GTL grading reports to describe stones, (of hues other than yellow and brown), of very low saturation and tone.

Brown and yellow diamonds of similar saturation and tone would receive a letter grade on GIA's colorless diamond scale. In the GIA lab, colorless diamonds tinted gray or brown from K color downward will also be given a written description along with their letter grade. K though M colors are designated faint brown, or gray, N-R very

85 John M. King et al., "Color Grading of Fancy Color Diamonds in the GIA Gem Trade Laboratory," *Gems & Gemology*, (Winter 1994), p. 222.

86 King, "Color Grading," p. 222. *Fancy vivid* refers to medium-toned gems (forty to sixty percent) with vivid saturation. *Fancy Deep* refers to colors of medium to dark tones with vivid saturation. For a full discussion of the relationship between hue, saturation, and tone see Chapter 3.

light brown or gray and S-Z light brown or gray. Note well, despite the written description, a diamond given a letter color grade K-Z is not considered a fancy color diamond.

A diamond with the faintest trace of blue or green hue face up will be graded "light blue" or "light green" on the GTL fancy grading scale. A yellow, gray, or brown stone of identical saturation and tone will be classified somewhere between S and Z on the GIA colorless diamond scale.[87] The latter grade classifies the stone as an off-color colorless diamond. Yet the yellow or brown stone has at least as much color saturation as the blue stone. However, though neither stone has much face-up color at all, GIA's methodology dictates that the yellowish or brownish stone receives a low letter grade on the colorless scale and the rare bluish and greenish stones receive the very desirable accolade *fancy light*.

1.07ct
Fancy Intense Green　　**1.07ct**
Fancy Intense Green　　**1.02ct**
Fancy Intense Green

Three fancy color green diamonds with identical GIA-GTL grades of Fancy Intense Green. Note the vast range of saturation and tone within this grade and the steely green hue. To call the 1.02 carat gem "intense" green is surely a reach. Photo: Stephen Hofer.

The author examined a diamond accompanied by a GIA-GTL grading report designated the stone as fancy blue. The stone appeared to have a primary gray hue of twenty percent tone, with a five to ten percent blue secondary hue. In short, the diamond was a light-toned slightly bluish gray. Since gray is not considered a hue in the GTL grading system, it was partially ignored, but only as a hue. If the gray primary hue had not been present, the stone probably would have been graded faint blue, but the lightness or darkness, in short the tone of the gray, was factored in since it contributed to the overall saturation and tone, justifying a grade of fancy blue. Thus a visually light bluish gray diamond becomes a fancy blue diamond.[88]

In the certificate-driven world of fancy color diamonds, the lab grade will

---

87 S-Z are called "very light yellow" on the GIA colorless diamond grading scale. Stones graded at this level will show very little color face up because the hue lacks sufficient saturation and tone to be considered more than yellowish. Blue diamonds of a similar tonal level face up would be graded as blue. King, "Color Grading," p. 237, states: "Because yellow is by far the most common, a greater depth of this color is required for a stone to receive a "fancy" grade. In contrast colors such as pink and blue are both relatively rare and occur in much narrower (lower) saturation ranges. Thus, a *fancy* grade is given for a paler stone."

88 John M. King et al., "Characterizing Natural-Color Type-IIB Blue Diamonds," *Gems & Gemology*, (Winter 1998), p. 262. In this later article the authors note that blue diamonds take up a much-compressed area of color space when compared to yellow diamonds.

be a primary determinant of price. Diamonds in the very rarest hues (red, pink, green, blue, violet, purple, and orange) will be classed as fancy color diamonds if they exhibit even the faintest, barely discernable, tints of these hues. Under such circumstances, connoisseurship falls by the wayside, and the beauty of the gem has little to do with its price in the marketplace.

Master stones used by the Gemological Institute's Gem Trade Lab (GIA-GTL) to grade fancy yellow diamonds. From left to **right: Fancy Light, Fancy, Fancy Intense, and Fancy Vivid.** © *The Gemological Institute of America*

Furthermore, according to GIA's John M. King "Most gray diamonds are those that are so low in saturation that no hue is readily perceived, only the light-to-dark tonal changes."[89] This point may seem obscure but its effect is easily seen when the color observed in the diamond is compared to the tortured language of the certificate. It appears that the saturation and tone of the gray in the rarer hues of fancy color diamond is added to the tone of the chromatic hue, in effect pumping up the grade. The result: stones that are visually grayish blue, gray blue, or bluish gray become fancy intense blue; greenish gray gems are described as fancy intense green. In such cases, the extreme rarity — a superheated market coupled with a too theoretical approach to grading — distorts the fundamentals of connoisseurship. The connoisseurship equation is distorted

when rarity supplants beauty as the primary determinant of quality.

Of the nine grades used by GIA-GTL to describe the saturation and tone of all fancy color diamonds.[90] Each of these descriptions describes not a single point on a scale but a range of saturation and tone. Thus, for example, not all fancy intense yellow diamonds are created equal. One may appear a bit more vivid or a little darker in tone than another. This is a good reason to avoid what dealers call "buying the cert." Twenty-three categories are used to describe colorless diamond; it is easy to see why six descriptions are not adequate to accurately describe the range of saturation and tone in any hue of fancy color diamond.

Some have called fancy color diamonds the "ultimate gemstone," but ultimate

---

89 *ibid.*

90 As a rule of thumb, the aficionado should avoid stones carrying the certificate grades of faint, very light, and light. These grades exist mainly to fill in the cracks between the GIA-GTL's colorless and fancy color diamond grading scales. The saturation and tone of stones in these categories are simply too pale and too light to qualify them for the contemplation of a true connoisseur.

in what sense? Is it the most beautiful gemstone? Or, given the breathtaking prices paid for the finest examples, are we simply confused and made giddy by its price in the marketplace?

Alan Bronstein, the man who put together the Aurora collection (the finest collection of fancy color diamonds in existence), admits that the most expensive gemstone on earth, red diamond, is not the paradigm of red gemstones. Nor is the blue diamond the ultimate in blue gemstones. According to Bronstein, the only fancy color diamond that can claim that title is the yellow, which, he maintains, with some real justification, is the most beautiful of all yellow gems.[91] A red diamond, no matter how exceptional, can never stand toe to toe with an exceptional ruby. The same is true of blue diamond compared to sapphire. Yet, in modern times, ruby has never approached the price paid for a red diamond, nor has any sapphire come close in price to that paid for a blue diamond.

The major premise of this book is that beauty is the ultimate criterion of connoisseurship and that rarity is a secondary concern. However, in the marketplace, beauty drives demand and rarity drives price. The gem market follows this credo. That is, the more beautiful the gem, the higher the price it will command. In the fancy color diamond market these criteria are reversed: value is linked to rarity. Rarity drives price. This leads to odd situations where rather unattractive gems will sell for astronomical prices. However, once rarity is established, beauty then

becomes a factor. All things being equal, a beautiful pink diamond will command a higher price than an ugly pink diamond. Although the comfort of having price closely associated with beauty is removed, discrimination still does have a significant part to play.

In the final analysis, collectors wishing to specialize in fancy color diamonds must understand that these stones live in a sort of parallel universe where beauty is relative to the possibilities available in that universe. Blue diamonds are compared to other blue diamonds. An aficionado looking for the ultimate in blue must turn to sapphire or perhaps spinel.

Despite the realities of the market, the aficionado should beware of getting too bogged down in certificates and should avoid, as mentioned above, "buying the cert." The fact is, most of GIA-GTL's clients are diamond dealers. The language of the certificates issued by the lab, in fact GIA's entire grading methodology, is more about marketing than accuracy. The aficionado should base a buying decision on the beauty of the stone, not on the language of the certificate. This discussion was meant to alert the collector to the realities of the market, not to suggest a different acquisition strategy for fancy color diamonds. Market realities change, beauty is eternal.

The aficionado should beware of entering the fancy color diamond marketplace unless armed with a good understanding of the importance of rarity and a solid grounding in the arcane world of diamond grading reports.

91 Alan Bronstein, personal communication, March, 1999.

As the reader will note, the criteria used by the Gemological Institute of America's Gem Trade Lab (GIA-GTL) for evaluating fancy color diamonds is unique and differs in some respects with the system used generally throughout this book. The role of gray has been discussed; the expanded range of possible hues, thirteen in all, is explained in Part II in the introduction and overview to fancy color diamonds.

GIA does analyze key color and uses the terms hue, saturation, and tone to describe a diamond's face-up appearance. In fact, much of the terminology used on certificates is the same as has been used throughout the book. An orange diamond with a slight pink modifying or secondary hue would be described on a GIA-GTL report as pinkish orange, the "ish" indicating a slight modifying hue. A stone with a stronger though not dominant (less than fifty percent) secondary hue is termed a pink-orange. Due to the stone's rarity, descriptive language takes on a level of precision rarely seen in other colored gemstones. GIA recognizes that fancy color diamonds may have more than one secondary hue; thus, descriptions such as brownish pink-orange and pinkish-orangy-yellowish brown appear routinely on fancy color diamond certificates.

Multicolor Effect

Multicolor effect is a characteristic of all gemstones, including fancy color diamonds. A given stone may show tonal variations of one hue or occasionally two different hues when viewed in the face-up position. Since diamond is singly refractive, the two-color effect is the result of cutting or, more rarely,

For the purposes of grading, the GIA Gem Trade Lab describes the hue that shows on a majority of the face of a stone. GIA terms this the "characteristic color." The major part of the face-up mosaic of the gem is fancy dark orangy brown. The lighter-toned roughly hourglass-shaped area toward the center of the diamond is not considered when writing the description that will appear on the grading report. © The Gemological Institute of America.

color zoning (or some combination) in the stone. GIA-GTL uses only two terms on its grading reports, even and uneven, to describe this phenomenon. The primary hue – that is, the hue that covers the largest percentage of the face-up surface of the gem – will be listed as the primary hue on the grading report.

Oddly enough, GIA-GTL does not use the terms pinkish red or reddish pink in its grading reports, even though stones of this description do exist. Since pink is a lighter, paler red, the lab considers the term redundant, like calling a stone reddish red.[92] However, GIA-GTL does use the terms brownish orange and orangy brown even though the terms would seem to be similarly redundant, since orange is a lighter, paler brown.

92 John M. King et al., "Characterization and Grading of Natural-Color Pink Diamonds," *Gems & Gemology*, (Summer 2002), p. 134.

GTL also uses the term "uneven" to describe a stone that shows extinction, which is really a part of the stone appearing dark gray to black due to lack of brilliance.[93] As currently defined, the GIA-GTL category of uneven is unequal to the task of describing the visual scene, what this book terms the face-up mosaic of the gem (see Chapter 4). I examined a fancy purplish red diamond which exhibited a halo of vivid pink flashes circling the edge of the crown and surrounding the purplish red that was concentrated toward the center of the crown. This is an example of prominent multicolor effect. The gem was purplish red on the grading report, but purplish pink red to the eye. The written description did not describe the appearance of the gem.

Light and Color

Fancy color diamonds are graded face up. GIA-GTL uses a very specific neutral gray-sided grading box designed by the MacBeth Corporation, together with a 6,500-kelvin MacBeth fluorescent daylight-equivalent lamp. This lighting is roughly equivalent to daylight at noon (5500-6500k , which is the standard grading light for all gemstones. As with all other gems of color, connoisseurship requires that fancy color diamonds be examined, and their beauty evaluated, in both incandescent, LED and in natural daylight.

93 *ibid.*, pp. 233-234.

Clarity in Fancy Color Diamonds

In 1987, the Hancock, a ninety-five point round diamond graded fancy purplish-red sold at Christie's auction house for the astronomical price of $880,000, or $926,315 per carat, a world record. The Hancock is visibly included, meaning that on the GIA diamond grading scale the stone would be "I2" or "imperfect." Clarity plays a much less important role in the evaluation of fancy color diamonds. In contrast, colorless diamonds are clarity-graded using the loupe standard; that is, a stone is considered flawless if no inclusions can be seen under 10x magnification. As the above example suggests, the loupe standard is not particularly relevant in establishing the value of the rarer varieties of fancy color diamond.

The rarer hues of fancy color diamonds are graded using a de facto two-step scale. As with other colored stones, the first and most important step is at the point where the gem is eye-clean or eye-flawless (GIA grade SI1-I2). A stone showing no visible inclusions in the face-up position will command a substantial premium over a stone that is visibly included (GIA grade SI1-I3). An additional premium will be charged for a loupe clean stone that is, flawless under 10X magnification (GIA grade IF).[94] Fancy color diamonds are rare;

94 Stephen Hofer, personal communication, 2002. Clarity grades will affect prices in the more common brown and yellow fancy colored diamonds. See also John King, et al., "Exceptional Pink To Red Diamonds: A Celebration of the 30th Argyle Diamond Tender," *Gems & Gemology*, (Winter 2014), p.278.

therefore, there is no need to adhere to the overly stringent standard used for colorless diamonds. Unlike the situation with colorless diamonds, there is little need to create an impression of rarity where none exists.[95]

Grading under magnification is the second step in the two-step scale. According to dealers I interviewed, stones with grades above SI2 with no eye visible inclusions would carry some dollar premium over the stone that was visibly included. As to how high the premium, there was no consensus. Some dealers maintained that there were only two grades: eye-clean and imperfect (eye-visible inclusions), but that is not true. One thing is clear: the higher the degree of rarity of a particular color, the less important clarity becomes. It is fair to say that in what might be called the "connoisseur categories" – fancy intense, fancy vivid, fancy dark, and fancy deep – fancy color diamonds are so rare that those graded VS2-VVS1 carry only the smallest dollar premiums over eye-clean stones. Gems graded IF would carry an additional premium.

Treatments

Given the prices asked for fancy color diamonds, it is not surprising that methods have been found to induce or enhance diamond color artificially. Treated

stones sell at dramatically lower prices than those of natural color. The two most common types of treatment are irradiation and high temperature/high pressure (HPHT). Diagnosing the origin of color often requires the use of sophisticated instruments and techniques not available to the jeweler-gemologist. The aficionado is well advised to consider only fancy color diamonds evaluated by GIA-GTL or other competent independent laboratories and that have "natural color" clearly stated on the grading report. For important purchases, there is no substitute for a GIA grading report.

There is a great deal more to say about fancy color diamonds. Specific colors are explored in depth in Part II, Chapters: 43-49

Grading Pearls

"It is amazing, after a day's buying is done, to see how the value of the pearls increased once we had bought them. Sheikh Mohamed would bring them out, gloat over them, weigh and grade them properly instead of haphazard as he did when buying. . ."

–Alan Villiers, 1940

To begin to understand cultured pearls you first must bracket much of what you already know in general about the qualities of gemstones. It takes some people a long time to warm up to pearls. Pearls, it seems, lack that old flash and dazzle. Faceted gems trumpet their charms in a burst of visual pyrotechnics. Diamond dazzles us

---

95 One prominent industry pricing publication, *The Guide*, has developed standardized pricing grids for the more common fancy color diamonds at FL-VVS, VS, SI, and I1 graded under 10X magnification. These categories show percentage increases as high as fifty percent between FL-VVS and SI grades for certain colors (blue and pink). Are such percentages justified by the realities of the market? I believe they are not.

A fine cultured white South Sea pearl. Photo: Tibor Ardai, courtesy Assael International.

with brilliance; ruby drenches us in color. Not so the pearl. No, pearl is more refined and far subtler. Which leaves us precisely where? Well, setting poetry aside, the point is that grading pearl is a different sort of experience.

A pearl's chief asset is its skin. Faceted gemstones have no skin. They possess a cold surface polished to a hard luster with rows of precisely arranged facets. The skin of a pearl, by contrast, is soft and glows rather than gleams.

The uncritical dismissal of the pearl is really due to lack of attention to the subtleties of the gem. John Singer Sargent, painter of wealthy belle époque beauties, required seven separate brushstrokes, each stroke using pigment of a different hue, to

create a realistic picture of a single round white pearl. No, the pearl is not simple, but to appreciate its beauty requires close attention and serious effort.

In modern times we like to use scientific rather than poetic language when we talk about gems. Scientific language is comforting in its precision. Thus we talk about the pearl's surface rather than its skin. However, the older term is more accurate because it is more evocative and more precisely connotes the feeling one gets viewing a beautiful pearl. Like the skin of a beautiful woman, the skin of a pearl can be silky smooth or blemished, translucent, luminous and alive, or opaque, chalky, and lifeless.

Pearl grading uses a specific set of terms which are different from, though somewhat analogous to, those used to grade gemstones. Once again, beauty is the bottom line. The criteria used to grade pearls are as follows: *body color, luster, orient/overtone, translucency, nacre thickness, symmetry,* and *texture.*

Body Color

Cultured pearls today come in many colors. Saltwater akoya pearls may be gray, yellow, or white. South Sea pearls may be white, cream, golden, or steel gray to black. Chinese freshwater pearls come in plum, bronze, peach, and champagne. In pearls, the chromatic colors—red, orange, yellow, green, blue, and violet—as well as the non-chromatic hues of white, black, and even brown and gray are included as possible body colors. Pearls, like faceted gems, can be said to have hues made up of primary

and secondary components; thus, plum may be more precisely described as violetish red. Natural pearls come in a similar range of colors and the qualities discussed in this section apply to the natural product as well. Natural pearls will be discussed in greater detail in Part II.

The actual body color or hue is not part of the connoisseurship equation; no particular hue is more beautiful than another. As discussed a bit further on, compatibility with the skin of the wearer or what I call *simpatico* is the more important criterion. In South Sea and akoya white pearls, a green tint in the body color is considered a fault. With gray it is a bit more complex: Is it grayish and dull or silvery and bright? With gray it's really a question of whether the gray acts as a mask. A grayish pearl with a dull luster is less than desirable, a grayish pearl with good luster appears more silver than gray and is much sought after.

The pearl's body color by itself is insufficient to define the beauty of the gem. It is the combination of body color, luster, translucency, overtone and/or orient which characterizes a fine pearl. Think of the beautiful glowing skin of a healthy baby or the dewy complexion of a beautiful young woman. These are excellent analogies because healthy skin has qualities similar to the skin of a pearl, and it is these qualities that give it its sense of life.

Orient and Overtone

A fine pearl exhibits that subtle misty iridescence that connoisseurs call *overtone*, a glow that seems to emanate from inside the

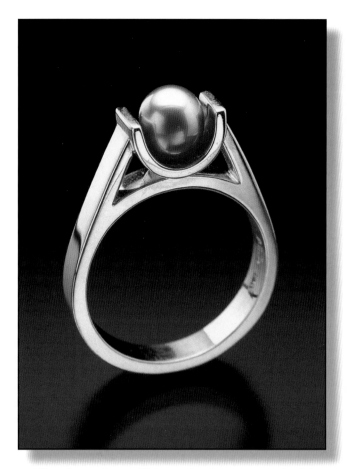

Ring featuring a Chinese freshwater pearl with an almost metallic luster and a distinct pinkish monochromatic overtone hovering around the edge of the pearl and reflecting off the polished gold. Photo: Jeff Scovil; courtesy of R.W. Wise, Goldsmiths, Inc.

pearl, hover above and cling to its skin, like sunlight through an early morning fog. This quality that led the English Renaissance poet Thomas Campion to enthuse: "looked like rosebuds fill'd with snow."[96]

Orient is derived from the Latin word *oriens*, meaning "the rising of the sun." Luster is caused by light reflection; orient is the result of light diffraction. Nacre or pearl essence is composed of translucent layers of aragonite, a type of calcium carbonate made up of tiny polygonal crystals mortared together by conchiolin, a protein which

---

96 Richard W. Wise, "A Meditation on Pearls," *Pearl World*, vol. 9, no. 3, (July-September 2001), pp 10-11.

to cling to some or the entire pearl surface, and is normally monochromatic. Orient occurs most often in baroque pearls where light interference caused by light reflecting off an uneven surface adds to the effect.

Chinese freshwater pearls exhibiting natural pastel colors and a mirror-like sub-metallic luster. Pearls reflecting in pearls. These gems were hand selected from kilo-sized parcels. Photo: Jeff Scovil, courtesy R. W. Wise, Goldsmiths

acts as a binder. In some types of mollusk the structure of nacre includes a series of tiny grooves, as many as three hundred to the millimeter, which acts as a reflection (diffraction) gradient, breaking light up into its constituent colors and creating iridescence or the rainbow effect. The more grooves the more distinct the orient.[97] The number of grooves and the spacing determine the colors we see. Some species and varieties of mollusks have this groove structure; some do not.

The terms orient and overtone are often used interchangeably, but they refer to slightly different effects. Orient is multi-colored and effervescent. It occurs in a number of rainbow colors at once, its effect is akin to dispersion in diamond, and it seems to scintillate off of the pearl's surface. Overtone occurs in large patches, appears

Orient and overtone combined with luster are the qualities which give the pearl its air of mystery and its life; it characterizes the finest of pearls. Overtone may occur in any spectral hue or mixture of hues. For instance, the finer examples of black pearl will normally appear greenish, bluish, pinkish, or, in rare cases, purplish against a tonal continuum (body color) of gray through black. A pearl may exhibit contrasting, overlapping overtones on different sections of the gem. The apparent color of the pearl, the color we see, is a combination of the gray-black body color and the chromatic overtone. White South Sea pearls may appear pinkish. Overtone will appear in cultured pearls only with relatively thick nacre coating. Akoya pearls are rarely left in the shell long enough to build up sufficiently thick layers of nacre to exhibit true orient.

Body color and overtone can be separated by a fairly simply procedure. Place

97 Yan Liu, J.E. Shigley, and K.N. Hurwit, "What Causes Nacre Iridescence?" *Pearl World*, vol. 7, no. 1, (December 1999), pp. 1, 4.

the pearl under the light of an incandescent bulb. The portion of the surface which directly reflects the bulb's image will have the orient color. In a round pearl that will be the center of the orb. The area outside and surrounding the reflection will exhibit the body color. This procedure works particularly well with black pearls. With Chinese freshwater pearls, however, this method may yield exactly the opposite result. Concentrated incandescent light will often bleach out the color toward the center of a round pearl turning it brownish or grayish, leaving the orient color in the surrounding halo. Chinese freshwater pearls look their best in natural light. Another test is to immerse the pearl in a shallow white bowl and the distinction will become evident.

Luster

Luster is the reflection of light off a surface. Luster in a pearl is analogous to brilliance in a transparent gemstone. In a fine pearl, the reflection off the pearl's surface is crisp, sharp, and well defined. As quality decreases, the reflection becomes fuzzy and dull. The proper way to evaluate a pearl's luster is to position a pearl under an incandescent light bulb. If the pearl is particularly lustrous you will be able to see the shape of the light bulb clearly mirrored in the reflection off the pearl's surface. The more clearly defined the image, the higher the luster of the pearl. If the bulb is clearly visible and well defined, the pearl is of the highest possible luster. If you can read the trademark on the bulb you really do have a gem!

Nacre Thickness

A major difference between natural and cultured pearls is the pearl's nucleus. There are two types of cultured pearls, tissue activated and bead nucleated. The former is pure pearl, the latter is the result of a bead, normally a spherical bead cut from a clam shell, which is implanted into the pearl mollusk. The mollusk secretes nacre over the bead, creating the pearl. A natural pearl may have a nucleus, but it is usually quite small while the bead inserted into the mollusk to create the cultured pearl is normally quite large, often ninety percent of the pearl's volume. If the nacre coating is too thin, the dull surface of the implant will show through the translucent nacre layers and the pearl will be little more than a nacre-plated bead.

Pearls with distinct orient will naturally have thick skins; however, treated pearls, such as the Japanese akoya, may display a distinct pinkish overtone which is the result of dyeing.[98] Experienced dealers will often examine the nacre thickness by shining a light through the drill hole and peering through the hole using a 10X loupe. A simple method to check the thickness of the nacre in a strand is to roll the pearls on a white surface under strong lighting. Thin-skinned bead nucleated pearls will *wink*; they will brighten and darken as the pearl is rotated. This winking effect is simply the mother-of-pearl bead showing through the

---

98 One exception to this is the so-called late harvest pearl. Akoya pearls are often harvested in November when the waters off Japan are at their most frigid. The nacre of late harvest pearls is dense and closely grained. After processing (bleaching and dyeing) the surface will appear opaque despite the fact that the nacre is quite thin. This technique has been known to fool experts. Sidney Soriano, personal communication, 1999.

nacre.

Another method is called candling. The pearl is held over a concentrated light source or placed on the lens of a flashlight. The concentrated beam of a Maglite is perfect for performing this test. Since nacre is semi-translucent, this will often allow the bead to be visible. In thick-skinned cultured pearls the bead will appear as a roundish dark smudge. In thin-skinned pearls the bead will exhibit a series of parallel stripes. These stripes are the growth layers of the bead insert itself, which is normally cut from a clam or oyster shell. Pearls which show these striations have very thin nacre coatings and should be avoided. Black pearl, because of the color and opacity of the nacre, will not show through. Thin-skinned black pearls will often exhibit a muddy brownish secondary hue.

Artists have a great appreciation of pearls. Iowa artist Sara Bell professes an ongoing love affair with the pearl. Bell devoted a year of study prior to executing her recent oil painting aptly titled *Pearl*. "My mother wore a natural pearl engagement ring that contained all the colors of the rainbow," she said. "I was frustrated trying to find pearls in jewelry stores. The ones I saw were mostly plain opaque white. They didn't look at all like my mother's."[99]

The precise realism of a painting entitled *Girl with a Pearl Earring* by the 17th

century Dutch master Vermeer fascinated Bell. Vermeer's technique of painting a pearl closely mirrors the process by which nature itself creates the pearl. The paint is built up in thin layers using multiple colors. This is the same technique painters use to give a quality of life to a subject's skin. In her own work Bell found she needed mixtures containing the entire range of primary and secondary colors to make a realistic rendering of the simple white orb. Her finished painting bears an uncanny resemblance to the real thing.

Water is an ancient term, but one which aptly connotes the combination of luster, translucency, body color, and overtone, which characterizes the finest of pearls. Real translucency, the ability to see light through a pearl, is non-existent in bead nucleated cultured pearls. However, tissue activated Chinese freshwater pearls may exhibit translucency to a very limited degree.

Simpatico

The pearl has another quality, unique among gems: its beauty can be enlivened or subdued by placing it in contact with a woman's skin. The Spanish word *simpatico* is a term I have chosen to describe this somewhat mysterious quality. A pearl, no matter how beautiful in itself, will seem to come alive if it is placed next to skin of a certain color, texture, and tone with which it is simpatico.

Pearls' simpatico can be determined with a simple test. The pearl or pearl strand should be placed against the inside wrist of the intended wearer. The inside wrist is a protected area of the skin which rarely tans

99 The artist is simply reacting to the look of the newer thin-skinned akoya pearls. Two decades ago pearls were left in the oyster for longer periods, normally exceeding two years. Thicker-skin Akoya pearls will exhibit the attributes the artist describes.

and is the same color and texture as the area around the throat and ears, the likely area where the pearl will be worn. The simpatico between the skin and the pearl becomes readily apparent when several pearls of different hues are compared in this way. Simpatico is a test of compatibility, not quality.

Symmetry

Other factors are also necessary qualifications for a fine pearl. Symmetry historically has been of great importance. A perfectly round pearl was the most desired of all. In the world of natural pearls, round pearls barely exist (Chapter 24). However, since the introduction of the cultured pearl in the late 1920s, cultured round pearls are not particularly rare in the marketplace though they constitute less than two percent of the crop of the average pearl farm. The thin-skinned akoya pearl has accustomed consumers to expect a spherical perfection of form. Few can resist the perfection of an oval or egg-shaped pearl or the sensuous curves of a pear shape. Ah, but still, there is something so elegant about a round!

The closer the pearl comes to achieving perfect symmetry, the greater its value. In the marketplace, round is the most desired shape, followed by pear, oval, and button. All other shapes are classified as baroque. Prices of baroque pearls can vary widely. The nuclei implanted in cultured pearls are perfectly round and blemish-free; however, nature has no commitment to symmetry

The Canning Brooch. A natural baroque pearl forms the upper body of the merman figure. Photo © Victoria & Albert Museum, London.

and can play many little tricks during the pearl's growth. As a result, the average harvest yields only a very small percentage of perfectly round pearls.

Designers often find inspiration in off-shaped pearls or baroques. Several historically important jewels such as the Canning brooch were crafted from freeform pearls whose shape suggested a sculptural form.

Size

All factors being equal, the larger the pearl the more valuable it will be. Natural pearl prices vary widely and the factors affecting value will be discussed in the appropriate chapter. Cultured pearl prices increase in an arithmetic progression up to seven and one half millimeters. From eight millimeters the price increases become larger. South Sea cultured pearls, both black and white, are seldom seen under eight millimeters. This is because the mollusk itself is larger than the oyster which produces the akoya and freshwater pearl, and is capable of accepting implants of up to ten millimeters. In the warm waters of the South Pacific, nacre will accumulate at roughly five times the rate that it accumulates in the relatively cool waters off the coast of Japan. The physical and economic conditions in the South Seas are right for the production of larger pearls. For this reason akoya pearls above 8mm will increase dramatically in price whereas South Sea pearls will not increase until they reach 12mm; at that size, pearl prices take another giant leap. At 16mm a cultured pearl is considered to be quite rare; a fine round pearl over 20mm is a museum piece.

Texture

A pearl's surface may be bumpy or smooth or textured. Texture on the skin may be interpreted as coarse and, if it detracts from the pearl's beauty, a distinct negative. But many bumpy baroque pearls, particularly freshwater pearls from China, exhibit excellent luster coupled with rainbow-like orient caused by light rays refracting and colliding with each other as they bounce from the pearl's surface. Perhaps this is a visual *mea culpa* to make up for the baroque's lack of formal perfection. This phenomenon is distinct from orient, and is caused by light entering the translucent crystalline layers of nacre. Technically described as light interference,

this rainbow iridescence is rarely found in a smooth skinned, symmetrical pearl.

Bumps, blemishes, tiny pits, or anything intruding on a perfectly flawless skin is considered something of a negative. The skin should be smooth and silky. Most pearls will have slight imperfections visible somewhere on the skin or surface of the pearl; the issue is how much they disturb the eye. In the case of symmetrical pearls, connoisseurs are most concerned with how the pearl will "face up," that is, if imperfections will still be visible when the pearl is set. The best face-up position is found by rotating the pearl. In baroque and button shapes imperfections visible on the backside are of much less concern. However, pearls with surface cracks anywhere on the skin are considered almost worthless.

It may be that we are entering another golden age of pearls. In Roman times, pearls were ounce for ounce the most valuable thing on earth. The fashion for pearls has waxed and waned over the centuries, at times eclipsing the value of diamonds and other crystalline gems. In the twentieth century, pearl production and demand reached its height in the 1920s, only to plummet in the early 1930s under the twin pressures of the stock market crash and the uncertainty generated by the introduction of the cultured pearl.

Today there are greater quantities of more varieties of pearl available than at any time in history. The American freshwater pearl, the Chinese freshwater pearl and the black South Sea pearl are available in quantity. South Sea pearls from the

Philippines and Indonesia are new to the market. Burmese and Baja California pearls are reentering the market. Pearl sales have increased dramatically in this decade and the consumer has become more accepting of variations in shape, size, and color.

The pearl horizon seems sunnier than at any time in the past sixty years. However, a few dark, dirty little clouds lurk just at the edge of the horizon. Pearl treatments have also multiplied. In addition to bleaching, a venerable treatment stretching back into history, dyeing, irradiation, waxing, coating, heating, and other treatments are proliferating as the demand for the pearl increases. Although treated pearls are of little interest to the connoisseur, treatments are acceptable so long as they are disclosed to the buyer. But, in many cases, these treatments are not disclosed. Cultured pearls are rarely sold with laboratory reports, natural pearls are rarely sold without one. If the aficionado is contemplating a significant purchase, it would be wise to insist upon a report from a well-known independent lab.

Hints on the Appreciation of Pearls

As one well-known dealer who grew up in the Japanese pearl business once pointed out, perfectly round cultured pearls are either an extreme rarity or a highly processed fraud.[100] The point is that most pearls are not round, not white, nor are

they perfectly matched. Pearls, even the cultured variety, are a natural product. We have become so enamored of the highly processed, thin-skinned Japanese akoya pearls that we are unable to appreciate the true virtues of unenhanced pearls.

Baroque pearls are not to be disparaged. Symmetry is only one part of the value equation. As with all gemstones, beauty is a balance of factors. In the case of pearls these factors include hue, orient (translucency), luster, texture, and shape.

There is much more to say about pearls. In Part II (Chapters 24-28) we explore the characteristics and idiosyncracies specific to various types of pearls both cultured and natural.

---

100 Fuji Voll, personal communication, 1988.

Chapter 6

Caveats:
Gemstone Enhancement

"...the stones are heated in the crucibles which are specifically designed for heating gems. This process is continued for a period sufficient to melt a mithqal of gold. A poultice is applied to the stones for cooling them. The stone finally crystalizes as a clear and transparent gem and fetches a higher price.[101]

–al Beruni 1052 AD

High tech ovens used in the heat treatment of Montana sapphire. Except for precision control of temperature and duration, the technique differs little from that described by Al Beruni. Photo: R. W. Wise.

The treatment of gemstones to enhance their visual appearance has a history going back at least three thousand years.[102] Excepting the ability to precisely control heating and cooling, the technology described above in the 11th century quotation differs little from treatment processes in use today. Advances in technology have added a few new tricks to the repertoire. Dyeing, oiling, heating, irradiation, and lasering are some of the better-known treatments.

Historically, a distinction has been made (in the trade) between treatments: those which should be disclosed to the buyer and those that need not be. Treatments, which are not permanent, such as dyeing, were to be disclosed. Permanent treatments, such as the heat treatment of ruby and sapphire, and those done ubiquitously – that is, to all gems of a particular type (such as oiling of emerald) – needed not be disclosed. The rationale for this was difficult to justify or understand. Happily, this situation has changed and complete disclosure is now the standard. Unfortunately many jewelers are still relatively unsophisticated in the area of treatment, and treated gems set in jewelry are frequently sold without any disclosure whatsoever.

Terminology is important. Some organizations such as the American Gem Trade Association, an organization of gemstone wholesalers, make a distinction between treatment and enhancement.

101 Al Beruni, The Book Most Comprehensive in knowledge on Precious Stones, (New Delhi, Adam Publishers and Distributors, 2007), p. 37

102 Kurt Nassau, *Gemstone Enhancement*, (London, Butterworths , 1984), p. 7.

AGTA prefers the latter term because it sounds better. From the perspective of the aficionado the terms are interchangeable. Something has been done, other than cutting and polishing, to change the look of the gem. That something should be disclosed in writing on the sales document.

Recent revisions of guidelines published by the Federal Trade Commission in the United States have begun to clarify this issue. According to the new guidelines, disclosure of a treatment is required if it has "a significant effect on the stone's value."[103] The issue is somewhat muddied by the fact that some types of treatments have a real effect on the value and price of a gemstone in the marketplace and others do not. One example is lasering, the use of a laser to drill a tiny channel into a diamond so that acid can be introduced through the hole to bleach dark inclusions and thus improve the diamond's apparent clarity. Under previous guidelines, only nonpermanent treatments were to be disclosed to the buyer. Lasering is permanent and does affect price so under the new guidelines must be disclosed to the buyer.

Other types of enhancement, such as heat-treating of aquamarine and tourmaline, are both impossible to detect and have little or no effect on the value of these gemstones. Heat-treating of ruby and sapphire is detectable and will have a significant effect on price, particularly on larger, finer stones. In certain cases it's the treatment that makes the gem. Tanzanite, for example, in its natural state, is an unattractive dark gray to root beer brown. Heat treatment, baking the stone in an oven at approximately twelve hundred degrees Fahrenheit, drives off the brown, leaving behind a lovely violet to purplish blue. The types of treatments used on specific gem varieties and their effect on that gemstone's value will be discussed later in this book in the chapters dealing with the specific gem.

In the detection of treatments, gemologists are constantly playing catch-up. Treatments often are initially created to fool the prospective buyer. It is up to the gemologist to figure out first that something has been done and, second, how to detect it. With advances in treatment technology it has become almost impossible for the average jewelry store gemologist to keep up. This situation has led to the rise of the gem laboratory.

These labs are staffed by specialists and have the advanced equipment necessary to detect more sophisticated treatments. The aficionado would be wise to insist that any and all treatments be disclosed in writing on the sales document. In cases where the purchase is significant, the buyer should insist on a certificate from a recognized gemological laboratory with a specific statement regarding the presence or absence of any treatments or enhancements.

---

103 Chapter I of Title 16 of the Code of Federal Regulations, Part 23: Guides for Jewelry, Precious Metals, and Pewter Industries. The legal requirement to disclose a treatment turns on the definition of "significant." The FTC guidelines do not require full disclosure. The FTC decided that there should be a practical, commonsense limitation on when disclosures should be made. The federal register's notice, which goes along with the regulation and further defines how it is to be interpreted, stated: "Disclosure of permanent treatments is necessary only where the treatment's effect on the value is likely to affect a consumer's purchasing decision." Thus, if the treatment has no monetary effect on value it would seem that no disclosure is necessary.

Chapter 7
New Sources

"The list of precious and semi-precious stones must be kept open sine die, like the lists of heavenly bodies, of the elements, the planets, the species of insects. New discoveries are constantly being added."

–Louis Kornitzer, 1930

Africa: Cradle of New Gemstones

In the 1960s one completely unknown gemstone and a host of new varieties of known gemstones were discovered in East Africa. Tanzanite, a normally blue but sometimes green variety of the mineral zoisite, was discovered near Merelani in Tanzania. Tsavorite, a rich green variety of grossularite garnet, was also first found in Tanzania near the village of Komolo, not far from the Kenya border.

Since that time various unusual garnets as well as sapphire, ruby, spinel, and alexandrite have been unearthed in East Africa, all within a geological formation known as the Mozambique belt. Consisting of high-grade metamorphic rocks, this geological formation stretches three thousand miles up the east coast of Africa: from Mozambique in the south-southeast through the countries of Tanzania, Kenya, Ethiopia, and Somalia and extending into the Red Sea. The rocks of the Mozambique belt are a diverse mix of volcanic rocks, ancient sediments, and intrusions which have endured several metamorphic phases,

Masai warrior takes a break along the long dusty road to Tsavo. Photo: R. W. Wise.

entirely altering the original character of the rocks.[104] In most cases, gemstones from these new locations have visual characteristics which are different from the same gemstone types found at traditional sources.

Other parts of Africa have also yielded

---

104 N.R. Barot, personal communication, 1998.

new finds of known gemstones. Zambia produces fine amethyst and along with

Eighty-five years old, this venerable **gariem-piero** *is still working the mines; Minas Gerais, Brazil. Photo: R. W. Wise.*

Zimbabwe is producing emerald with visual characteristics differing from each other and from those found at traditional locations in Colombia. Recently, demantoid, green tourmaline and mandarin garnet have been found in Namibia; red and green tourmaline and spessartite garnet strikes have been made in northwestern Nigeria. Opal was first discovered in Ethiopia in the early 1990s with a major discovery in Wello Province in 2006 and at Wollo in 2013. As

of this writing (2016) Ethiopia has become a major opal producer.

Two of the most exciting new areas are Mozambique and Madagascar.   In recent years a huge variety of gemstones have been found in East Africa and on the Spice Island. Sapphire, ruby, emerald, alexandrite, tourmaline and an array of interesting new varieties of garnet, all with unique visual characteristics, are currently being mined.

Brazil: Gem Paradise

Since 1554 when the Portuguese explorer Francisco Spinoza first discovered green tourmaline in the foothills of the Brazilian state of Minas Gerais, Brazil has consistently yielded new surprises to titillate and amaze the gem world. Tourmaline, along with spinel, is the most consistently underappreciated gem of all time. Originally termed "Brazilian emerald," it took two hundred years and the birth of modern gemology before tourmaline was recognized as a distinct gem species.

In addition to tourmaline, Brazil is the world's largest producer of an abundance of gems, including aquamarine, amethyst, citrine, emerald, garnet, opal, and diamond. A twenty square-mile area surrounding the town of Ouro Preto, in the state of Minas Gerais, is the sole major producer of commercial quantities of topaz. Two hundred miles north, near Hematita, lies the site of the single major find of gem-quality alexandrite.

In the early 1980s a new source of emerald was located in the state of Goiás at Santa Terezinha. Perhaps the most exciting

Remains of 19th century gem washing rig at Montana's Gem Mountain. Photo: R. W. Wise.

North America

While certainly no gem paradise, North America or, more specifically, the United States, produces its share of beautiful gemstones. America's oldest gem mine is located in the state of Maine. Tourmaline deposits were first discovered at Mount Mica, outside Paris, Maine, in 1820. This deposit, despite several interruptions, is currently in production.[105] In 1972 a huge pocket of gem quality tourmaline was located thirty miles away at Plumbago Mountain, in Newry, Maine. These two mines are part of a single geological formation called a pegmatite runing in a straight line from Brunswick on the Atlantic coast through Paris and Newry to the New Hampshire border. Approximately seventy-five percent of the gemstones mined at Plumbago are pastel pink to violetish red gems. The remainders, along with the majority of production from Mount Mica, are the lovely mint green stones for which Maine is justly famous. Mount Mica and the Havey Quarry are currently producing tourmaline.

California is known for its tourmaline. The major tourmaline producing area is found at the Pala and Mesa Grande districts in San Diego County. These areas were major producers of tourmalines in the

discovery of all was near the border of the western state of Rio Grande do Norte, just outside the village of Paraíba. A new variety called cuprian tourmaline was found, which derived its incredibly saturated color from trace elements of copper and gold. Marketed as Paraíba tourmaline, this gem set the gem world on its ear as the price of Paraiba tourmaline escalated to heights previously unheard of for this lowly gem species. More recently, other deposits of cuprian tourmaline have been found in Nigeria and Mozambique.

Very little gem material is currently being produced in Brazil. Surface deposits are worked out and the increasing capital outlays for deep underground mining coupled with environmental regulations have discouraged investment in gem mining.

105 Richard W. Wise, "Oldest Mine in the U.S. Reopens," Colored Stone Magazine, (July/August 1992), cover, p. 8. See also John Sinkankas, Gemstones of North America, vol. 3 (Tucson, Arizona: Geoscience Press, 1997), pp. 468-470.

late 19th and early 20th centuries. The Mountain Lily, Carmelita and Ocean View Mines are currently being worked. The historic Tourmaline Queen, Pala Chief, and Elizabeth R. mines are still worked sporadically.

In ancient times, certain agates, specifically carnelian, were ranked among the precious gemstones. If beauty is any criterion, history is about to repeat itself. Along with carnelian, which is found in Washington State, gem chrysocolla, found mainly in Arizona, deserves a place among the new precious stones.

Australia: New Wins From Down Under

While no entirely new species of gemstone has been found in Australia, a number of new varieties of known gems have been located in the land *Down Under*. Perhaps the most famous of these is opal.

Generally speaking, Australia produces the finest opal in the world. In fact, until the discovery of opal in Ethiopia, just about all the opal in the world market came from Australia. Before the Australian discoveries the only known source of gem opal were the diggings southwest of the northern end of the Carpathian Mountains in what is now Slovakia. This source of opal may have been known in Roman times, and produced opal with a milky white body color.

Opal was first discovered in the Australian state of Queensland at Springshure in 1872. Opal types are classified by body color, into six basic types: black, white, gray or semi-black, boulder, crystal, and fire opal. Five of the six basic opal types, excluding the orange-based fire opal, are found in Australia. Two other precious stones, sapphire and diamond, are also found in Australia.

Chrysoprase, the apple green variety of chalcedony, has been discovered at two locations in Australia. One deposit was found at Marlborough Creek in Queensland and another near Yerilla in Australia's Northern Territory.

Boulder opal rough for sale in the Queensland outback, Australia Photo: R. W. Wise.

PROBES: A NEW LIST OF PRECIOUS GEMSTONES

The charge legitimately can be made that the selection of stones I have designated as precious and included in Part II simply represents the author's opinion and is entirely subjective. In reply, I plead *nolo contendere*. The term *precious* is a qualitative designation and, as has been demonstrated earlier in this volume, a somewhat arbitrary label. Simply put, the gems selected are, in my opinion, the most beautiful and the most important gemstones available today.

As for the aesthetic judgments: these should be seen as what the late philosopher Marshall McLuhan called probes. They are attempts to clarify; they represent the author's best shot at the truth. If the reader simply accepts uncritically the author's conclusions, this book will have failed in its primary objective to aid the reader in developing true connoisseurship. A fine gemstone is like a fine wine. It must be held up to the light, tasted, rolled around on the tongue, savored, and finally judged. Not all wines appeal to all palates — so too gemstones. An honest disagreement based on reasoned contemplation is the beginning of true discrimination.

Many of the essays in Part II, particularly those on gemstones discovered in the last fifty years, contain the first critical

discussions of the aesthetics of these gem species and varieties written for public consumption. They are, therefore, a jumping-off place, a beginning, a point of departure. Other opinions will certainly follow, some of which will challenge the author's conclusions. So be it! Or, as the Roman emperors are reputed to have said upon entering the Coliseum: "Let the games begin!"

In my opinion, the finest — the most beautiful — examples of each gem variety may be termed precious. As a practical matter, it is not possible to discuss in detail every variety of every gemstone. Naturally, essays on traditional gemstones are included here, but an equal emphasis is placed on the newer precious stones. Some of these new precious gems are recent discoveries, some are simply newer sources of traditional favorites, and some are gem species such as tourmaline that have been kicked around for centuries and are just now beginning to be appreciated. The primary criterion I have chosen is beauty, not tradition.

In one case, akoya pearl, the ubiquitous white Japanese saltwater pearl that is part of every woman's wardrobe, has been excluded from the list. This is because the culturing

process is so short and the pearl has such thin skin and is so highly processed (bleached and dyed) it no longer possesses the quality characteristics that make for a true natural product.[106]

Preciousness is an evolving concept. Thus this book's list of precious stones specifically includes, for the first time in over a thousand years, four varieties of chalcedony and two varieties of feldspar: Holley/Mojave blue agate, gem chrysocolla, carnelian, chrysoprase, moonstone and sunstone. These gem varieties were selected because of their breathtaking beauty. No one who looks at these four gems with an unbiased eye could possibly call them semi-anything.

Durability and rarity, as limiting factors, have also played a part in the selection. A gemstone should be durable enough to be set in jewelry. In fact, it was partly the amazing durability of gemstones that first brought them to man's attention. Therefore faceted gemstones with a hardness less than six and one half on the Mohs scale, stones that are softer than steel, have been excluded from the list of the precious stones included here. Although advances in lapidary technique have made it possible to fashion many exotic materials, this does not automatically qualify them as precious gemstones.

Extremely rare materials, those not available in sufficient quantity to make a market, have also been excluded. Gemstones such as red beryl, which is every bit as beautiful as its siblings, emerald and aquamarine, is not covered simply because there is too little of the material around, and space dictates that a line be drawn somewhere.

Due to their extreme rarity, fancy color diamonds exist in a world of their own. They are without question one of the rarest, most costly substances on earth. They are, in fact, the only diamonds that are truly rare. Connoisseurship, or at least the current value of fancy color diamonds, is based on the rarity of a given hue. Beauty is, at best, a secondary consideration. Despite this fact, fancy color diamonds are of great interest to collectors, and for that reason have been included in this volume.

Part II (2nd edition) adds eleven *additional* gems; jade, natural nacreous pearls, conch pearls, moonstone, sunstone, cobalt blue spinel, red/pink spinel, peridot, demantoid garnet and Golconda (type IIa) diamonds for a total of the forty-five gems I believe should be included in any contemporary list of precious gemstones. In this section the principles discussed in Part I are applied to these specific gem varieties. The reader should note that the essays in Part II assume that the reader has studied and understands these principles.

106. "When Mikimoto first cultured his pearls, he left his oysters in the water for four to six years, a time span that produced a luxurious nacre coating that would last for generations. During every decade since 1960, most Japanese pearl farmers clipped another year off culturing times. In the 1990s we first saw pearls that had been in the water six months or less. The nacre was so thin you could flick it off with a fingernail." Fred Ward, "The Wisdom of Pearls," *Pearl World, The International Pearling Journal*, vol. 11, no. 2, (July/August/September 2002), 11.

Chapter 8
Alexandrite

The Ceylonese alexandrites are, on the whole, finer than the Uralian, the columbine red colour seen in artificial light being especially beautiful. . . Its colour by daylight was a fine sap-green with a trace of red, while in candlelight it appeared a full columbine-red, scarcely distinguishable from a purplish-red Siamese Spinel."

–Max Bauer, 1909

Ask the mirror on the wall which is the rarest gemstone of them all? If the answer is not alexandrite, trade in the mirror. With the exception of some varieties of fancy color diamonds, alexandrite is indeed the rarest of them all. Alexandrite is a variety of chrysoberyl which changes color when exposed to different light sources. Daylight and incandescent are the traditional light sources for evaluating color change.

Alexandrite was discovered in 1830 in an emerald mine near the Tokovaya River in the southern part of Russia's Ural Mountains. The stone was discovered by the mine's manager, Yakov Kokovin.[107] Later, at the suggestion of Finnish geologist Nils von Nordenskjold, the stone was named Alexandrite in honor of Alexander Nikolayevich, soon to become the ill-starred Czar Alexander II. Later, small deposits of alexandrite were located on the island of Ceylon. For the next ninety-some years these were the only known sources of this rare gemstone.

What is truly remarkable about

alexandrite is not that it changes color; several gemstone varieties, including sapphire and some members of the garnet group, will shift color when exposed to different light sources. Most other gemstones which exhibit color change will shift only to the adjacent hue on the color wheel; for example, from red to violet or blue to purple. Alexandrite fairly leaps across the color wheel, shifting from hues of green to blue green in daylight to red purple in incandescent light.

In the mid-1980s the first and thus far only major new strike of alexandrite was made in the Brazilian state of Minas Gerais at Hematita. In this case the term "major" should be put in perspective. Once news of the find leaked out, three thousand garimpeiros, the Brazilian term for independent prospectors, descended on a small valley, five hundred feet wide by six hundred fifty feet long, and began digging.

The run lasted approximately twelve weeks, April to June of 1987, with an average of one death by gunshot per week until the bloodshed resulted in a government order to shut down the mining area. By this time the area was essentially

107 Schmetzer, Karl. *Russian Alexandrites*, (Schweizerbart Science Publishers, 2010), p.25.

mined out. The estimated production from this strike was two hundred fifty thousand gem carats in the rough.[108] Figuring a weight loss of forty percent in cutting the stones, this translates into one hundred fifty thousand carats of cut gemstones. This constitues a very small amount when compared, for example, to diamond, with a production of over one hundred million carats per year. Alexandrite is a very rare gem.

Before this strike, small amounts of alexandrite had been found a bit further north in Minas Gerais at Malacacheta. Since the big strike ended in June 1987, production from Hematita has been sporadic. Small amounts of alexandrite have come from Tanzania and Madagascar; very limited amounts have begun coming out of Russia since the fall of the Soviet system and more recently from India.

Color Change: The Key Factor

Alexandrite is defined by its color change. If it doesn't change color from red to green, it's not alexandrite; it is simply chrysoberyl or if it changes to any other color, color change chrysoberyl. The traditional view is that the best alexandrite shifts from an emerald green to a ruby red. This is a bit wide of the mark. A majority of gems from Hematita show a teal or greenish blue to a slightly yellowish green in daylight, turning a violetish red to purple in evening light. The strength of the color change is the key element in the judgment equation.

Two views Realistic color change in a typical fine 3.51 carat alexandrite from Brazil, purple in incandescent, slightly grayish green in daylight. The gem would appear more reddish in candlelight. Photo: Jeff Scovil, The Edward Arthur Metzger Gem Collection, Bassett, W. A. and Skalwold, E.A.

To be considered a fine stone, the color shift must be distinct and dramatic with little of the former color left after the shift. Generally speaking, stones with a ninety percent shift should be considered fine. For example, a stone which is teal green in daylight should become purplish red under the light bulb, with no more than a tiny bit of green left over after the lighting environment has been switched.

108. Keith Proctor, "Chrysoberyl and Alexandrite from the Pegmatite Districts of Minas Gerais", *Gems & Gemology*, Spring 1988, pp. 26-28.

Judging the Change

Although gems should always be observed in a variety of lighting environments, when viewing alexandrite it is especially important to pay attention to the light source. True daylight and old-fashioned light bulb lamps (or light from a candle) established the color standard.[109] Beware the dealer's lamps: they most always are carefully chosen to show the stone to best advantage. Lamps marked "daylight" fluorescent come in a variety of kelvin temperatures, and new halogen incandescent lamps coming on the market can generate light close to the daylight range (see Chapter 4). As lighting technology improves, there no doubt will come a time when dealers will be able to "tune" the light and thereby alter the apparent color of an alexandrite to show the stone to its best advantage. In fact, that can be done to some extent today. Meanwhile, the aficionado is advised to observe the stone in true daylight and then in incandescent light.

Hue

In alexandrite, a stone which shows a pure hue — that is, a shift from a relatively pure red to a relatively pure green or greenish blue — is the most desired. Ruby red to emerald green stones are reputed to exist, but if they do I have yet to see one, except in a famous photograph (pictured below) by the late Tino Hammid. According to Mary Murphy Hammid, a well-known gem connoisseur and owner of the stone in the photograph, the green has absolutely no gray mask and is a pine green; the red has a definite purplish secondary hue.[110]

Pioneering gemologist Max Bauer, writing in 1904, describes the Uralian stones as being green in the daylight and "red to violet" by night, "an emerald by day and an amethyst by night."[111] He also describes the daytime color as "grass green to emerald green" and the nighttime color as "columbine red inclined to violet."[112]

Part of the color confusion is due to the fact that very little Russian alexandrite has been seen in the West since the Russian Revolution of 1917, the major part of the strike having been exhausted some years before. Still, an aura of legend clings to the appellation Russian alexandrite. The color range described by Bauer is remarkably similar to the color range for Brazilian alexandrite described in the following discussion of hue. Bauer even uses the term "raspberry red" to describe alexandrite

109 . In the mid–19th century, prior to the invention of the incandescent bulb, candlelight was used to describe the color change. A letter from Count L. A. Perovskii to Finnish geologist von Nordenskiold confirms that this was the method used by the discoverers. Cf: Schmetzer, Karl, *Russian Alexandrites*, Schweizerbart Scientific Publishers 2010, pp. 25-29. At 1,500 kelvin, candlelight enters the red range of color as perceived by the eye. The color temperature of the average light bulb, by contrast, is 2,700 kelvin, distinctly into the yellow range. This is, no doubt, part of the reason why alexandrite that turns the mythic ruby red night color is almost impossible to find.

110. Mary Murphy Hammid, personal communication, 2002. Ms. Hammid further states that she purchased the stone in rough specimen from the estate of Ralph Holmes, late professor of mineralogy at Columbia, in 1977, for the princely sum of $75. The finished stone weighs 1.29 carats and was cut by the late Reggie Miller. The refractive index establishes Russia as the country of origin.

111. Max Bauer, *Precious Stones*, trans. L.A. Spencer (New York: Dover Editions, 1968) vol. 2, p. 305.

112. Ibid.

The Myth: the famous Hammid alexandrite. This altered image of a Russian alexandrite appeared in many articles for more than a decade fueling the legend that the best color change was a pure red to a pure green hue Photo: Tino Hammid.

from Sri Lanka. However, the colors of the original Uralian material were sufficiently like the Russian Imperial Military colors, red and green, as to make the gem extremely popular among the Russian aristocracy.

The color range of alexandrite in the world market today varies somewhat according to source. Brazilian stones will exhibit a reddish-purple to purple-red in incandescent light. Some exceptional stones from Hematita will show a blue primary hue with as little as fifteen percent greenish secondary hue when exposed to daylight. The Brazilians call this color *pavão*, which

translates to *peacock blue*. The evening color of Tanzanian and Madagascar stones is more to the purple.

Brazilian alexandrite is more likely to show a bluish green to a blue green in daylight. Stones from Tanzania normally show a purer green daylight color. Tanzanian alexandrite will sometimes exhibit both day and night colors simultaneously when viewed under the light bulb, particularly when some diffused daylight is present. While this bicolor pyrotechnical display may be technically a fault, it is certainly a most fetching one.

The green daylight color of gems from Hematita is rarely found without some admixture of yellow, which visually dilutes the green. In such cases, the less prominent the yellow secondary hue shown in daylight, the better. The finest daylight color of Brazilian alexandrite is a rich bluish eighty-five percent green with perhaps fifteen percent blue. The market loves visual purity of hue, especially green; however, the teal blue pavão daylight hue is, to my eye, more desirable.

The nighttime colors of Brazilian alexandrite will normally vary from violetish pink through pinkish-purplish red to a purplish red. The red night color of Brazilian alexandrite will always have a strong purple secondary hue. Of the three ranges, pinkish is the least desirable. The most beautiful evening color of the Brazilian alexandrite can be described as reddish purple or raspberry. It is a red more purple than pink. It's the color that stains the mouth and hands while picking wild berries in the sultry heat of a late summer

afternoon. The evening color of Tanzanian alexandrite is normally more toward the purple with just a bit of reddish (ten to fifteen percent) secondary hue.

Saturation and Tone

Except for the very finest examples, alexandrite is not a particularly beautiful gemstone. Although gray is predominant, brown can be found as a saturation modifier or mask in alexandrite. Brazilian and Madagascar gems tend toward a distinct grayish mask; stones from Tanzania tend to be brownish. This tendency increases with size. As a result, stones over five carats which do not show a mask are extremely rare.

Most stones from all sources occur in paler tones of forty to sixty percent with weak color change. Those with a strong color change often will have tones of seventy-five to eighty-five percent plus which, coupled with poor crystal, give them a murky overcolor appearance. Day and night tones will usually correlate. That is, a light daylight green of fifty percent tone will shift to a light pinkish color of fifty percent tone in incandescent.

Crystal

The best tone for alexandrite falls between seventy and seventy-five percent. Eye-clean stones with tonal values closer to seventy percent are more transparent, have better crystal, and will exhibit color shifts that are truly breathtaking. The more transparent the gem the more visible the color shift. Darker-toned Brazilian stones rarely show good crystal. The best of the gems from Tanzania have greater

transparency and exhibit better crystal than those from Brazil. Although currently less highly valued in the marketplace, the finest of the Tanzanian alexandrite is among the best in the world.[113]

Cut

Due to alexandrite's extreme rarity, almost anything found will be cut! In order to retain as much weight as possible, alexandrite often is offered highly included, flawed, and very poorly proportioned. Well-cut stones are the exception rather than the rule and the serious collector should be prepared to make some compromise if the stone is a genuine beauty. It is possible to find stones that are well cut or at least face up well; the serious aficionado will not be deterred by the hunt can be the most exciting part of collecting.

Brazil Versus Tanzania and India

Brazilian alexandrite is the current market standard bearer. Excepting the Russian stones, this means that the characteristics typical of fine stones from Hematita, at this writing, command the highest prices in the market. Thus, an eye-flawless bluish green to raspberry (purple red) gem of eighty percent tone with an eighty-five percent color shift will command the top price. Tanzanian gems with their green to reddish purple color shift, their superior crystal, and their charming tendency to exhibit both hues in mixed lighting, will carry a substantial discount. What's a collector to do? Indian stones, as stated above, tend to have a purer green day

113. Allen Kleiman, personal communication, 2002.

color. With the new federal standards for incandescent lighting effective in 2012, the colors you see may be quite different than what you saw under the old light bulb.

As discussed in Part I, gemstones from newly discovered areas often are compared to those from traditional sources, usually to the detriment of the newer source. This is an expression of the innate conservatism of the gem market and was precisely the position that Brazilian stones occupied in the early 1990s when Russian stones were considered the standard bearers.

Today Brazilian stones have what is considered to be the classic look. Remember, in the final analysis, gems have no pedigree. The idea that one gem is better than another simply because it comes from a famous source is both a snare and a delusion. The current low price for Tanzanian gems with their superior crystal presents the aficionado with a buying opportunity that should not be missed.

Synthetics and Treatment

Alexandrite is not usually subjected to color or clarity treatments. Synthetic alexandrite exists and is particularly difficult to separate from natural. The separation requires equipment not normally available to the jeweler-gemologist. The aficionado should insist on a grading report (certificate) from a well-known independent gem laboratory clearly specifying natural origin when purchasing alexandrite.

Cat's-Eye

Alexandrite is a variety of chrysoberyl that can also occur as a cat's-eye

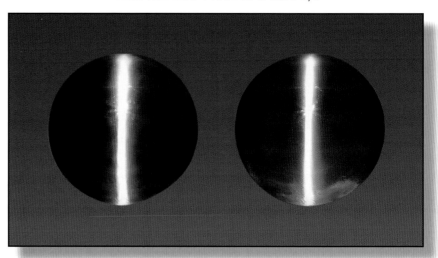

World-class cat's eye alexandrite from Brazil showing both its green daylight and nighttime colors. Fig. Figure 50: cat's-eye Alexandrite. Photo: Harold & Erica Van Pelt, courtesy of Kalil Elawar.

stone. Alexandrite cat's-eye is simply a chrysoberyl cat's-eye which changes color in the prescribed manner. Chrysoberyl is not the only gem exhibiting a cat's-eye phenomenon. Gem varieties as diverse as emerald, tourmaline, and opal can also be cut as cat's-eyes if the requisite inclusions are present within the rough gem.

Because alexandrite cat's-eyes, like all others, are cut en cabochon, light must be transmitted through the stone to determine if it can be classified as alexandrite. The stone may be a cat's-eye, but to be considered alexandrite it must have the sort of color change discussed above. For the stone to be properly evaluated, the following must be considered: the four Cs, the strength and percentage of the color

shift, and the position, strength, and visual characteristics of the cat's-eye. Chapter 5 contains a detailed discussion of the grading of cat's-eye gems.

Translucency and Crystal

Crystal, the fourth C, plays a defining role in the evaluation of alexandrite and all other cat's-eye stones. Phenomenal stones are often dark toned and only semi-translucent. Stones of this description may have a very distinct eye. Once the quality of the phenomenon is determined, the next criterion in order of importance is transparency or crystal; that is, the more translucent and limpid the stone, the more visually pleasing it will be.

An opaque stone, even one with a sharp, well-defined eye, is less desirable. A high degree of translucency is the demarcation between the very good and the very finest qualities in the connoisseurship of alexandrite cat's-eye, as well as other cat's-eye gems. The finest cat's-eye stone is first of all a fine cabochon, one that glows, has a limpid, transparent body color, and, in addition, shows good color change and a bright well-formed eye.

As with alexandrite generally, cat's-eye alexandrite is usually either very light or very dark in tone. Darker stones in this case are more desirable because they will exhibit a much stronger eye. Though much rarer than faceted alexandrite, the cat's-eye type does not command nearly as high a price. Rough stones with a high degree of clarity, crystal, and color change are almost always faceted to bring the highest price. Therefore the most available cat's-eyes are stones with

tonal values of at least eighty-five percent. In daylight these stones will appear almost black, and it is necessary to shine a strong light directly into the stone to observe the color change.

The Rarity Factor

In 1990, shortly after the big strike at Hematita, I transported a large, fine alexandrite cat's-eye about the size of a small walnut to a local dealer in the Brazilian town of Teofilo Otoni. The dealer was seeking a single stone to complete a matched necklace suite. After coffee and a pleasant chat, the gentleman brought out a large box containing a dozen matched cat's-eyes. The suite was magnificent! The dealer was disappointed because after a brief examination it was clear that the fifty-carat gem I had brought was just too small to complete the suite. Does this mean that large alexandrite cat's-eyes are not rare? No, but it does point out the sort of temporary glut that can occur when a large pocket of even a very rare stone is found. Fine alexandrite is rare in any size. Rarity increases at three carats and dramatically over five carats; the next jump in rarity is at ten carats. Large fine alexandrites are true museum pieces.

Polishing a Brazilian amethyst. Photo: courtesy Nomads.

Chapter 9
Amethyst

"The Mursinsk (Siberian) amethyst at times is very dark violet blue, surpassing that from Ceylon, but mostly it is pale violet-blue or spotted and striped violet-blue and colorless."

–Gustav Rose, 1837

Quartz is earth's most plentiful mineral; it is the primary component of dust, making up twelve percent of the earth's crust. It is surprising, therefore, that the finest examples of amethyst, the purple variety of quartz, are so rare and difficult to find.

Hue and Tone

Amethyst occurs in a continuum of primary hues from a light-toned slightly pinkish violet to a deep Concord grape purple. Amethyst is expected to be eye-clean and, given its relatively low cost, should always be finely cut.

Amethyst may exhibit one or both of two possible secondary hues—red, blue or both. A light rosy red, what we think of as pink, is the usual secondary hue found in lighter-toned stones. The ideal tone for amethyst is between seventy-five and eighty percent; at this tonal level the secondary

Figure 51: 34.67 carat custom cut deep Siberian color Brazilian amethyst, cut by John Dyer. Photo: John Dyer.

hue, if there is one, may be red, blue or both. The color in amethyst may occur in overlapping zones of purple and blue. When the stone is viewed face up, the color will be a slightly (ten to fifteen percent) bluish purple. The blue adds a velvety richness to the purple hue.

Multicolor Effect

Multicolor effect is a key element in the connoisseurship of amethyst. The finest gems show a rich grape-juice

Figure 52: A 48.85-carat amethyst gem sculpture by Michael M. Dyber. The stone is a visually pure purple (Siberian color) of about eighty percent tone coupled with excellent crystal. Photo: Jeff Scovil; courtesy of R.W. Wise, Goldsmiths, Inc.

purple body color with deep red flashes in incandescent lighting. As shown on the color wheel, mixing red and blue make purple. Amethyst is a nightstone. When the stone is faceted, the relatively yellowish light of the incandescent bulb draws deep red flashes of brilliance from the heart of the stone. This describes the finest color in amethyst and is called *deep Siberian* color. The red flash is the chief difference between deep Siberian and the number two or Siberian color. Under the light bulb, Siberian stones will show a deep purple hue but no red in its key color. This difference is visible only in incandescent light.

Saturation

The usual saturation or intensity modifier in amethyst is gray although occasionally brown is also seen. In darker-toned stones the mask may be very hard to see, but pure chromatic hues are always vivid. Dark purple is normally a rich warm hue. Gray will dull the hue and give it a cool aspect. Brownish stones appear muddy and overdark. The finest stones will show little or no gray or brown mask.

The Finest

To reiterate, the finest amethyst can be described as an eye-clean stone with a primary purple hue between seventy-five and eighty percent tone with perhaps fifteen to twenty percent blue and (depending upon light source) red secondary hues— the deep Siberian quality.[114] Following this is the second quality, or Siberian, which has a slightly bluish purple key color. Deep Siberian color is exceptionally rare. I have spent weeks searching for this quality in Brazil and in Africa and often returned home empty-handed.

---

114. As indicated by the quotation at the beginning of this chapter, the term deep Siberian is more than a bit of a misnomer. A personal examination of the gems on display at Moscow's *Diamond Fund* (2010) confirmed this fact. Very little amethyst from the Siberian source ever achieved the finest quality. If there is a best source for amethyst it is — or rather, was — the West African country of Zambia. Gems from this strike were prevalent in the market in the late 1980s and early 1990s. I bought several kilos of rough Zambian amethyst in Nairobi in 1990. This material contained beautiful dark blue zones juxtaposed against purple which, when faceted, added a bluish secondary hue to the purple. This mixing of purple with a bit of blue lent the Zambian amethyst a lovely rich hue. Very little material has been available from this source since the mid-1990s.

Commercial grade amethyst occurs in tonal variations of violet, from ten to sixty percent tone. Lighter-toned gems (thirty to forty percent tone) with a pinkish violet hue are called *rose de France*. Gems of this type with a high degree of transparency (good crystal) are very reasonably priced and quite beautiful.

Beware of Synthetics

The collector must beware of synthetic amethyst. Synthetic amethyst, like most synthetics, is made to imitate the finest grades of the natural gemstone. Unfortunately, synthetic amethyst is impossible to differentiate with complete certainty from natural without the use of sophisticated gemological instruments normally unavailable to the jeweler-gemologist. Testing services are available but, given the relatively modest prices of even very fine smaller stones, the cost of testing may exceed the cost of the gem. The collector should buy only from dealers willing to certify natural origin on sales slips. An independent gem lab should test larger stones.

The Rarity Factor

Fine amethyst is rare in any size. Amethyst can be found in extremely large sizes. Stones tend to decrease in value on a per carat basis over twenty-five carats.

Fine aqumarine. Photo: courtesy Nomads.

THE BERYL FAMILY

Our pousada (a Brazilian B & B) sits with its backside hanging over a cliff overlooking the Brazilian town of Nova Era. Puffy white cumulus clouds glide across an aquamarine sky. The rainy season has just ended. The foothills are clad in a coat of green velvet, the grass cropped close by the broad flat teeth of grazing cattle. Hill follows hill, rolling toward the horizon like a sea of verdant waves. In the valley which contains the town, a winding river sambas along the valley floor, the water, like the clay tiles on the village roofs, stained a deep reddish brown. Just above the river a dirt road carves a livid ruddy scar, paralleling both the river's color and course.

The color of the iron-rich lateritic soil is characteristic of the entire northeastern portion of the Brazilian state of Minas Gerais, and is similar to other gem-rich areas throughout the world, including central Thailand, upper Burma, and the opal mining areas of western Queensland.

Five hundred million years ago super-hot liquid magma injected itself into gaps or cracks in the pre-existing country rock. As these intrusions cooled, rare minerals such as lithium, boron, beryllium, and manganese concentrated in pockets, aiding and abetting the formation of gem crystals such as tourmaline, emerald, and aquamarine.

Our jeep bounces along a one-lane dirt track, following the course of the Piracicaba River as it works its way

The Posso Grande aquamarine mine outside the town of Nova Era, Minas Gerais, Brazil.

westward off the main highway. Our destination is the Posso Grande aquamarine mine. Brazilians are optimistic by nature. In Portuguese *posso grande* means "it could be big." And maybe it will!

Emilio Castillo, a jolly, rotund *pedraista* (gem dealer) who has lived for twenty-eight years in Nova Era, tells us that Philippe, the mine's owner, has hit pay dirt no less than four separate times in this area. Philippe is a seasoned *gariempiero*, or independent miner, and, like many of his brothers, has a sort of Zen attitude toward fortune. In short, he lives for the day. With each strike he has made a fortune, and each time he has managed to allow that fortune to slip through his fingers.

Unfortunately for Philippe his much-touted luck appears to have deserted him. Just as he hit a pocket deep in the hill, disaster struck —a mudslide has buried his tunnel. We arrive at the mine to find Philippe and his five-man crew at loose ends. The mine is several hundred yards up a steep hillside above the river. Two previous tunnels have proved unproductive, and it will take backbreaking weeks of labor to dig away the slide by hand. Still, Philippe is philosophical. He is cutting green bamboo with his machete, the heart of which will be sliced and fried up for lunch. Money is tight, food is scarce, and there is no capital available to purchase heavy equipment which could considerably speed up the excavation.

Back at his home, Emilio pours several kilos of previously won Posso Grande aquamarine rough onto a table. Most pieces weigh less than three grams, all a limpid blue crystal without a trace of green. Aquamarine rough is normally greenish. Both the green and the blue color are derived from iron —the blue from ferrous iron and the green from the yellow of ferric iron. Gentle heating will drive off the ferric iron, leaving just the blue, but in this case, heating will not be necessary. With Emilio's permission, we select several pieces that we will cut into one- to three-carat gems.

Rebekah Wise sorting aquamarine rough from the Posso Grande Mine. Note the pure blue hue. This aquamarine has not been heat-treated.

After a country lunch at Emilio's home, and a detour so that he can present us a gift of a bottle of cachaça, the fiery Brazilian cane liquor, we continue our journey. The sky is cloudless and the afternoon turns sultry. We find ourselves bouncing our way up yet another dirt road, this time on the north side of the mountain from Posso Grande toward Capueirana, a village in the Nova Era region.

Nova Era first drew the attention of the gem world because of its emerald, a light- to medium-toned slightly yellowish to grass green limpid stone discovered in the 1980s. Many of the mines that

produced Nova Era emerald are clustered around this small village.

"This is also the back road to Hematita," Paulo Zonari, our host and driver, remarks. "Twenty years ago this was one of the busiest roads in Brazil." Busy indeed! For three months, thousands of garimpeiros worked a small valley east of Nova Era. This was the first and only major strike of alexandrite anywhere. After three months, and a great deal of blood letting, the valley was mostly mined out. Fortunes were made and fortunes were lost—often in a hail of bullets.

This afternoon we are to visit an emerald mine, one of the few still working in the area. We pass through the small village. The tiny whitewashed houses and tin-roofed shacks flank both sides of the narrow dirt track. A group of men are gathered on the veranda of a local bar to exchange the day's gossip. Glasses and beer bottles sit on narrow homemade wooden tables.

The men are dressed in traditional miner's garb, torn shirts and dirt-soiled pants. Some are shoeless, their swarthy faces burned dark by the tropical sun. They gaze, eyes narrowed, in our direction – some curiously, others with suspicion. We pull up to a gate. A small concrete house sits several yards up a hillside. It looks more like a small fazenda, or farm, but we have arrived at our destination – the mine owned by Sergio Martinez.

About twenty-five yards above the house a square hole leads to a round shaft – similar to the lebin, the square reinforced vertical shafts dug for millennia in the ruby-bearing soils of the Mogôk Valley of upper Burma. This one is a bit more up to date. A tin roof covers the shaft and a motorized pulley system has replaced the simple hand crank used to lower the miners and retrieve the gem gravel. I am strapped into a leather harness rigged to a tripod above the shaft. Below us a gaping black hole, perhaps six feet in diameter, drops three hundred feet straight into hell. Paulo introduces us to the mine manager; he will be my guide into the depths of the mine.

Suddenly we are at the bottom of the shaft. It is quiet. The single sound is an echoing drip of water coming from below us. We rappel off the shaft wall into a horizontal tunnel. There is no reinforcing structure, no supports of any kind. The cavernous shaft that has been carved out of coal-black schist reaches back into the darkness like the view into the belly of a whale. Am I apprehensive, perhaps a bit scared? Hell, no – I love this stuff! A

Author prepares to descend a vertical shaft leading to an emerald mine, Nova Era, Minas Gerais, Brazil. Photo: Rebekah T. Wise.

look back into the shaft, and I see the dangling harness, swaying limply. I was the only volunteer!

Tiny flakes of mica twinkle like stars against the black schist as the manager's flashlight plays across the sides and roof of the tunnel. The floor is wet and slippery and slopes upward. The air is cool and damp as we work our way deeper into the shaft. Behind us the steady dripping of water seems impossibly loud. We are, it seems, at the very roots of the mountain. At Posso Grande we found large "books" of mica, as much as twelve inches in diameter; here tiny flakes mix into the crumbling black schist.

My guide takes me down several tunnels, pointing out areas where emerald was found. The work here was done mostly by hand, dug out with pick and shovel and transported by wheelbarrow to the tunnel's mouth, then raised the three hundred feet to the surface. There is no one working underground today, but in the unnatural silence, I hear the faint echo of rock against steel.

I am not sorry to find myself back at the surface. The sky is somehow bluer and the air seems sweeter. Three men are at work on a huge pile of schist, two breaking up the larger chunks and one washing the gravel on a large corrugated steel table — all looking for the telltale green flash of emerald.

Later in the trip, Haissam Elawar tells me the real story of the Martinez mine. "Some years ago," Haissam begins, holding up his index finger to emphasis his point, "about thirty people invaded this mine." Haissam is Brazilian, born of Lebanese parents. He is a strongly built fellow with a broad face and an infectious laugh. To make a long story short: the owner of the mine was naturally upset having these squatters setting up in his mine and refusing to leave. He called the police. The police, including a high-ranking colonel of police, entered the mine. A gunfight ensued and the colonel, along with several others garimpeiros and police, were carried out on a slab.

The beryl family consists of four varieties: aquamarine, emerald, morganite, and red beryl. I have included two of these, aquamarine and emerald, among the new precious gemstones.

Chapter 10

Aquamarine

"An aquamarine, particularly of good deep blue-green colour, is a stone of great beauty, and it possesses the merit of preserving its purity of tint in artificial light."

–G-F Herbert Smith, 1910

Aquamarine is one of the few gemstones which sometimes requires the hand of man to give nature a bit of a nudge to produce the most beautiful gems.

Although the English word aquamarine comes from the Latin for "sea water" denoting a greenish blue hue, the gem aquamarine is valued chiefly for the purity of its blue hue, a condition normally achieved by gently heat treating the gem to drive off the green.[115] Aquamarine contains two types of iron, ferris and ferric. The former is responsible for the blue, the latter the green. Low temperature heating drives off the ferric iron, leaving behind the blue. Far from having a negative effect on the value of this gemstone, heat enhancement puts the finishing touches on nature's efforts.

Heat treatment is not always necessary. Some aquamarine rough occurs naturally in a pure blue color. However, it is safe to say that the vast majority of aquamarine is heat-treated. This type of treatment is undetectable and, in fact, increases the

A 9.39-carat Mozambique aquamarine showing a visually pure blue primary hue with no secondary green; tone is sixty percent. Jeff Scovil, courtesy of Mine Design

value of aquamarine.

Hue/Saturation/Tone

In judging the desirability of aquamarine, the marketplace considers tone, not saturation or hue, to be of primary importance. In simplest terms, this means that the darker the stone, the more

---

115. Some connoisseurs prefer the greenish blue of natural untreated aquamarine. Natural pure blues do exist.

it will be valued. Unfortunately, darker-toned stones are almost always grayish. The reader will recall that color in gemstones breaks down into three components: hue, saturation, and tone (see Chapter 3). In evaluating most gemstones, hue is far and away the most important of the three components. In aquamarine, this situation is altered: the darker the blue the better the stone. Thus, tone is primary. The hue should be, of course, a visually pure blue.

In practice, even gems that have been heat treated will normally show a bit of green, usually between five and twenty percent. Although some connoisseurs prefer a greenish stone, a greenish secondary (what might be called a true aquamarine hue), is generally considered a negative. A pure blue hue is the most desirable. The greater the percentage of green secondary hue, the less desirable the gem.

Aquamarine tends to be pale (ten to thirty percent tone) and the key color of stones of this description will have a washed-out look. The ideal tone is between fifty and sixty percent. Aquamarine is never too dark; all other factors being equal, the darker the tone the better the stone.[116]

The collector should take care not to be seduced by the gem's body color. Aquamarine is normally light in tone and quite transparent, and it is easy to be drawn in by the stone's lovely internal glow. This can be a costly mistake unless the gem in question is cut en cabochon. The key color of aquamarine is often quite a bit paler of

hue and lighter in tone than its body color, so much so that a delicate light to medium blue body color will often translate into a light-toned pallid key color.

Although a slight brown mask may be found in unheated stones, gray is the normal saturation modifier or mask found in aquamarine. Darker-toned stones will often be distinctly grayish and have a somewhat dull appearance. A bit of gray (five to ten percent), according to some experts, is said to enhance or to darken the blue. But, while a bit of gray can enhance, a bit more makes for a stone that is distinctly grayish and therefore dull. In the market place, the aficionado will be charged a premium for dark toned stones which are distinctly grayish. However, as with other blue gemstones discussed in this volume, a pure hue is still the most desirable. That said, in darker-toned gems, the smaller the percentages of gray mask the more desirable the stone.[117]

Despite current market opinion, the author prefers a stone a little lighter in tone (fifty percent) having visually pure robin's egg, medium sky blue hue with no visible green secondary hue or gray mask. This hue coupled with fine crystal is extraordinarily beautiful, particularly when compared to darker (grayer and duller) examples.

Nightstone

Aquamarine is a true lady of the evening, a nightstone. The aficionado is advised therefore to pay particular attention to the gem in daylight. As a rule of thumb, if a night stone looks good by day it will

---

116. C.R. Beesley, *Colorscan Training Manual* (New York: American Gemological Laboratories, Inc., 1984, part 2, p. 1.

117. Ibid.

16.50 carat aquamarine exhibiting medium-toned, visually pure, blue hue and fine crystal from Padre Paraiso, Minas Gerais, Brazil. Photo: Jeff Scovil, courtesy R. W. Wise, Goldsmiths.

only improve under the light of the bulb.

Crystal

One rarely thinks of crystal in connection with aquamarine because the gem is consistently highly transparent. However, super-diaphanous aquamarine does exist. Some of the finest is found in the northern part of the Brazilian state of Minas Gerais near the town of Padre Paraiso and also further north and east near Medina. Exceptional crystal punches up the luster, brilliance, and saturation of the gem. Gems of this description are best described as icy blue and even lighter-toned gems with fine crystal can be strikingly beautiful.[118]

In the early 1980s a process was developed to artificially enhance colorless

118. Richard W. Wise, "Gariempiero Dreams," *Gemkey Magazine*, (May/June 2000), p.42.

topaz to create a pure medium-toned blue that, to the untrained eye looked identical to the finest aquamarine. The market was then flooded with inexpensive blue topaz, causing a precipitous drop in the price of aquamarine. It took almost a decade for aquamarine to recover, but recover it did and today fine aquamarine sells for yet higher prices.

Much of the finest aquamarine comes from Brazil, specifically from the state of Minas Gerais. Although in recent years Africa has eclipsed Brazil as a source for aquamarine, the standard is still Brazilian. In 1954 a particularly fine crystal weighing thirty-four kilograms was found on a fazenda north of the town of Teofilo Otoni. The crystal was named *Marta Rocha* after the woman crowned Miss Brazil that year.. The color of the crystal was reputed to perfectly match her eyes.. Experts at the time pronounced gems cut from the Marta Rocha to be the finest quality aquamarine ever found. The sultry Miss Rocha is no longer with us, but gems cut from the Marta Rocha crystal can be found at the Hans Stern Museum in Rio de Janeiro.

Treatments

As mentioned above, aquamarine is normally heat treated to drive off the yellow

component of its hue in order to enhance the blue. This treatment is applied at relatively low temperature and is currently not detectable by standard gemological testing. Heat treatment adds to the value of aquamarine. A slightly greenish unenhanced stone will sell at a lower price than a heat treated gem with a pure blue hue.

The Rarity Factor

Fine aquamarine is rare in any size. This gem is often available in very large sizes and prices tend to decrease on a per carat basis above twenty-five carats.

Chapter 11
Emerald

"The second-finest shade of green emerald is to be found in stones from the original Chivor mine. This shade tends to have more blue than the Muzo shade. To some neophytes the Chivor blue-green stones appear, at first sight, to have more warmth and fire. The Muzo stones often appear to be over-dark or to have a hint of yellow in them."

–Benjamin Zucker, 1984

Fine Colombian emerald. Photo: Tino Hammid.

Prior to the discovery of Colombian emeralds, some writers have held that the gem known as oriental emerald was green sapphire. The first source of true emerald known in the West was the famous Cleopatra's Mine located in the Sinai Desert.[119] Evidence suggests that small amounts of true emerald from Austria's Habachtal Mine were also available beginning in the mid 17th century.

The first European encounter with Columbian emerald occurred in 1526. While surveying the western coast of Central America, the Spanish conquistador Pizarro sent his pilot, Bartolomé Ruiz on a reconnaissance mission along the coast. Just south of the Equator, off the coast of Ecuador, Ruiz overhauled a balsa-trading raft and found emeralds among its trade goods.[120]

Two mines, Chivor and Muzo, both discovered — or rather stolen by the Spanish conquistadors from their Native American owners — are superior in all respects to the pale, highly included stones from the mines of ancient Egypt.[121] Of these two mines, separated by less than one hundred miles, Chivor was in production first, sometime before 1555. Muzo, a source of larger and even finer crystals, was located five years later. The chief difference between emeralds found at Muzo and Chivor lies in the

---

119 Located in Egypt's Eastern Desert Region southwest of Marsa 'Alam. Pliny mentions a source in "the mountains of the Scyths'," perhaps Pakistan. One European site, Habachtal, Austria, State of Salzburg, was worked as early as the 13th century.

120 Kris Lane, *Colour of Paradise, The Emerald in the Age of Gunpowder Empires*, (Yale University Press, 2010), p. 25.

121. John Sinkankas, *Emerald and Other Beryls* (Radnor, Pennsylvania: Chilton Book Company, 1981) p. 29.

Into the belly of the beast. A quarter mile in, then 300 yards down. Consorsio Mine, Boyacá, Colombia. Part of the La Pita complex. Approximately seventy percent of current production comes from La Pita. Photo: R. W. Wise

in Pakistan, Afghanistan, Russia, the Brazilian states of Bahia and Goiás, and lately in Hiddenite, North Carolina,[124] now produce stones which rival some of the best that Colombia has to offer.

Hue

Emerald, like ruby, blue sapphire, and tsavorite garnet, is one of the primary-color gemstones. That is, the most desirable color is the purist green possible. Although it is true that a small percentage of yellow (ten to fifteen percent) will enhance or frame the dominant green hue, the purer the green the finer the stone. Emerald personifies the pure idea of green. Close your eyes and picture the richest, purist green you can imagine: rich meadow grass growing by a shaded pool in high summer — that is the finest color in emerald![125]

Emerald is usually either a bit yellowish

secondary hue. Geologically, the Chivor deposit is older; these stones tend to have a slightly bluish secondary hue, while those from Muzo tend slightly toward the yellow.

Although some connoisseurs maintain that stones mined in the 16th and 17th centuries, called "old mine" emerald, were the finest of the fine, Colombia is still a major source of fine emerald.[122] About three years ago a new source was discovered in an area near Muzo called La Pita. Stones from this source are almost impossible to separate from Muzo stones.[123] La Pita, a complex of six mines discovered in 1997, is responsible for between sixty and seventy percent of current production in Colombia. Other recent discoveries

122 Zucker, *Gems and Jewels*, pp. 54-55. Zucker's excellent essay on emerald was written before stones from the Pakistan and the newer Brazilian deposits became available in the market and is therefore a bit dated.

123 Ray Zaicek, personal communication, 2003. La Pita emeralds are considered to be more brittle than stones from the older mines.

124 Discovered in 1875, the *North American Gem Mine* has produced over 3,000 carats of gem quality emerald since it began commercial production in 1998, including a number of fine gems such as the 8.85 carat *Carolina Duchess*.

125. The Arab scholar Ahmad ibn Yusuf al Tifaschi, writing in the 13th century, described this color as *zhubabi*, defined as "a very deep green without any other shade of color." He further described it as similar to the iridescent color of the back of a spring fly (not a house fly, mind you, rather the big juicy outdoor type). Multicolor effect, the tendency for a gem to show tonal variations of green in the face-up position (see Part I), along with any defect in the transparency (crystal) of the gem, were considered the other major faults in emerald. See Huda, *Arab Roots of Gemology*, p. 104.

Exceptional emerald crystal specimen from La Pita Complex, Boyacá Colombia. Photo: Jeff Scovil.

or a bit bluish; in practice, it may have a bit of both as secondary hues. A slightly bluish secondary hue (ten to fifteen percent) adds a richness and warmth to the overall appearance of the gem. For this reason it is a more desirable secondary hue than yellow. As always, the gem should be examined in all light sources. Incandescent light will bring out the blue. This is the primary reason why some connoisseurs prefer an emerald that is slightly yellowish in daylight. They feel that a bit of yellow balances against the tendency toward blue under the light bulb.[126] If the stone shows more than a fifteen percent blue secondary hue in incandescent lighting it is over-blue.

126. C.R. Beasley, personal communication, 2000.

African emerald from Zambia often suffers this fault.

Saturation and Tone

The color green achieves its optimum saturation, its gamut limit, at about seventy-five percent tone. Thus, emerald achieves its most vivid saturation at this tone. Gray is the normal saturation modifier in emerald. Gray will reduce or dull the hue. Gems with visible gray should be avoided. Sometimes the gray component will be very difficult to discern. Aside from a sense of dullness, the hue of a grayish emerald will appear distinctly cool in incandescent light.

Clarity and Crystal

Emerald is most often found with some visible inclusions. It is rarely eye-clean in sizes above one carat. Because of this a greater degree of tolerance should be exercised when judging the clarity of emerald. More attention should be paid to diaphaneity (crystal) than to strict flawlessness. The finest emeralds exhibit a wonderful clear crystal which gives the stone a marvelous inner glow. If the stone has this trait, a few visible inclusions — what experts call "jardin" (garden) — are easily forgiven. Such stones are more highly valued than those, which are strictly flawless but lack the limpid quality of good crystal.

Old Mine and Gota de Aceite; A Clarification of Terms

In the upper echelons of connoisseurship, the aficionado may hear terms such as *old mine*, *gota de aceite (drop of oil)* and *butterfly wing effect (efecto aleta de mariposa)*. These are all terms used by

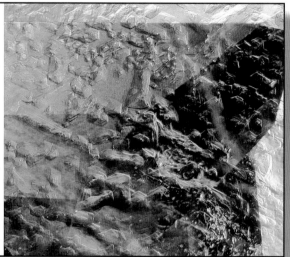

Emerald 10X magnification showing "honey-like" (gota de aceite) transparency considered a hallmark of old mine emerald. Photo: Ronald Ringsrud.

Transparent hexagonal growth structures are the cause of the honey-like or oily texture seen in the emerald pictured above viewed at 30X magnification. Photo: Ronald Ringsrud.

the cognoscenti to describe the look of a particularly fine emerald. *Old Mine* is the most confusing term because it is used in two distinct senses within the trade. In the first sense, old mine is used to designate a stone mined in the 16th, 17th or 18th centuries.[127] The implication is that stones mined long ago were of finer quality than emeralds mined more recently, a difficult proposition to validate. In the second sense, old mine is used to describe the specific look of these old stones. What separates the old mine look from the new? A specific sort of softened transparency described as a honey like consistency,[128] like a drop of oil (gota de aceite) or like looking through a transparent butterfly's

wing (efecto aleta de mariposa). Ronald Ringsrud, a dealer who is best described as besotted with emerald, specifically Colombian emerald, has done a great deal of research on the true meaning of these terms.

Old mine emeralds were originally thought to have originated in India, but as Ringsrud rightly points out, most were, in fact, Colombian emeralds.[129] The Spanish sent these gems from the Colombian mines to India by the Asian route across the Isthmus of Panama and from there by ship to India via the Spanish colony of the Philippines.[130]

127 Kris Lane, *Ibid*, p.28-29. Historically emerald was divided into Oriental and Occidental. The Oriental emerald was supposed to have been mined in Egypt and/or India and be superior to Occidental emerald, which was probably green sapphire. However, modern gemology has proved that the so-called Oriental stones were, for the most part, mined in Columbia and more specifically Muzo.

128 Benjamin Zucker personal communication 2007.

129 Ronald Ringsrud, *Emeralds: A Passionate Guide*, (Green View Press, Oxnard, Ca., 2009), p.228.

130 Jean Baptiste Tavernier, *The Six Voyages, Vol. II*, pp. 82-83. Writing in 1689, Tavernier notes the confusion among European jewelers and was well aware of the early trade route from South America to India via The Philippine Islands. Tavernier uses the similar term "old rock" *(la vieille roche)* rather than old mine to describe the finest emerald and turquoise.

The other two terms: drop of oil and butterfly wing do describe a very special sort of muted or honey-like transparency (crystal) with a texture which is greatly appreciated by emerald connoisseurs. Ringsrud has been able not only to describe the effect but also to isolate its physical cause.[131]

The effect is the result of light refracting through transparent partially crystallized hexagonal growth structures. These structures scatter light and produce the thick oily transparency which softens the look of the crystal. This effect compares to the look of light passing through a drop of colorless oil or honey. To describe stones with the gota de aceite effect, Indian merchants use the Sanskrit word, *snigdha*. The word translates as "smooth" or "tender."[132]

All other factors being equal, a very fine emerald that exhibits this honey-like crystal face up should bring a premium. However, when I first began researching the finer points of emerald connoisseurship, many dealers and most jewelers seemed confused about the meaning of these terms. As always, superior knowledge may provide a buying opportunity for the savvy aficionado. Ringsrud recommends that the term old mine be restricted to genuinely old stones. However, the term is used most often to describe stones that exhibit the effect.

131 Ringsrud, *ibid.*, pp. 214-215.

132 *ibid.* p. 219.

Emerald Cut

Most aficionados prefer emeralds cut in the traditional step or emerald cut. This cut has seventeen long, narrow, step-like facets. A majority of emeralds are cut in this style, both because it accentuates the warm satiny hue of the gem and, by happy coincidence, it is usually the most efficient use of the rough material. Emeralds cut in the brilliant style exhibit a great deal more scintillation, but lack that satiny look.

The emerald's satiny brilliance, like the luster of a satin ribbon, gives emerald a look of softness, which contrasts with the crisp brilliance of tsavorite garnet, its only rival for the title of "greatest of the green." This is partly a function of emerald's relatively low (1.57-1.58) refractive index and partly a result of cutting style.[133] A slightly bluish green emerald with excellent crystal appears to glow with an appealing richness and warmth which is alien to tsavorite. Emerald is soft where tsavorite is hard. To fully understand this quality the budding connoisseur must educate his eye by comparing a large number of emeralds. Stone to stone comparisons between emerald and tsavorite can also be useful.

Treatments

Emerald is a gemstone which the collector must approach with fear and trembling. An overwhelming majority of

133. Refractive index measures the degree to which light is bent as it enters a substance. Emerald is a beryl colored green by a combination of chromium and vanadium. Tsavorite is a garnet colored green by chromium and vanadium. Tsavorite garnet has a refractive index of 1.74 against emeralds of 1.56-1.58. Tsavorite is also much more dispersive (see Chapter 12, tsavorite).

Very fine old mine Muzo emerald with no enhancement in antique platinum setting. Photo: Jeff Scovil. Cora N. Miller Collection, Yale Peabody Museum of Natural History. Courtesy R. W. Wise, Goldsmiths

fissures without sophisticated testing. In addition, green dye may be introduced into the polymer filling, which can improve the apparent color of the stone dramatically.

Current industry opinion is that oiling and, to a lesser extent, the use of unhardened polymers is acceptable, though dyeing is not. However, in practice, many dealers turn a blind eye to both types of enhancement. Given the high prices of fine emerald, collectors are advised to insist on full written disclosure backed up by an expert laboratory analysis before finalizing the purchase of an emerald.

the emerald available on the world market is treated with a variety of substances to enhance the clarity and sometimes the color of the stone. Rough emerald is often highly fractured. Since earliest times, various oils have been used to hide cracks to improve the clarity of the gem. This practice has been going on for so long it has become accepted and is rarely disclosed to the buyer.

In recent years, polymer plastics have begun to replace the traditional oiling. Opticon is the brand name of a popular polymer with a refractive index virtually matching that of emerald, making it impossible to detect surface breaking

Recently many labs around the world have adopted a uniform seven-step classification to describe emeralds which have been treated with colorless oils or polymers. (Chapter 4, Certificate Games) The classifications are as follows: *none, insignificant,*[134] *minor, faint to moderate, moderate to strong, strong to*

134 Emerald clarity grading is not consistent and may differ markedly from lab to lab. The author submitted four emeralds cut from crystals in his presence in Bogotá to each of the major labs and found a great deal of disagreement. The rough gems were not treated by the lapidary and were represented as coming directly from the mine and being totally untreated. Only one received a grade of *none* by all labs. A gem graded *minor* by one lab received a grade of moderate from another. All labs use minor, but the grade *insignificant* is used by only a few. Some will not issue the grade *none* if there is the slightest microscopic residue of oil inside a fissure. Others will use *none* to describe a stone with a tiny bit of oil residue that has no visual effect on the stone's clarity. The aficionado should bear in mind that lab grades are opinions only.

prominent, and prominent. Stones that fall into the first four classifications, none to insignificant to faint to moderate, are rare and worthy of consideration. Regardless of appearance, stones that fall into the last two classifications, moderate to prominent, are best avoided. Exceptional stones that are both eye-flawless and untreated (none to minor) will command premiums of fifty percent and more in the marketplace. The careful buyer will require a grading report from a recognized gem lab as part of the purchase agreement.

The aficionado should bear in mind that these are grade levels of treatment, not clarity. That is to say, an emerald graded none is not necessarily a flawless stone; it is simply a stone which has not been oiled or treated with polymers. An untreated emerald can still look like a piece of a broken Coke bottle. Conversely, a stone may appear flawless to the eye and still be found to have a prominent level of treatment. This means that the treatment has effectively covered up many sins.

Beware Synthetics

Synthetic emerald can also be a problem. There are a number of types of synthetics currently available in the market. The aficionado's best defense against both treated and synthetic emerald is a certificate from a recognized independent gemological laboratory.

The Rarity Factor

As with all rare gemstones, fine emerald is rare in any size. Prices tend to increase at one, five, eight and ten carats and level off at twenty.

Sunstone country! Plush, Oregon. Photo: courtesy Desert Sun Mining.

FABULOUS FELDSPARS, SUNSTONE/MOONSTONE

Feldspars are a group of silicates, which taken together, are the earth's most common minerals, making up almost sixty percent of the earth's crust. Most rocks contain feldspar. Gem varieties such as corundum (ruby and sapphire) are siblings with identical chemical and crystallographic makeups which differ only in the types and amounts of the trace elements responsible for color. Groups, by contrast, are sometimes distant cousins with widely varying structure and chemistry. The feldspars are divided into roughly two types, those rich in potassium and those containing sodium and aluminum. Potassium-rich

The many colors of sunstone. Photo courtesy Dana Schorr and Desert Sun Mining Co.

feldspars are known as orthoclase. Sodium-aluminum to calcium-aluminum rich feldspars are known as plagioclase.

Gem quality moonstone is a translucent to semi-transparent, colorless variety of orthoclase known as adularia which, when properly oriented and cut en cabochon exhibits a milky-misty-pearly, billowing blue sheen known as *adularescence* appearing to hover or float above the rounded dome of the fashioned gem. In recent years a plagioclase exhibiting a strong "sapphire blue" adularescence has been found in Northwestern India in the State of Bihar outside the city of Patna.

Sunstone was known and treasured by Native Americans for its medicinal qualities. In the early 20th century, Tiffany & Co. held claims near Plush, Oregon and coined the term "Plush Diamond" to market the gem only to eventually give up and divest itself of its mining claims in 1911[135]. Sunstone rose once again to prominence in the early 1990s with the beginning exploitation of large deposits of the material on the volcanic highlands of southeastern Oregon.

As there is a broad range of feldspars of varying chemistry called "sunstone" the gem is best evaluated by appearance. The original sunstone, discovered in the 18th century on Sattel Island off the coast of the Russian city of Archangel[136] and more recently Northwest of Arusha, Tanzania, is an albite rich plagioclase feldspar. When cut the material from Tanzania produces a spectacular multi-hued spangled "confetti" effect when exposed to light[137]. This phenomenon is caused by light refraction through tiny concentrations of ultra thin semi-translucent hematite (micacious iron) inclusions. This effect differs from adularescence and is known as *schiller*[138].

The second type, known as Oregon sunstone, is a type of aventurine (plagioclase) which derives its schiller effect from the refraction of light from a constellation of tiny flakes of included native copper lined up in serried zones like a dusty milky way. The material from Tanzania has yet to make much of a splash in the gemworld and appears to be mined out in sizes over five carats. Oregon sunstone is the only sunstone of importance in the gemstone market today.

135 Alexandra Arch, "Sunstone Mining in Oregon", 1859 *Oregon Magazine*. http://1859 oregonmagazine. com/sunstone-mining-in-oregon.

136 Bauer, Max, *Precious Stones*, p.428.

137 Although this chapter will concentrate on Oregon sunstone, Tanzanian sunstone with the distinct confetti like spangling can be very beautiful and smaller stones are currently a bargain in the marketplace.

138 Another species of plagioclase containing slightly less albite is known as andesine. Beginning in 2002 large quantities of vivid translucent *andesine* was introduced into the gem market and became a boutique gem promoted by TV retailers. Originally deemed natural, in 2008, J.G.G.L. a Japanese gemological lab, determined that translucent red and green Mongolian andesine was color treated by copper diffusion.

Chapter 12
Adularia, Queen Of Moonstones

"This stone is a strange one as the overwhelming color is white with an astonishing radiance coupled with a delicate transparent clarity, but a bluish spot can be seen inside it, similar to a cat's eye..."

–Ahmad ibn Yusuf Al Tifaschi, 1253

The term moonstone is often used quite loosely. Many translucent to semi-transparent minerals exhibiting a milky to pearly glow have been called moonstone. This includes several varieties of feldspar and quartz such as the well-rounded alluvial quartz pebbles littering the mainland beaches along the shores of Block Island Sound, Rhode Island. True moonstone, however, is a type of feldspar, a term which encompasses a group of very common minerals with varying chemical formula and, in the case of feldspar, crystal habit as well.

Moonstone has been known since very early times. Though mentioned by Pliny and Al Beruni[139], the first truly accurate description of adularia moonstone is found in the 13th century writings of the Egyptian gemologist Al Tifaschi. who specifically discusses stones cut to produce the narrow chatoyant band known

Fine Indian adularia moonstone. The camera lights tend to bleach out the adularescence. Note the blue surrounding the white reflection of the light source. Photo: Jeff Scovil. Courtesy R. W. Wise, Goldsmiths, Inc.

as a cats-eye (see Chapter 5)[140]. However, it is fair to say that although cats-eyes are greatly appreciated, moonstones today are

---

139 The Arab term is *hajar al qamar* (moonstone), see Al Beruni, *The Book Most Comprehensive in Knowledge on Precious Stones*. S. Sajid Ali for Adam Publishers, New Delhi, 2007. pp. 66.

140 Huda, S. N. A., *Arab Roots of Gemology, Ahmad ibn Yusuf Al Tifaschi's Best Thoughts on the Best of Stones*, The Scarecrow Press, Lanham Md., 1998. pp. 213-214

purposely cut flatter so as to exhibit a broad billowing blue across the face of the stone rather than the narrow   band characteristic of the cat's-eye.

Traditional sources of adularia moonstone include Sri Lanka, Burma and India. The finest moonstones, those described by Al Tifaschi, undoubtedly came from Sri Lanka. Unfortunately production from the ancient mines located at Meetiyagoda and Tissmakarama in the southwestern part of the island slowed to a trickle and stopped producing entirely by the late 1990s[141]. Burma also produced some exceptional quality adularia, but production from that deposit ceased in the late 1970s[142]. Current production is centered in the northwestern Indian state of Behar near the town of Patna.   Unlike the Sri Lankan moonstones, which are orthoclase feldspar, these gems are a type of plagioclase[143].

Hue/Saturation/Tone

Al Tifaschi goes on to say that "the best quality is a strong, white (colorless) transparent one with a clear spot... showing great brightness." This is precisely correct. Adularia may occur with blue and reddish-brown body colors, but the crème de la crème of adularia moonstones are colorless. Fine moonstones will exhibit a vivid medium-to-medium dark toned blue adularescence with no visible gray mask. This phenomenon characterizes the best

141 Manu Nichini personal communication, August 2014.

142 Ibid..

143 "Gem News", *Gems & Gemology*, Summer, 1997, pp. 144-145.

of adularia. According to moonstone specialist Manu Nichini, the finest phenomenal hue can be described as a "sapphire blue" and though this term is a useful analogy, beware of pushing it too far[144]. Indian adularia will fluoresce a weak to an occasionally moderate blue under long wave ultraviolet light, which may enhance the visible saturation.

Adularescence

Adularescence or the *moonstone effect* is a result of light refraction off tiny serried cleavage planes that are oriented parallel to the curvature of the cut gem together with layers of slightly varying chemistry. Other feldspars both orthoclase and plagioclase may exhibit adularescence in a variety of hues. Rainbow moonstone, a newly discovered variety of plagioclase found near the Indian city of Patna, will sometimes exhibit a distinct adularescence composed of a rainbow of colors consisting mostly of mottled blue with yellow. Adularia, however, is the only member of the group which consistently exhibits the beautifully distinctive billowing blue effect.

The moonstone effect is ethereal, like a thick translucent floating blue cloud. The effect shows a cool aspect like the light of the moon. Blue is normally a cool color in gems except in fine sapphire where the addition of medium dark purplish secondary hue warms it. Some gems will exhibit a blue brush effect, as if a paintbrush was drawn once across the surface leaving evidence of its bristles. This is often due to visible cleavage and reduces the desirability of the stone.

144 Manu Nichini, *ibid.*

Clarity

The finest examples of adularia are eye-flawless and this is the basis for judgement. In most gems, the inclusions responsible for the adularescence are planes easily identifiable as razor thin very straight cracks in the finished gem. Gems will usually disclose inclusions under magnification.

Crystal

Adularia is the most transparent of all gems described as moonstone. Though normally described as translucent to semi-transparent, the best examples exhibit a lovely clean, crisp, colorless, dew-like crystal which gemologist Max Bauer described as "perfectly transparent."[145]

Size & Rarity

With only a single location producing adularia, the gem is quite rare. Stones are normally available in sizes to one carat. Fine stones above five carats are rare and prices double at this size. Gems above ten carats are extremely rare with per carat prices doubling again at the ten-carat mark.

---

145 Bauer, *ibid.*, p.428

Carved sunstone pendant. Photo: courtesy Desert Sun Mining.

Chapter 13

Sunstone

"In my opinion once the clarity of the stone is graded correctly, then people will decide what color(s) they prefer and make their choices accordingly. Yes, some colors are considered more valuable than others, and in general the more intense and bright the color of a gemstone the more valuable"

–Jeff R. Graham, 2007.

Hue

Green, one of the rarest colors of Oregon sunstone though vivid red will command the highest price. Photo courtesy Dana Schorr and Desert Sun Mining Co.

Sunstone occurs in a range of hues: blue, green, yellow, pink, brown-orange, green and what is called "red." (see Introduction & Overview) Most reds are a distinct reddish-brownish-orange to orangy red although occasionally an example of true red will be found. While orangy brown to brownish orange is the characteristic color, blue, green and red are the rarer hues of Oregon sunstone. All may exhibit a fairly consistent sugary-metallic sunstone effect known as *schiller*. Aside from a gritty texture, schiller differs in little but name from the billowing phenomenon of adularescence found in moonstone.

Saturation/Tone

Generally speaking, sunstone hues are not highly saturated. Most contain a tinge of brown or gray mask rendering the hue either muddy or dull when compared to most other gemstones. That said, the aficionado will occasionally run across a highly saturated gem if he is willing to pay the price.

Clarity/Crystal

Dense concentrations of microscopic copper inclusions reduce the transparency of most Oregon sunstone rendering the majority semi-translucent at best. However, it is these very inclusions coupled with a strong underlying transparency which creates the schiller phenomenon. Fine,

"Dreamscape", a 9.48 carat Oregon sunstone gem carving by award winning gem sculptor John Dyer. This is the classic reddish-orange hue. Note the artistic use of the parallel bands of tiny copper inclusions. Photo: Lydia Dyer. Courtesy John Dyer Gemstones.

highly crystalline eye-flawless gems also exist and these normally will command the highest prices. So what might be called a double standard exists between phenomenal and non-phenomenal sunstone. Other gem varieties produce redder reds and more vivid greens. In this writer's opinion, the Oregon sunstone's uniquness rests in the subtle combination of color, crystal and phenomenal effect. The aficionado is invited to form his or her own opinion.

Cut

Faceted, cabochon, fantasy, gem sculpture; Oregon sunstone is cut in so many different styles that it is difficult to get a handle on any sort of quality hierarchy. Given the lack of pure blues and greens, a faceted, eye-flawless visually pure red hue will command the highest price, but the gem must be cut en cabochon to exhibit the schiller effect.

The Rarity Factor:

There are large reserve deposits of sunstone in Oregon and speaking generally, the gem should not be considered rare[146]. Blue is the rarest color and I have not seen a pure blue. Pure green is next and normally shows a distinct gray or brown mask; red rates third in rarity. It is available in a visually pure hue, but is again quite rare[147]. Sunstone is available in a full range of sizes. Prices jump at one, two and five carats. The yellow material is quite inexpensive. Carvings, sculptures, fantasy cuts (whichever term you choose to call them), are priced per piece and prices reflect the skill and reputation of the cutter.

146 Ponderosa, the largest mine, contains reserves estimated at a trillion carats of usable cutting rough. Cf: Duncan Pay, Robert Weldon, et al., "The Occurrences of Oregon Sunstone". *Gems & Gemology*, The Gemological Institute of America, Fall 2013, p. 3.

147 Dana Schorr, personal communication, 2015

CHALCEDONY:
ARISTOCRATIC AGATES

Chalcedony has been an important gem material since the very earliest times. It may be the oldest mineral used as a gemstone. Chalcedony is both hard and dense. Beginning in prehistoric times, flint, a fairly uninteresting variety of chalcedony has been used to make pointed weapons and cutting tools. It is not surprising that our ancient ancestors, while seeking material for practical uses, would chance upon beautiful examples of chalcedony which perhaps caught a young girl's eye and found uses other than practical. Flint knapping was, in fact, the earliest lapidary technique.[148]

Fine quality agates have been found at a number of archaeological sites throughout the Mediterranean world. Agate beads dated to 7000 BC have been discovered at Catal Huyuk in Anatolia. The bead cloak of Queen Pu-abi, found in the royal tombs of Ur in Mesopotamia and dating from 2500 BC, contains hundreds of carnelian and agate beads. At Aidonia on the Greek mainland, agate seal stones of exceptional quality and workmanship have been unearthed from Mycenaean shaft graves dated back to 1500 BC. Carnelian, the orange variety of chalcedony, as well as other agates, were much sought after and were surely among the precious gems of antiquity.

In recent times the trend towards all things faceted has relegated chalcedony to the semi-precious backwaters. For most of the twentieth century, agates have been seen as a curiosity and not taken seriously as a gem material of real rarity and value.

Thus far the terms chalcedony and agate have been used interchangeably. Chalcedony is the gemological term used to describe a type of quartz with a micro or cryptocrystalline structure, as opposed to single crystal quartzes such as amethyst and citrine. Agate is a nontechnical term used as a common name for chalcedony, but is more often used to describe chalcedony gems with visible bands of color. However, jasper, carnelian, and sardonyx are varietal names also used to describe specific types of banded chalcedony. Here the term agate is used in the more universal sense as a common name for all varieties of

---

148 Rhodes, J.G., *Flint Rings and Egyptian Armlets. Antiquity Journal* **44**, Issue 174, June 1970, 145-146 doi:10.1017/S0003598X00104752. The earliest known flint jewelry are one-piece knapped armlets from the pre-dynastic period excavated from a cemetery at Tarkhan near Kafr Ammar in 1911.

chalcedony.

Unlike crystalline materials such as amethyst, which is a single crystal, this species of quartz is composed of tiny interlocking microscopic fibers giving rise to the term *cryptocrystalline* (hidden crystal) quartz.

Recently, several varieties of this chalcedony have become popular. Gem Chrysocolla, Holley and Mojave blue agate, carnelian, and chrysoprase are increasingly in demand for jewelry. Why? because new sources of these agates are producing gems of such surpassing beauty that they have simply become impossible to ignore. American and European lapidary artists have made extensive use of these materials in gem carvings. A few years ago *Town & Country* magazine featured an article on blue agate attesting to the fact that chalcedony is enjoying a new coming of age. The best examples of the first three types mentioned are found in the United States. The last two are found in Africa and Australia. All are from sources discovered in the past thirty years.

Treatments

Since at least 2000 BC, chalcedony has been subjected to heat treatment to create carnelian. Heat treated agates were among the treasures unearthed from the tomb of King Tutankhamen (1300 BC).[149] Most of the carnelian currently on the market has been heat-treated. Agate is quite porous and can be soaked in acid to create a dark outer layer in preparation for carving en cameo. Porosity also makes chalcedony a good candidate for dyeing. Dyeing will sometimes produce a very highly saturated unnatural-looking hue. For example, "green onyx" is colorless chalcedony which has been dyed a rich dark green; it looks simply too green to be true and it is! Black onyx is also universally dyed. The blue varieties of chalcedony are also subject to dyeing. Under magnification, natural color agate will usually show evidence of thin color bands. Specialists call these bands *fortifications* of color. Heat treatment will burn out the bands. Acid treatment produces a dark, dense, "burnt" look in carnelian. Many first- to third-century Roman intaglios, engraved seal stones, fit this description perfectly. Because agates are relatively reasonable in cost, treatments are rarely disclosed to the buyer.

This section will consider the five most beautiful and commercially important varieties of chalcedony: gem chrysocolla, Holley blue and Mojave blue agate, carnelian, and chrysoprase.

---

149. Nassau, *Gemstone Enhancement*, p. 25.

Chapter 14
Gem Chrysocolla

*"Gem chrysocolla is in high demand in the Fareast.
In Taiwan they call it "blue jade" but to my eye it has a
far more vivid hue."*

–Michael Randall, 1992

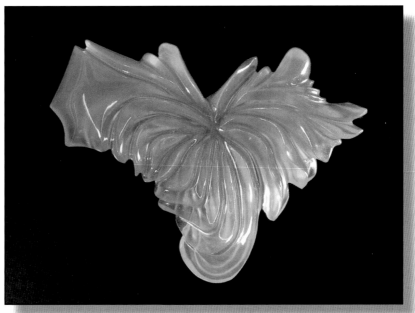

Gem carving (36.62 carats) in exceptionally fine translucent gem chrysocolla. The gem exhibits a highly saturated blue primary hue of sixty percent tone. Carving by award-winning gem sculptor Glenn Lehrer. Photo, Glenn Lehrer.

The rare rich sky-blue variety of chalcedony has been referred to variously as *silicated chrysocolla, gem silica, agated-chrysocolla*, and *gem chrysocolla*. The issue lies with the term chrysocolla. True chrysocolla is a soft, non-crystalline mineral which owes its vivid color, ranging from turquoise to sky blue, to the staining effect of copper oxides. gem chrysocolla, by contrast, is a relatively hard cryptocrystalline silicate which has been stained blue by the same copper oxides.[150]

Given the vast difference in hardness, the aficionado should have little difficulty in separating the two materials by appearance alone. True chrysocolla is opaque, with sub-vitreous luster, and has a hardness of two to four on the Mohs scale. Materials of this hardness normally can be scratched with a copper penny. Gem chrysocolla may be opaque to translucent and have a vitreous, or glassy, surface luster and a hardness of seven, which makes it harder than steel. Standard gemological tests can easily separate the two materials. True chrysocolla may have all the beauty of gem chrysocolla, except the luster, but it lacks the requisite hardness and durability necessary to qualify it as a gemstone.

Much of the finest gem chrysocolla is found in the copper mines of Pinel County, Arizona. The *Inspiration Mine* was

150. Glenn Lehrer, personal communication, 1998.

a legendary source of some of the best. Located in Miami, Arizona, about eighty miles east of Phoenix, this mine, specifically the *Live Oak Pit*, produced a particularly dense material which was unaffected by and did not lose color in the dry desert environment. This source was more or less played out in the 1990s. Another source, located east of Inspiration, The *Ray Mine*, produced particularly pure hued, limpid material well into the 1980s. In the 1960s a deposit of the gem was found on the island of Taiwan. In the East, gem chrysocolla is sometimes referred to as "blue jade," an unfortunate term which further adds to the linguistic confusion. A new site located in Young, Arizona, was recently reported[151].

Bacan Doko," highly translucent green gem chrysocolla from Doko village, Kashrut Island, Indonesia. Gems from this location run the gamut from sky blue to green. Photo: Ibnul Ghufron, Ibsy Gemstone.

Deposits of fine blue to green chrysocolla are being exploited in the Arequipa region of southern Peru and on Kashrut Island in the Indonesian archipelago. Known as *Bacan stone*, Indonesian chrysocolla varies from a translucent green reminiscent of jade to a true blue.

Historically, gem silica was a byproduct of copper mining and miners who found particularly beautiful bits of rough carried it out in their lunch boxes. Strip mining, with its use of chemicals to leach out the copper put an end to that and little new material is available on the market.

Hue, Saturation, Tone

Vividness of hue is the quality that assures gem chrysocolla a place among the precious gemstones. The hue varies from a green to a slightly greenish medium dark-toned (fifty to seventy percent) turquoise blue to a similarly toned visually pure sky blue. In Arizona stones the green secondary

151 Thomas Stricker, personal communication 2014

hue rarely exceeds ten percent. As with all azure gems, the smaller the percentage of secondary green, the more desirable the gem. Gem chrysocolla rarely shows any evidence of either a gray or brown mask, though gray may occasionally be present. Thus, it has a consistently blue hue which varies in saturation or quantity of color from pale to vivid. Fine stones can be described as a visually pure vivid blue of approximately 50% tone. Tonally darker gems tend towards the opaque. A fine stone treads a tightrope balanced between tone and transparency (see below).

Color Fading

It is estimated that over eighty percent of gem chrysocolla is subject to a degree of color fading when exposed to a dry environment. This is particularly true of the Mexican material. As the stone dries out, the translucency and the color saturation of the gem are both diminished. According to Chris Boyd, a dealer who spends summers in upstate New York and winters in Arizona, and who has worked with gem chrysocolla for many years, fading is an issue in areas with an average relative humidity below fifty-five percent. His stones look much better in New York than in the Arizona desert. Fading can be reversed when the gem is rehydrated by directly exposing it to moisture or to a moist environment for a short period of time.

Crystal

Given the consistency of hue, saturation, and tone, it is the degree of transparency which defines the various grades of gem chrysocolla. This gem is never completely transparent. It is the degree of translucency that is the defining factor. If held under a strong light, a pencil placed behind the stone will appear as a distinct shadow in a fine example of this material. It is fair to say that crystal is at least co-equal with color in the connoisseurship of gem chrysocolla, as well as in the other agate varieties to be discussed in this section. Visually pure blues with a high degree of translucency are the most prized. A greenish blue gem with a high degree of translucency is more desirable than an opaque pure blue. Green chrysocolla is priced lowest, but some of the Indonesian material is very jade like and exhibits finer crystal than most chrysoprase, which can appear milky by comparison.

Cut

Gem chrysocolla is often cut in freeform (nonsymmetrical) shapes. Shape has little effect on value except that the more interesting free-forms may command a premium. Particularly translucent gems will sometimes be faceted.

Clarity; Inclusions and Value

Drusy, tiny colorless quartz crystals growing on the gem, are the most sought after inclusions in gem chrysocolla. Although perfection in all characteristics tends to be the way the finest gems are defined, the aficionado should be alert to pleasing compositions of gem chrysocolla with lovely inclusion patterns of drusy quartz. Malachite, a vivid green/black mineral is a common though less desirable inclusion, although an artful juxtaposition of color and pattern in a vivid blue gem

can be quite beautiful. It is the skill and sensitivity of the lapidary that makes all the difference is such cases.

The Rarity Factor

Gem chrysocolla with the qualities discussed is extremely rare. I might see a piece or two every few years usually in the hands of dealers. If a fine piece is on offer, the aficionado must be prepared to accept the price, or not have a second opportunity for many years. Size is rarely a factor in the price.

Chapter 15
Chrysoprase

"Of all the so-called jaspers none were so highly valued (in ancient times) as those of a green color . . . but there is every reason to suppose that the true jade was always more highly prized than its jasper substitute for it was easily distinguishable, by its translucency, from jasper of a similar color."

–G.F. Kunz, 1904

hrysoprase is a green chalcedony which owes its "apple green" hue to the presence of trace amounts of

of as much as 2.35%, shows the purest green hue and is, generally speaking, the finer of the two[152]. Other sources include

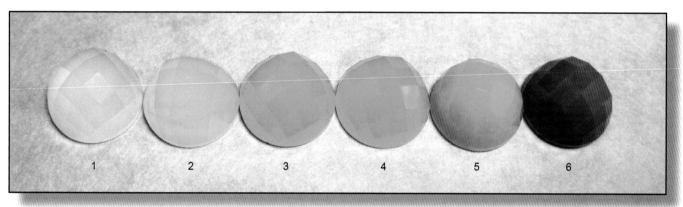

Chrysoprase Grading set: Moving from left to right, a tonal run of chrysoprase from Marlborough Creek. Number 4 exhibits a lovely combination of color and crystal. Number 5 shows finer color, but lacks transparency. Number 6 is both overcolor and opaque. Photo courtesy Candala Chrysoprase

pimelite, a type of nickel. Chrysoprase is very close-grained microcrystalline quartz.

Two sources of chrysoprase were discovered in Australia in the late 20th century, each producing, in substantial quantities, most of the material available in the market today. The first and most important source is a vast deposit found in 1965 at Marlborough Creek, two hours out of Rockhampton, Queensland. The second was unearthed in Western Australia in the Yerilla District in 1992. Material from Marlborough Creek with a nickel content

Brazil, and Kazakhstan in Central Asia, (a legendary find), much of which later was determined to be green opal. Opal of a similar color is often associated with chrysoprase. In earlier times it was referred to as *chrysopal*.

Hue, Saturation, Tone

The finest chrysoprase is usually described as apple green. It is sometimes called *imperial* chrysoprase, an obvious

152 Pay, Duncan. "Fine Australian Chrysoprase Rough and Carvings – Tucson 2014". *Gems & Gemology*, Spring 2014,

attempt at a comparison to jadeite. It is a vivid visually pure green between sixty-five and seventy percent tone. The material from Marlborough Creek normally ranges from a visually vivid pure green to a slightly bluish green. Gems from Yerilla may show a yellowish secondary hue. This yellowish secondary hue is considered a fault. Gems with even the slightest trace of a visible yellow secondary hue are far less desirable.

Gray is the normal mask or saturation

Chrysoprase sculpture by gem sculptor Steve Walters. Note the particularly fine medium toned slightly yellowish green hue and milky-translucency (crystal). A fine example! Photo: Steve Walters.

modifier in chrysoprase. The best of the Australian material shows no gray at all and can be best described as a vivid apple

green that is arguably every bit as beautiful as jadeite of a similar hue. The general run of chrysoprase from these two sources is remarkably consistent in hue, saturation, and tone.

Another variety called lemon chrysoprase is found in Western Australia. It is a pale opaque lime green color and is not a chalcedony at all but a nickeloan magnesite.

Crystal

Given a consistent hue, saturation, and tone, the relative degree of translucency defines the quality grades in chrysoprase — the more translucent the material the higher the price. In fact it can be said that *crystal* is really the first C of connoisseurship in grading not only chrysoprase but also the other varieties of agate discussed in this section. The material from Marlborough Creek has a distinct translucency which contradicts the statement made by G. F. Kunz at the head of this chapter. From the perspective of beauty, the finest qualities of chrysoprase will stand toe to toe with all but the finest qualities of jadeite and chrysocolla. Translucency tends to decrease with size and thickness. As we shall see (Chapter 34) jade, by comparison, may be semi-transparent. Chrysoprase possesses a milky crystal and is, at best only semi-translucent.

Clarity

Ironstone is the normal inclusion found in chrysoprase. However, with the relative abundance of material, chrysoprase is normally cut flawless. Stones from Western

Australia will sometimes show small black dendritic inclusions. Any visible inclusions disqualify the stone from being considered top quality.

Color Fading

Chrysoprase, like gem chrysocolla, has a reputation for drying out and fading when exposed to heat or a dry environment. This is certainly true for material from older European sources, chiefly Silesia. This seems not to be the case with material found in Australia; gems from New Marlborough and Yerilla are stable. However, it is still advisable to keep chrysoprase away from long-term exposure to direct sunlight and other forms of extreme heat.

The Rarity Factor

As the new millennium begins, high quality chrysoprase is available in quantity and at very low prices. I learned long ago that abundance in gemstones is will-o'-the-wisp — here today, gone tomorrow! Whenever a large deposit of any gemstone is found, the material floods the market, temporarily reducing prices. The operative word here is temporarily! Deposits, even large ones, are quickly depleted. A successful collector is an accomplished opportunist and should not be put off by the relatively low price. Even though it is inexpensive, a fine example of chrysoprase is well worth collecting.

Carved carnelian (with druzy) gem sculpture by Glenn Lehrer. Photo: courtesy Glenn Lehrer

Chapter 16

Carnelian

"Fine dark stones of uniform colour and free from faults are described as "carnelian de la vieille roche," or as masculine carnelian. By transmitted light they appear a deep-red colour, and in reflected light a blackish red shade . . . all carnelians, whatever be their color, are strongly translucent."

–Max Bauer, 1907

arnelian was one of the most sought after gems of antiquity. The Egyptians, Sumerians, Hittites, Babylonians, and Mycenaeans adored this gemstone. The Roman historian Pliny said that "among the ancients there was no precious stone in more common use." A majority of classical Greek and Roman seal stones were carved in carnelian or dark toned brownish-red sard. Sardonyx is a chalcedony with alternating bands or *reinforcements* of white and orange which was favored for the making of beads.

The ancients differentiated between five types of *sardion*. Of first importance was the translucent male "blood red" stone; second came the paler yellowish orange, which was considered female. Third was the darker-toned brownish orange that we call sard. The fourth classification included agate with alternating layers or stripes of brownish orange and white. The fifth quality had blood-red stripes alternating with white. The first three types are to be found in all the best tombs.

Hue/Saturation/Tone

Figure 73: Although not the legendary blood red, this natural carnelian cabochon is a fine example of the translucent visually pure orange "citrus agate" found in Namibia. Photo: Jeff Scovil; courtesy of R.W. Wise, Goldsmiths, Inc.

Regarding the components of color – hue, saturation, and tone – the modern value system almost mirrors the ancient. The dark red hue is still the most desired, although it is difficult to find a true blood red. Almost all carnelian will show some evidence of an orange secondary hue. Most of the fine carnelian beads unearthed from the royal tombs at Ur are a medium to dark orange to reddish orange. Thus, the finest quality might be described as a medium to dark tone (eighty to ninety percent) red primary hue with an orange secondary hue (ten to twenty percent). Carnelian can also occur in a medium-toned (fifty-five to sixty-five percent) highly saturated visually pure orange hue. These gems, sourced in Namibia in modern times, are highly prized. Stones of this description are at least as desirable as the so-called blood reds.

Crystal

As with all chalcedony, diaphaneity plays a defining role in the quality equation. An orange stone with good crystal is more desirable than an opaque red stone. *Crystal* is at least co-equal with *color* in the connoisseurship equation. Recently some medium-toned (sixty percent) highly translucent orange carnelian has come on the market. This material, reportedly from southern Africa, has been nicknamed "citrus agate" due to its pure orange primary hue and its high degree of translucency. Stones of this description will command the highest prices, despite the fact they are not red or even reddish.

Treatments

Most of the carnelian currently on the market has been heat treated. Since ancient times, chalcedony has been subjected to heat treatment in order to create carnelian. Under magnification, natural color carnelian will usually show evidence of the thin bands called fortifications, even in stones which seem uniformly colored to the naked eye; heat treatment will burn out the bands. Acid treatment produces a dark,

dense, "burnt" look in carnelian. Treated stones are less translucent than natural gems.

Most of the gems currently available are heat treated or enhanced in some fashion. Much of the material available today at low cost, in calibrated sizes in a consistent opaque orange hue, is undoubtedly treated. Calibrated stones are cut to precise proportions, e.g., round stones are precisely 5mm, 6mm; ovals, cushions, emerald cuts, and pear shapes will be exactly 7x5, 8x6mm, etc. The aficionado should always be suspicious of a calibrated stone.

The Rarity Factor

The real challenge is finding natural stones. The same problem is encountered with turquoise. Fine natural carnelian is available; however, it's much easier to treat low-grade material than to bother with the natural. This reflects one of the ruling dynamics of the mass market: that which can be reduced to a commodity will be reduced to a commodity. In the marketplace, natural gemstones are inconvenient because they don't occur in precisely uniform colors, shapes, and sizes to be placed into pre-made settings which are produced by the thousands.

Distortions in the market can work for the connoisseur. Uncalibrated stones, those not of standard measurements (7x5, 8x6) are often available at reduced prices. Natural carnelian can often be bought at a better price in carvings. If the collector is attentive, a gem of exceptional quality often can be picked from lots of mixed sizes at a lower price than treated stones in calibrated parcels.

Chapter 17

Blue Agate

"Think of quartz and words like "commonplace" and "abundant" come to mind. That any member of this broad gem and mineral family could be a rarity of keen interest to connoisseurs and collectors (not to mention designers and manufacturers) seems incongruous and far-fetched."

–David Federman, 1992

Like many another gemstone, blue agate is marketed under a variety of aliases: damsonite, Mojave blue, and Holley blue agate are the most common. All are chalcedony from a variety of locales, with some small differences in color and transparency.

Mojave blue is a trade name describing a slightly grayish blue opaque to semi-translucent stone found in the Mojave Desert, about a hundred miles south of Los Angeles. A similar quality material is found in Namibia. Damsonite is an opaque violetish material found in central Arizona. Holley blue is an opaque to translucent, purplish or amethystine-hued variety which has been found in and around Sweet Home, Oregon, and was named, apparently, for the Holley School near one of the sites. Another variation, Ellensburg blue, is found as glacial till scattered about the Washington state township from which it draws its name. Ellensburg blue agate often has a purer blue hue.

Hue/Saturation/Tone

Regardless of source, blue agate is chiefly valued for its translucency and purity of hue. The general mine run of

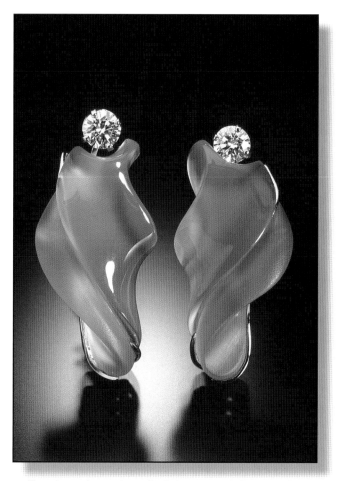

Earrings featuring slightly grayish "Mojave" blue agate show exceptional crystal for this gem variety (blue agate is never more than translucent). Carvings are by Steve Walters; the agate sculptures are accented with a pair of D-color ideal-cut diamonds. Photo: Jeff Scovil; courtesy of R.W. Wise, Goldsmiths, Inc.

stones from California and Namibia tend toward a slightly gray, medium-toned blue hue, sort of a stormy-sky blue. In

this system of evaluation, gray, along with brown, are not regarded as hues but are treated as saturation modifiers which, when present, dull or muddy the saturation or mask the intensity of the hue. A gray mask is nearly always present in this variety of blue agate. Like a dirty film on your windshield, gray dulls the hue, reducing the saturation or brightness of the color. The gray may be difficult to detect but pure hues are always vivid, so the presence of gray may be inferred from a cool dullish blue. Obviously, the brighter the blue the better the stone.

Purple is the normal secondary hue and, when it is present, adds dramatically to the beauty and price of the stone. Holley blue is famous for its amethystine secondary hue, but a hint of purple can also be found in damsonite and even in stones from other locations. Occasionally a stone will be found which has a purple primary hue, but these are exceedingly rare. Stones of this color will have a blue secondary hue; that is, they will be bluish purple rather than purplish blue. The market prefers bluish purple, purplish blue, and grayish blue, in that order. The greater the percentage of purple and the smaller the percentage of gray the better the stone. I have seen only three examples of a pure blue hue without measurable gray.

Crystal

As with all varieties of agate, it is diaphaneity or what is called "good crystal" which is a primary determinant of value in blue agate. *Color* and *crystal* are more or less co-equal in the valuing equation. This gem is normally found with a milky or cloudy body color. Larger, thicker cabochons are barely translucent. Translucent stones are rare and expensive. Although agates are normally cut *en cabochon*, especially transparent or limpid examples will occasionally be faceted.

Clarity

Blue agate is almost always cut flawless. Stones with inclusions other than drusy, a carpet of tiny quartz crystals growing on the surface, are almost without value as gemstones. Drusy is quite desirable, particularly when the cabochon is cut in a freeform shape. In this case it is a question of composition, a pleasing proportion of drusy balanced against the size and shape of the stone. Designer goldsmiths particularly seek stones of this sort. Gems with banding, what agate lovers call *fortifications*, can be quite pleasing, but will not command the price a cabochon of perfectly uniform color will fetch.

Treatments

Both irradiation and heat can alter the color of blue chalcedony. As with amethyst, prolonged exposure to high heat will turn some stones an orangy yellow. Irradiation can be used to enhance the purple secondary hue. Prolonged exposure to the sun or dry environments may also bleach some of the color out of the stone. Soaking the gem in water can restore the color.

The Rarity Factor

As we plunge ever deeper into the new millenium, blue chalcedony is only getting rarer as it becomes more desirable in the marketplace. The popularity of the

gemstone has increased markedly in the past few years. Increasing market rarity coupled with actual rarity means that prices of the finest material are destined to rise. Blue chalcedony is normally sold by the piece rather than by the carat. Pieces too large to be used in jewelry are rarely found.

Ideal cut diamond. Photo: Michael D. Cowing, ACA Gem Lab..

BLUE WHITE DIAMONDS

"...the never failing test for correctly ascertaining the water is afforded by taking the stone under a leafy tree and in the green shadow one can easily detect if it is blue."

–Jean-Baptiste Tavernier, 1689

Historically the finest colorless diamonds have been known as *blue whites*. Beginning in the 1970s diamonds so described began to fall out of favor. This was due to a misunderstanding in the trade as to the precise definition of the term and the difficulties involved in grading gems so defined.

In modern times, a blue white diamond has come to be defined as a colorless diamond which fluoresces blue in ultra violet light[153]. Historically this was not the case. The original blue white diamonds were the legendary Golconda diamonds of India. Unlike most diamonds available today, many of these gems fell into a very rare category of diamond known as type IIa. Type IIa diamonds are characterized by a lack of nitrogen impurities in the crystal lattice, a

9.07 carat Type IIa diamond with a mysterious blue body color. Note the large window beneath the table. This stone was identified as a type IIa diamond and graded D IF with no fluorescence by GIA-GTL. Photo: Jeff Scovil. The Cora N. Miller Collection, Peabody Museum, Yale University. Courtesy: R. W. Wise, Goldsmiths.

---

153 According to Federal Trade Commission Guideline, it is an unfair trade practice to characterize any diamond with any trace of color other than blue or bluish as blue white.

white to bluish body color and the fact that they do not fluoresce in ultra violet light. One percent of the world's diamonds are type IIa and the other ninety nine percent are type Ia.

Type Ia diamonds do contain nitrogen, the cause of their slight yellowish body color; fully one third of type Ia gems fluoresce blue under UV light.

In the mid to late 17th century, Golconda was a small sultanate located in in the modern Indian State of Andhra Pradesh about seven miles from Hyderabad. Golconda was a diamond-trading city. The closest diamond mines were situated five days by horseback at Rammalakota, a village about twenty miles south of Karnul. Another mining center, Kollur, was a leisurely seven-day journey east of Rammalakota. The diamonds traded at Golconda could have come from either of these mines or from a half dozen or so mining locations further northwest.

Ruins of the fort at Golconda. Photo: Bernard Gagnon, Wikimedia.

No one knows what percentage of old Indian diamonds traded at Golconda were type IIa, but many of the celebrated colorless diamonds surviving in museum collections with demonstrated Golconda provenance, including the *Koh-e-Nor*, *Regent* and, *Beau Sancy*, are type IIa.

Not all type IIa diamonds were sourced from India. Type IIa diamonds have been found in Brazil and continue to be mined in southern Africa. The diamond mines of India and Brazil were worked out in the 17th and 18th centuries respectively and most diamonds in the market today hail from Africa, Canada and Russia. Ninety-nine percent of these are the more common type Ia.

Due to the paucity of nitrogen, type IIa diamonds exhibit a snowy white and some exhibit slightly bluish body color. [154] Gems of this type are also known for their exceptional transparency or crystal. Given the lack of UV fluorescence, the source of the bluish body color in type IIa diamonds is not well understood[155]. Historically, highly

154 Frank B. Wade, *Diamonds, A Study of the Factors That Govern Their Value*, Putnam & Sons, N.Y., 1915, P.16. Frank Wade was one of the first American diamond experts. He was a member of the GIA Student Advisory Board. In 1936 Wade became one of the first honorary members of the Institute. In 1947 GIA founder Robert Shipley inducted Wade into the "Committee of 100 World Gem Authorities.."

155 Lenzen, G., Lapworth P. B. trans., *Diamonds and Diamond grading*, Butterworths, 1983, p.12.

transparent diamonds with white and bluish white body color were considered gems of the finest water, a rare quality known as *river*.

In the late 19th century, fully one hundred years after the Indian mines were tapped out and with production from Brazil slowing to a trickle, diamonds from southern Africa began to flood the market. Some of these were blue fluorescent diamonds. According to the legendary gemologist, Frank Wade, writing in 1915, those that did not bleed color, that is, did not appear yellowish face up in the low ultraviolet of incandescent light were added to the river category[156].

Blue and yellow are complimentary colors. Complimentary colors nullify each other. Blue ultraviolet fluorescence will cancel out the yellow in a diamond's body color, resulting in a stone which appears "whiter" (less yellow) in natural daylight. Unfortunately the vast majority of blue fluorescent gems will exhibit a slight yellowish body color when viewed in incandescent or LED lighting. The color grade is dependent upon the type of lighting used. For this reason blue fluorescent diamonds cause a great deal of trouble for the professional grader.[157]

During the hard asset investment craze of the late 1970s, investors insisted that diamonds they purchased come with a laboratory report from the Gemological Institute of America (GIA-GTL). Thousands, sometimes tens of thousands of dollars, rode on the color grade given a diamond on the laboratory report. Investors paying high prices often insisted on resubmitting their stones and nervous graders subsequently downgraded strongly fluorescent stones which faced up better than their facedown grade.[158] This set off something of a controversy within the trade.

GIA formerly taught that diamonds should be graded under low UV lighting. That opinion was reversed in 2008 when the institute decreed that diamond grading should be carried out in light with some ultraviolet because ultraviolet is a constituent of natural

---

156 Wade, *ibid.*

157. Under FTC guidelines, the use of the term "blue-white" is considered an unfair trade practice if the stone "shows any color or any trace of any color other than blue or bluish" under north daylight lighting (23.14). Thus it would seem that any diamond graded below F on the GIA-GTL colorless scale legally cannot be termed a *blue white*.

158. Diamolite, the light tube used since the 1950s for diamond grading by GIA, was designed to have low ultraviolet emissions to avoid stimulating fluorescent stones. However, due to a change in manufacturers, bulbs made since the early 1990s actually emit a good deal of ultraviolet light. Fluorescent diamonds graded between 1990-2000 are likely to have received a higher color grade than the same stone graded a decade earlier. Grading experiments using filters to screen out ultraviolet produced by Diamolite bulbs have resulted in downgrading of color on an average of between one half and one and one half color grades. See Thomas E. Tashey, "The Effect of Fluorescence on the Color Grading & Appearance of White and Off White Diamonds," *The Professional Gemologist*, vol. 3, no. 1, Spring/Summer 2000, p. 5.

light[159]. Others, most notably The Accredited Gemologist's Association (AGA), maintained that light with UV filtered out should remain the standard particularly when grading blue fluorescent diamonds. Don't let your D become a G. In experiments conducted by AGA, fluorescent diamonds have been shown to bleed, adding enough yellow to reduce the color by as much as four grades when evaluated in low UV lighting[160].

In 2000 GIA-GTL adopted a new grading methodology and a new grading environment using two Verilux F6T5 fluorescent tubes in a small box called DiamondDock. They also changed the grading methodology, increasing the viewing distance between the gem and the light source. This substantially reduced the effect of the UV in the lamp on the diamond being graded, effectively reducing overgrading of blue fluorescent diamonds from four to two color grades.

Following GIA's lead, most laboratories have adopted similar lighting, which still results in consistent "overgrading" of strong to very strongly fluorescent blue diamonds. As a result blue fluorescent diamonds continue to be priced at a discount in the wholesale market.

What does this mean for the connoisseur? Type IIa diamonds can be easily separated from type Ia by a gemologist using a simple test. Strongly fluorescent diamonds should be avoided. However, some blue fluorescent diamonds do not bleed color when the lighting environment is shifted from low UV fluorescent grading light to natural daylight. These together with type IIa's are the true blue-white diamonds. Paradoxically, due to commercialization coupled with the evolution of the term blue-white, dealers rarely test smaller diamonds in their stock. River quality stones are very rare, but these original blue white type IIa's and blue fluorescent diamonds that hold their color are well worth the seeking and likely to be priced well below their true value.

159 John M. King, et al., "Color Grading D-To-Z Diamonds at the GIA Laboratory", *Gems & Gemology,* The Gemological Institute of America, *Winter 2008*, p. 304. GIA spokesmen have claimed that a low UV environment "doesn't really exist anywhere," Of course it does, every night when the sun goes down and we turn on the interior incandescent and/or LED lighting. See also: AGA Executive Summary, *Lighting and Its Effect of Color Grading Colorless Diamonds...2009. http://accreditedgemologists.org/lightingtaskforce/ExecSummary.pdf*

160 Michael D. Cowing, "The Over-Grading of Blue Fluorescent Diamonds: the Problem, the Proof and the Solutions", *The Journal of Gemmology,* 2010, Volume 32/No.1-4. p.45. According to Cowing, the new methodology has reduced overgrading from four to two grades in very strong blue fluorescent diamonds and from one to two grades to a half to one grade in medium blue fluorescent stones.

Chapter 18
Colorless Diamonds

"As regards the water of the stones, it is to be remarked that instead of, as in Europe, employing daylight for the examination of stones in the rough…The Indians do this at night; and they place in a hole which they excavate in a wall, one foot square, a lamp with a large wick, by which they judge the water and the cleanness of the stone"

–Jean Baptiste Tavernier, 1675

The Four Cs of Connoisseurship in Diamond

A round brilliant cut diamond (RBC) exhibiting exceptional dispersion (fire) in natural sunlight. Photo: Michael D. Cowing. Courtesy: ACA Gem Lab.

*H*aving dispensed, in the Introduction, with the peculiarities of historical diamond connoiseurship, we are now ready to take a more holistic approach to diamond connoisseurship. The discussion that follows applies equally to both type Ia and type IIa colorless diamonds. Type IIa colorless diamonds will be considered in depth in the Chapter 18.

Cut: the First C

The form, the outward shape of a diamond, has from earliest times been an important factor influencing the gem's desirability. The *Ratnapariksa*, a Sanskrit text written sometime before the 4th century BC, sets forth the standards for evaluating a diamond. According to the text the most important thing is that the diamond have the perfect bipyramidal crystal form — "the six sharp points, the eight identical plane facets, the twelve narrow, straight edges are the basis of the natural qualities of the diamond."[161] Of course in those days the technology necessary to cut a diamond did not exist and gems of perfect crystal form were also the most brilliant and dispersive. Since the invention of faceting, lapidaries have worked to perfect the form of the diamond to maximize the optical qualities of brilliance, dispersion, and scintillation for which the gem today is so highly valued.

Since the 15th century when the

---

161. Lenzen, *History of Diamond Production*, p. 16.

technology was finally developed to cut the world's hardest substance, lapidaries have been focused on developing a set of proportions that would both retain maximum weight and bring out the gem's brilliance, fire and sparkle. The initial focus was on faceting. The first "cuts" were simple modifications of the gem's bipyramidal crystal shape. It was not until the late 16th century that the fifty-seven facet brilliant cut was invented.

The American Ideal Cut Round

In 1860 a Boston jeweler, Henry Dutton Morse, opened the first American diamond cutting shop in the city of Boston. At a time when most lapidaries concentrated on weight retention, Morse was a maverick, and, as is often the case, one blessed with a great talent. *Shopping for diamonds by weight, is like shopping for a racehorse by the pound,*[162] Morse proclaimed. He was interested, first and foremost, in maximizing the beauty of diamonds. He experimented with various angles and by 1870 had developed a set of mathematical proportions he judged would maximize the beauty of a round brilliant diamond.

Thirty years later, in 1919, Marcel Tolkowsky, the son of a master cutter, and working on an advanced degree in England described a mathematical set of proportions and angles which maximized the brilliance of a round diamond. Tolkowsky observed and measured a large

number of diamonds[163]. He then reduced the proportions of those with the greatest brilliance and dispersion to a mathematical formula which has come to be known as the *ideal cut*. The proportions of these stones closely matched the proportions developed by Henry Morse. Tolkowsky, however, wrote the book. He titled it *Diamond Design* and over the years Morse's proportions became known as the Tolkowsky ideal cut[164]. Since that time there have been many subtle changes to the Tolkowsky formula, depending upon who was calling the diamond "ideal ."[165]

Before proceeding further it is necessary to define our terms. *Brilliance* is the total quantity of light reflected by the gemstone back to the eye of the viewer. *Dispersion* (fire) is the property of breaking white light up into its component colors; this is

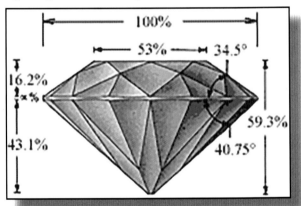

the property (illustrated above) observed as a rainbow when light is passed through

163Gilbertson, ibid. p.

164 Marcel Tolkowsky, *Diamond Design, A Study of Reflection and Refraction in Diamond.* E & F.N. Spon., London, 1919. Tolkowsky was well aware of Morse, in fact footnotes him. Tolkowsky's father followed Morse's ideas about proportions and most likely cut the diamonds he measured.

165. Tolkowsky's 1919 model diamond was an old European cut, the precursor to the modern brilliant.

162 Al Gilbertson, *American Cut, The First 100 Years*, The Gemological Institute of America, 2007, p. 30.

a prism. Dispersion is a property of all transparent gemstones but it is usually masked by the color in colored gems. Dispersion is a defining criterion in diamond.

Scintillation (sparkle) describes the breakup of light into individual points or flashes. Scintillation also requires movement of the gem, the observer, the light source or all three. Absent motion, the diamond has a face-up mosaic with of areas of light contrasted against areas of darkness.

A diamond's sparkle is directly related to the number of facets. Scintillation is not the production of light (brilliance); it is the sparkle, the breaking up of the light refracting from inside the stone into segments (scintilla). This would seem to lead logically to the conclusion that the more facets a gem has the better. This, however, is not the case. The more facets cut, the smaller each must be. Eventually a tipping point is reached where too many very small facets produce tiny little points of light creating a visual fuzziness which detracts from the beauty of the stone. In short, there is a point where too much scintillation can be too much of a good thing, hence the fifty-seven-facet brilliant.

In recent years, diamond-cutting standards have reached a level of precision rarely approached in colored gemstones. Diamond is singly refractive and is generally not color zoned. As we know, color is the prime criterion in judging the beauty of colored stones. Colorless diamond, since it has no color, is appreciated first and foremost for its brilliance (light return),

dispersion (fire), and scintillation (sparkle). These qualities are a function of cut. Cut is the critical issue and is, without question, the single most important of the four Cs in the evaluation of colorless diamond. Cutting is responsible for the brilliance, dispersion, and scintillation—in short, the beauty of a diamond.

Some experts have argued that crowning one set of proportions as ideal and labeling all others as less desirable is wrong. For years a debate over the concept of "ideal" proportions raged inside the diamond trade. Just as the space between any two existing points on a line is infinitely divisible, it is possible to proportionally alter the ideal measurements in an infinite number of ways to produce an infinite number of formulae which should maximize the brilliance of a diamond. Thus, by using computer modeling, it is at least theoretically possible to define an infinite number of ideal cuts. This, at least, was this author's argument in early 1998.[166]

For many years the pricing of a round brilliant diamond has been based on the Tolkowsky model. Gemologists were taught that there were four classes of cut. At the top of the heap sat the Tolkowsky ideal. Any variation from these proportions was discounted. As of this writing, there is as much as a seventy percent price differential between round diamonds carrying identical color and clarity grades solely on the basis of the stone's proportions. Previously, this had been the jeweler's fudge factor. So long as cut grades were kept off grading

166. Richard W. Wise, "Diamond Cutting: New Concepts, New Millennium," *The Guide, Gem Market News*, March-April 1998, p. 7.

certificates, sellers could meet and beat competitor's prices and retain their profit margin simply by offering the client the same color and clarity and a poorer cut.

Toward the end of the 1990s, under increasing consumer pressure, laboratories began issuing cut grades for round brilliants on their diamond grading reports. These grades were loosely based on traditional Tolkowsky proportions. Diamond dealers who fiercely resisted this trend feared

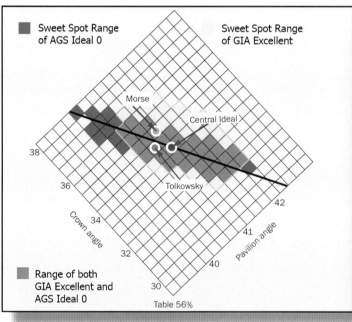

Cowing's Central Ideal of "Sweet Spot" and its relationship to Ideal proportions as defined by the Gemological Institute of America and the American Gem Society. Graph: Michael D. Cowing, Courtesy: ACA Gem Lab.

that once the buyer was able to obtain a certificate which graded all four Cs, the diamond would be reduced to the status of a commodity. As it turns out, with the rise of the internet, their fears were well founded.

Multiply tiny variations in proportions by a factor of ten thousand and you have

a general idea of the method used in a landmark study of diamond cut and brilliance by the Gemological Institute of America. The GIA computer modeled ten thousand variations in an attempt to determine, once and for all, the best proportions for a round brilliant cut diamond.

In the first edition of this book, I wrote following the GIA study: *"the best proportions for dispersion do not overlap with the best sets of proportions for brilliance."*[167] Happily, more recent studies call that assumption into question. According to researcher Michael Cowing, another industry maverick[168], it is possible to cut a round diamond to proportions which will maximize both brilliance and dispersion which will fall within GIA and American Gem Society's (AGS) ideal range. Taking an analogy from the sport of tennis, there is, says Cowing, a "sweet spot" or an *super-ideal* cut round brilliant diamond where maximal brilliance and maximal dispersion overlap. The

167. Hemphill et al., "Modeling the Appearance of the Round Brilliant Cut Diamond: An Analysis of Brilliance," *Gems & Gemology*, winter 1998, pp. 158-183; Reinitz et al., "Modeling the Appearance of the Round Brilliant Cut Diamond: An Analysis of Fire, and More About Brilliance," *Gems & Gemology*, Fall 2001, pp. 174-197.

168 Cowing, Michael, *Accordance In Round Brilliant Diamond Cutting*, www.acagemlab.com/news/ 2000, http://www.acagemlab.com/articles/SweetSpot.htm. The ideal parameters set by Cowing are: 1. Pavilion main angle = 41°, 2. Length of pavilion halves = 77%, 3. Crown main angle = 34°, 4. Table size = 56%, 5. Star Length = 55%, 6. Girdle size = thin to medium, 7. Culet size = small to none.

The author had an opportunity to examine a modern brilliant cut with the proportions Cowing described. The stone exhibited superior optical performance when compared to other ideal cut diamonds. Unlike most well cut modern diamonds which maximize brilliance and scintillation at the cost of dispersion, this stone exhibited a lovely balance of brilliance, fire (dispersion) and sparkle (scintillation). Thus, there exists an balanced "Ideal" and the connoisseur interested in purchasing a round diamond is advised to closely examine the diamond grading report and to seek out a gem with the angles suggested by Cowing[169].

Comparing dispersion (fire). The dispersion or fire in a 1 carat round brilliant ideal (left) compared to a 2.5 carat Old Mine or rectangular baroque brilliant cut (right). The visual pyrotechnics in the OEC are the result of shorter, wider pavilion main facets. Images approximately 30X. Photo: Michael D. Cowing courtesy ACA Gem Lab.

In theory diamond cut can be graded much as cut is graded in colored stones; i.e., by visually measuring light return to the eye (see Chapter 3, "Performing the test"). However, there are a couple of problems. Diamonds are relatively well cut when compared to colored stones. A ruby with eighty percent brilliance is considered a very well cut stone. A diamond with eighty percent brilliance is a very poorly cut stone. An exceptionally well-cut diamond can be as much as ninety-nine percent brilliant.[170] Detecting the difference between ninety-five and ninety-nine percent light return is very difficult, if not impossible, for any but the most experienced.

Buying for Beauty: a Contrarian Approach to Round Diamonds

There is much to admire about the Baroque Brilliant or Old European Cut. An earlier version of the brilliant cut popular between 1810-1920 which preceded the modern brilliant, some of these gems were cut the proportions and pattern that Morse perfected and Tolkowsky described. The Old European has same number of facets with a higher crown, smaller table and a slightly different pavilion. In the modern brilliant, the pavilion halves or lower girdle facets are larger and the pavilion main facets elongated to maximize scintillation. This pattern robs the stone of much of its fire (dispersion). In the single-minded pursuit of brilliance, the modern ideal cut has maximized total light return at the expense of dispersion (fire). Why was this done? In a word, lighting.

169 Several diamond companies have branded their own version of the ideal cut. These patented cuts sell for premiums of 40% and more verus unbranded gems of similar or superior quality. With a bit of study, the knowledgeable aficionado can realize significant savings on the finest ideal cut round diamond.

170. Wise, "Diamond Cutting," p. 7.

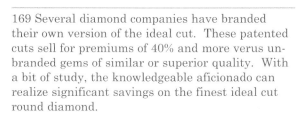 

Modern jewelry stores use spot and flood lights which light up the store from every angle. Older brilliant cuts were developed in a different environment under natural light, candlelight and early electric point light such as that generated by a single bulb in an overall darkened environment. That is why clients are often heard to remark that the diamond they purchased looked so much better in the jewelry store. They do, because they are cut specifically to look their best in this type of lighting.

A well-cut Baroque brilliant with shorter, thicker pavilion main facets can be very strong in dispersion. The big hearty shards of colored light seem to leap from the heart of the stone, but they must be

not feel wedded to the ideal or for that matter to any round diamond. A properly proportioned emerald cut is a vast cathedral of light—the oval brilliant is more elegant and the pear and the marquise brilliants are arguably more feminine than the ubiquitous round. In the creation of an symmetrical brilliant, oval, pear and marquise the elongation of the shape requires a pair of larger facets at the center of the stone. These larger facets often create a visible phenomenon known as a bowtie, a black area shaped like an hourglass across the short axis of the gem. Well-cut gems will not show this effect.

Color in Diamond

Diamond is a special case! Diamond is colorless so there is rarely a need to discuss

Colorless diamonds set facedown for grading. Tonal variations of yellow in the body color illustrate the range of colors from D to Z. Courtesy of J. landau, Inc.

seen in natural and point lighting. Under such conditions, the modern brilliant, even when ideal cut, seems anemic by comparison. Those cut within the parameters of Cowing's sweet spot, not so much. The problem is that very few Baroque brilliants were cut anywhere close to Morse's proportions. Enjoy the hunt.

Other Designs

Unfortunately most gem labs do not grade cut in shapes other than round brilliant. Even so, the aficionado should

either hue or saturation. Color in colorless diamond is normally limited to tints, tonal variations of yellow and less frequently gray or brown and on rare occasions, blue. The word tint is used to discuss hues which are so light in tone, a stone would be termed yellowish or brownish rather than yellow or brown; the tonally darker the tint the less desirable the stone. If the diamond does have a well-defined hue then it is a fancy color diamond, and therefore in an entirely different category and evaluated like any

other colored stone.[171]

Color in colored gemstones is judged face up, color in colorless diamond is judged face down.[172] Body color, not key color defines the grade. Turning the stone face down eliminates the distraction of brilliance and allows the grader to observe the minute tints (tonal variations) of yellow, gray or brown, which define relative colorlessness. The objective is the total elimination of color. The body color should be snowy white or occasionally slightly bluish[173]. The color may vary depending upon the viewing angle so the stone should be viewed face down, through the girdle and through the crown. Diamond colors are graded on an alphabetical scale from D to Z; there are no A, B, or C grades. D is totally colorless. Z is discernibly yellowish.

The letter grades indicating the tonal variations in a face-down color are first grouped in threes. D through F are called *colorless*. That is, there is no discernable tint of yellow when the gem is viewed face up. G through I are termed *near colorless*. J through L show a *faint yellow* when viewed face up. From N to Z the groups get larger; N through R are termed *very light yellow*, S

through Z are called *light yellow*.[174]

This grading method sometimes can lead to seeming contradictions. Dealers are sometimes heard to say that the stone is slightly yellowish, *e.g.* J color, but it faces up white, meaning that despite the very slightly yellowish body color, the stone's key color appears colorless when the diamond is viewed face up. The dealer's point here is that the stone appears whiter (less yellow) and is therefore worth more than its color grade would indicate. This visual discrepancy may be a function of cut or more frequently a byproduct of ultraviolet fluorescence.

As discussed in the introduction, diamonds are color graded in a special lighting environment. Given the variability of natural daylight, experienced graders prefer diffused fluorescent lighting with a kelvin temperature rating between 5500-6500.[175] This type of lighting is difficult to find in a normal jewelry store or at a gem show. Jewelers prefer to show stones under incandescent spotlights or the new LED spotlights which are rapidly replacing them because they maximize the diamond's brilliance. This is perfectly legitimate; remember, gems are creatures of the light.

171. A tint can be thought of as a discoloration of the stone which interferes with its crystalline colorlessness.

172. The diamond is placed table down at a forty-five degree angle in a white grading tray under a specially designed fluorescent light source (GIA Gemolite). The point here is to judge the amount of yellow in the transmitted color as seen through the pavilion of the stone.

173 Diamonds with bluish body color is discussed in detail in the section on fluorescence (Chapter 5) and the chapter on Golconda diamonds.

174. Despite terms such as *light yellow,* diamonds graded S-Z are not considered to be yellow diamonds, they are off-color colorless and these terms do not appear on the grading report. According to the GIA standard, to be called yellow, a diamond must have a more saturated yellow hue than the Z master stone viewed face up. Fancy color diamonds are graded face up, as are all colored gemstones.

175. Since 2000, The Gemological Institute of America has used the *DiamondDock* at a grading distance of seven inches (see Chapter 4). Gemologists should use two 17" F15T8VLK Verilux full spectrum fluorescent tubes as equivalent to the Diamond Dock.

However, jewelers specializing in gems will often have special grading lamps in their offices or at gem shows.

In 1997, The Gemological Institute of America conducted a color preference test comparing fluorescent blue diamonds against diamonds of comparable color which were inert under ultraviolet light.[176] In this test, professional graders expressed a clear preference for the appearance of blue fluorescent stones. Surprise, surprise! In the world of colorless diamonds, blue fluorescence will improve the look of a diamond. As discussed previously, not only will the blue nullify some of the yellow in the body color, but often the fluorescence often will lap over into the visible spectrum, punching up the saturation of the stone and giving it a whiter-brighter appearance. The two downsides, as discussed in the Introduction, are overgrading and reduced transparency, poor crystal. The aficionado is advised to review these sections and as

---

176. Thomas M. Moses et al., "A Contribution to Understanding the Effect of Blue Fluorescence," *Gems & Gemology*, winter 1997. pp. 244-259.

A 2.24-carat round brilliant-cut diamond with a GIA-GTL clarity grade of SI1 viewed under magnification. 1: Invisible to the eye, grade-making inclusions are barely visible under 10X magnification. 2: Side view magnified to 20X. Inclusions of this size have no appreciable effect on either the beauty or the durability of the gem, but will substantially lower the price. Photo: Gary Roskin.

always, compare, compare, compare.

Clarity Grading Diamond

Diamonds are graded for clarity using the loupe standard; this means the stone is examined using 10X magnification. This approach is distinctly different from the standard for colored stones; in that case the naked eye is the standard for judging clarity. The dual standard exists for a number of reasons. A fine diamond is relatively colorless and has exceptional diaphaneity; that is, it is normally crystal clear. The

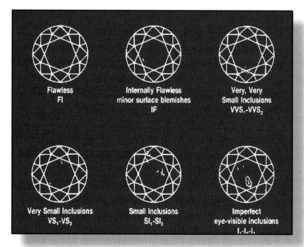

GIA Diamond grading chart. Assumes 10x magnification.

existence of visual inclusions, therefore, becomes relatively more important.

The adoption of a microscopic standard (or what dealers call a loupe-clean standard) for diamonds also had something to do with the *appearance* of rarity. Before the invention of the magnifying lens, visual appearance was, by default, the only standard for judgment. As South African diamonds began to flood the market in the early 20th century, it was necessary to create a means whereby apparent rarity was increased. What better way than to introduce an absurdly stringent standard of judgment? Why "absurdly" stringent? Because, by definition, the first five categories of diamond clarity — *internally flawless* (IF), *very very slightly included* (VVS1 and VVS2), *very slightly included* (VS1 and VS2), and *slightly included* (SI1) — are visually identical.[177] By definition, **the inclusions in the first six clarity grades, *internally flawless* (IF) through SI1, do not materially affect the beauty or durability of the stone.** Only in gems graded SI2 and *imperfect* I1, 2, and 3 do the inclusions become flaws. The first six clarity grades **are** visually identical. Why then should the aficiaado pay the thirty to forty percent premium demanded in the marketplace for

VVS or even VS gems?[178] Clarity grading in colorless diamond has more to do with the politics of beauty than with beauty itself. This fact is not completely lost on the market!

An examination of the very accurate wholesale diamond price list issued by *The Guide* shows that the visual standard of clarity matters most in the market. Except for the highest grade, the legendary *D-Flawless*, diamond prices in all color grades increase in proportionate increments from SI1 to IF. The largest percentage jump, seventy-three percent, is between the grade SI1, which is the lowest eye-clean grade, and *imperfect* (I1), the first grade in which inclusions are always visible when the diamond is viewed face-up without magnification.

What then is the difference between a D-color diamond graded flawless and a D-color diamond with a clarity grade of SI1? The answer, aside from the 40% premium, none. From the perspective of beauty, assuming that both gems are eye-flawless, there is no difference at all. Diamond salesmen will sometimes suggest that a perfect diamond, like the perfect relationship is one without flaw. Leaving aside the absurdity of that statement, the aficianado, and the bride to be, would do

---

177 Though I was taught that the SI2 was the first grade where inclusions were eye visible, apparently thare has been some evolution in standards. According to Gary Roskin; "In teaching the original system, the basic guideline was that all eye-visible inclusions were graded I1 at best. Today some larger step cut diamonds with eye-visible inclusions may be graded SI2 or even higher. Conversely a diamond can have inclusions that are not visible...and be graded I1." Roskin, G A., *Photo Masters For Diamond Grading*, Gemworld International, 1994, p1.

178. At what clarity is a diamond eye-clean? I was taught that any gem from SI1 up (SI1, VS, VVS grades and of course I-Flawless) were by definition eye-clean (no inclusions visible to the naked eye). It seems, though, that inclusions are allowed to be larger in larger gems. A five-carat diamond might very well have eye-visible inclusions and have a laboratory grade of VS2. I am referring to the eye-clean boundary, no matter at what letter grade. Cf. Gary Roskin, *Photo Masters for Diamond Grading* (Chicago: Gemworld International, Inc., 1994).

well to remember the distinction made between inclusions and flaws in Chapter 3. Inclusions become flaws only if the inclusion materially affects the beauty and/or the durability of the gem. Absent that, a diamond with tiny invisible inclusions is flawless.

Chapter 19
Golconda or Type IIa Diamonds

"Transparency is of great importance; but as few persons recognize the finer grades of transparency it is not ordinarily considered."

–Robert M. Shipley, 1936

As we have seen, colorless diamonds fall into two types, Ia and IIa. Ninety nine percent of the colorless diamonds readily available in the market are type Ia. Less than one percent of colorless diamonds are type IIa. Type Ia diamonds contain nitrogen, the source of the yellow, and type IIa contain no measurable levels of this element.

Historically, most, though not all type IIa diamonds hail from the old diamond mines of India and were sold in the gem souks of the ancient city of Golconda. It is important to understand that the terms Golconda and type IIa are not synonymous. Not all Golconda diamonds were type IIa and not all type IIa diamonds are from India. This rare type of diamond has also been found in Brazil and at South Africa's Premier Mine and other locations. Due to the lack of nitrogen, Golconda diamonds have earned a reputation as being both "whiter than white" (*i.e.* absent nitrogen they have no yellow tint) and having exceptional transparency (crystal).

Gem expert Benjamin Zucker suggests, "Place a Golconda diamond alongside a

The 76.02 carat Archduke Joseph Diamond. Graded D-Fl, note the exceptional diaphaneity and pure white colorlessness. In 2012 this diamond sold for a record $21.48 million, a record price for a Golconda type IIa diamond. Courtesy AP.

modern, recently cut D-colour diamond and the purity of the Golconda stone will become evident."[179] Mary Murphy Hammid, in an essay on Golconda diamonds written for Christie's auction house, maintains: "Golconda diamonds have a degree of transparency rarely seen in stones from other localities . . . it is variously called soft, limpid, watery or clear. It is not to be confused with clarity. . . . It is not to be confused with color grade. Rather, it is a quality in which light appears to pass through the stone as if it were totally unimpeded, almost as if light were passing through a vacuum."[180]

Legendary French gem merchant Jean Baptiste Tavernier at the diamond fields of Golconda from a 17th century engraving.

The Mysterious Blue Glow

Some non-fluorescent diamonds exhibit a slightly bluish body color. Some years ago, I acquired a 9.07 carat type IIa diamond graded D Flawless type IIa by GIA-GTL. The stone was cut in the old mine style and had a very large window beneath the table (pictured Chapter 20) leading me to believe that the stone was from the old Indian mine at Golconda. Upon examination in early June afternoon sunlight, the stone exhibited a blue glow which appeared to hover above the gem like an early morning mist[181].

Upon further research I discovered that a number of type IIa diamonds, including several with unimpeachable Indian provenance, also exhibited this phenomenon. Specifically, the *Regent, Queen of Holland* and *Orloff* are said to show a blue-to-blue green tint. According to diamond expert Ian Balfour, the 135.33-carat cushion cut Queen of Holland "does present a

181 Herbert Tillander, *The Hope Diamond and its Lineage,* Paper presented at the 15th International Diamond Conference, 1995. p. 7. The present owner states that the gem continues to exhibit a blue glow at times of its own choosing, Cora N. Miller, personal communication, 2001.

179. Benjamin Zucker, *Gems and Jewels: A Connoisseur's Guide* (New York: Thames and Hudson, 1984), p. 86.

180. Mary Murphy Hammid, "Golconda Diamonds," *Christie's Magnificent Jewels* (October 23, 1990, catalog), pp. 301-302.

definite blue tint rather like the color of cigarette smoke.[182]"

Several experts have suggested differing causes for this phenomenon. Tom Moses, GIA's point man on colored diamonds named the Tyndall effect; Dr. Emmanuel Fritch, University of Nantes suggests nitrogen levels below the detection level of the instrument used to determine diamond type may be the cause[183]. Prior to the adoption of the GIA colorless diamond grading system, D color diamonds, those with a snowy white to slightly bluish body color were classed as *river* quality gems (see Introduction).

Writing in 1915, the distinguished gemologist Frank B. Wade differentiated two types of *river* diamonds, those with and those without blue fluorescence. River gems, said Wade, had the additional quality of exceptional transparency. Following those of *river* quality are the *Jagers*, Wade continued, which exhibit a steely blue white body color. Most of these diamonds are from the old Indian and Brazilian diggings and a small percentage are from Africa[184].

Performing The Test

A simple test, developed by Wade may

aid the aficionado in determining if a diamond exhibits the ultra-transparency characteristic of *river* gems[185]. The test, echoing the 13th century writings of the Arab gemologist Al Tifaschi, requires two diamonds and an opaque white card placed directly in the sun[186]. Though Wade did not mention it, the two stones being compared must also have a similar polish, a fact usually noted on laboratory grading reports. The two stones, each held in diamond tweezers, are oriented so that a rainbow of dispersed light is projected on a white card. If one gem has the exceptional crystal of a river quality diamond, the prism projected will be the more vivid and "snappy." To compare color, both of these gemologists, separated by eight centuries, suggest comparing the diamond's color to that of rock crystal[187].

Since the first publication of this book (2003), there has been a dramatic upsurge in interest and in the price of Golconda Type IIa diamonds[188]. Several large diamonds of this type have been sold at the big auction houses and regularly command prices as much as fifty percent higher than type I diamonds of the same color/clarity grade. At a Sotheby's Magnificent Jewels

182 Balfour, Ian, *Famous Diamonds*, William Collins Sons & Co., London, 1987. p.179. Balfour goes on to describe the Orloff (p.20) and the Regent (p.64) as having a light blue tinge. See also Richard W. Wise, GemWise Blog, *Golconda Diamonds* 2007: http://gemwiseblogspotcom.blogspot.com/2007/06/golconda-diamond.html

183 Tom Moses, personal communications 2008 and Emmanuel Fritch, personal communication 2012. Moses has seen the phenomenon many times and states that it is not geographically specific.

184 Frank B. Wade, ibid. p. 16-17.

185 Wade, *ibid*.

186 Al Tifaschi, quoting certain Persian merchants, states that the Indians divide colorless diamonds into two types, *billawri* (pure white) and *zayti* (slightly yellowish). The finest, *billawri*, will project a rainbow on a wall. Indian dignitaries keep this quality for themselves. S. N. A. Huda, *Arab Roots of Gemology*, ibid. p. 118.

187 ibid.

188 Wise, Richard W., GemWise Blog 5/27/08. *Type IIa Diamonds Glow at Sotheby's and Christie's.* http://gemwiseblogspotcom.blogspot.com/2008/05/big-type-iia-diamonds-glow-at-sothebys.html

auction held in New York on April 17th, 2008, a 24.42 carat D color Internally Flawless diamond sold for 3.6 million or better than $148,443 per carat. On January 10th, 2013 a 52.58-carat Golconda stone sold for 10.8 Million or $207,600 per carat.[189]   These premiums are based on the fact that the best of these gems are both highly crystalline and whiter than white, that is, a D-Flawless type IIa will make an ordinary D-Flawless seem less transparent and slightly yellow by comparison.

Smaller type IIa diamonds are to be found in the market and can be separated by a simple test which can be performed by a seasoned gemologist.   They are often to be found in older pieces of jewelry. Gem dealers generally do not test their stock to distinguish diamond type. However, diamond type is normally noted on laboratory reports, particularly those issued by GIA-GTL. "Certified" type IIa diamonds will sell at a dollar premium of between ten and fifteen percent.

189 ibid.

THE GARNET GROUP

Picture a remote valley sandwiched within a range of humpbacked hills of northern Tanzania in the long shadow of Mount Kilimanjaro. The month is February, the end of the blistering African summer. A lone figure struggles through a copse of sansevieria, the wild African sisal, with sword-spike leaves that can slash a man's skin to ribbons. This is the domain of the tsetse fly, of kifaru the rhino, and of simba the hunting lion.

The man lifts his head and stares across the wide savanna; the dry, windblown brown landscape is dotted with the spindly umbrella-shaped acacia trees. He slowly walks in a grid pattern, head down, carefully scanning the parched earth. Then, he pauses mid-step like a water dog on point. The breath catches in his throat. A trained eye has detected the slightest hint of green twinkling in the equatorial sun. He removes his bush hat and wipes the sweat from his eyes with the back of his hand. Barely daring to hope, he squats down for a better look. "Yes! Here they are, shards of crystal, sparkling like tiny green diamonds!" The search is over, but the real work is about to begin.

Mining tsavorite garnet on the East African savannah; Scorpion Mine, Tsavo National Park, Voi, Kenya.

The garnet family is scientifically classified as a group. This means that although all garnets share the same atomic or crystal structure, they differ in chemical makeup. For this reason the properties of garnet, specifically hardness and refractive index, will vary greatly. Most people are familiar with almandine (also almandite), the

brownish red iron-aluminum garnet, and with pyrope, the deep red magnesium-aluminum garnet. These are only two of the seven varieties classed by chemical composition, which include pyrope, almandite, grossularite, andradite, spessartite, and pyrope-almandite.

To further muddy the waters, recent discoveries in Africa have turned up garnets such as malaya and tsavorite with chemical compositions that fall between the traditional classifications, rendering these classifications unwieldy and all but obsolete.[190]

From the connoisseur's perspective, this doesn't matter a great deal provided the stone can be identified. And in all cases positive identification is possible using the basic instruments and the services of a trained gemologist. It is only necessary to be aware that hardness and refractive index, and also rarity, differ markedly among the members of the garnet group.

In this volume the focus is on four of the most distinguished members of the garnet group: malaya, spessartite, demantoid and tsavorite. Although two of these, malaya and tsavorite, are among the latest discoveries, their outstanding visual characteristics make them irresistible candidates, demanding their presence among the new precious gems.

Working a reef. Tsavorite garnet mining, Scorpion Mine, Tsavo National Park, Kenya. Photo: R. W. Wise.

Treatments

Garnet is rarely subjected to any form of treatment. Heat treatment has thus far proved ineffective in most varieties, and irradiation has little effect on the gem. In the past few years, demantoid, specifically Russian demantoid, has been subjected to heat treatment at low temperature making detection of this treatment impossible at the current time (2016). However, the science of gemology is always engaged in a game of catch-up. When the guys bad guys discover a method of enhancing the look of a gem material, gemologists are left

190 Carol Stockton and Vincent Manson, "A Proposed New Classification for Gem-Quality Garnets," *Gems & Gemology*, (Winter 1985), 205. The authors suggest a new system of classifying garnet consisting of eight types: grossular (which includes tsavorite), andradite, pyrope, pyrope-almandine, almandine, almandine-spessartine, spessartine, and pyrope-spessartine (which includes malaya).

with the task of, first, figuring out that something is being done, and second, determining what is being done. As garnet prices rise, you can be sure that sooner or later someone will figure out a way to make an inferior stone seem superior.

Durability

For our purposes, garnet can be divided into two groups; hard and soft. Tsavorite, malaya and spessartite have a Mohs hardness rating of 7-7.25, are harder than dust and therefore suitable for daily wear; demantoid with a hardness of 6.5-7 is one of the softer members of the garnet group.

4.12 carat Demantoid garnet. Photo: Jeff Scovil, (2012) The Edward Arthur Metzger Gem Collection Bassett, W. A. and Skalwold, E.A.

Chapter 20
Demantoid Garnet

"Many collectors dream of having a demantoid because they know this is one of the rarest stones on Planet Earth. This 'Gem of the Czars' is not just one of the most precious members of the garnet family, but is one of the most precious gems of any kind."

Richard W. Hughes & Jonas Hjorned 2014

There are two varieties of green garnet of interest to the connoisseur, tsavorite and demantoid. Though they are cousins, they have very different properties. Demantoid is the softer, less durable of the two, but all things being equal will deliver superior pyrotechnics.

Demantoid was first discovered in 1853 in Russia's Central Ural Mountains near the village of Nizhniy (Tagil Nizhniy). Children playing near the village discovered some rounded green alluvial pebbles which, due to their yellowish green color were first thought to be peridot. In February 1864 Nils von Nordenskiold, the same Finnish mineralogist who was involved in the naming of alexandrite, identified the stone as green andradite colored by chromium, and he proposed the name demantoid (diamond-like) due to the gem's high diffraction and dispersion.[191]

A magnificent medium dark toned 4.55 carat Russian demantoid garnet. Some aficionados prefer a gem of a slightly lighter tone. Photo Courtesy Pala International.

Demantoid garnet occurs in a number of localities including, Sri Lanka, Pakistan and Italy, the other commercially exploited location is in northwestern Namibia. The Green Dragon Mine, the only commercial mine, is located between the Spitzkope and Erongo mountains. Approximately 14km from Erongo and 25km from Spitzkope

191 William R. Phillips and Anatoly S. Talantsev, "Russian Demantoid Garnet, Czar of the Garnet Family," *Gems & Gemology*, The Gemological Institute of America, Summer, 1996, pp.100-101.

the first known specimen of demantoid garnet from Namibia was found in 1936 and is in the collection of the British Museum.[192]

Hue/Tone

Demantoid can run the gamut between near colorless to a rich forest green hue. Most demantoids are a light to medium-toned green with a strong yellow secondary hue. There are two opinions regarding the finest color. Some aficionados prefer a medium dark toned (80%) visually pure green not unlike the color of tsavorite garnet. In fact, demantoid derives its rich color from chromium, which along with vanadium is also responsible for the color of tsavorite. In fact, a 75-85% toned rich forest green is the hue preferred by collectors in the United States. However, some connoisseurs prefer a stone of a slightly lighter tone (70-75% tone) which allows the gem's characteristic dispersion (rainbow effect) to show through. The lighter the tone, the more prominent the dispersion.

Saturation

Brown is the normal saturation modifier or mask found in demantoid. It is ubiquitous enough to be a characteristic of Namibian gems. Russian gems can be brownish and are sometimes slightly grayish.

192 Christopher T. Johnston, personal communication, 2015.

Dispersion and Brilliance

Demantoid garnet boasts the highest dispersion among the garnet group. At .055 it has the highest dispersion of all the precious gems eclipsing even diamond (.044). Dispersion is rarely a part of the connoisseur equation in colored gemstones because the effect is largely masked by the gem's color. In demantoid it is irrepressible and a scintilla of pure red emerging from the heart of a cool green gem can be disconcertingly beautiful. Dispersion is best observed in incandescent lighting.

A classic horsetail inclusion under magnification. Photo courtesy Pala International.

The high dispersion coupled with its equally elevated index of refraction causing light to bend dramatically when it enters the gem, yields a particularly brilliant and colorful face-up mosaic. Demantoid is normally cut in the brilliant style so as to maximize the gem's most obvious charms. Round brilliants are particularly desirable.

Clarity

Gem aficionados normally prefer eye clean gems. Some more mineralogically minded collectors prefer gems which exhibit the classic horsetail inclusion when viewed under magnification. These horsetails (fibrous serpentine) inclusions are present, whole or in part, in most cut gems. However, the lack of inclusions has little effect on price amongst gem aficianados.

Treatments

Garnet is known to be particularly resistant to treatment. However, some Russian stones are treated with low temperature heat to both lighten the tone and drive off the brown mask. This treatment is sometimes detectable depending upon the heating temperature and as always, the aficionado is advised to require a gem grading report from a recognized gemological laboratory.[193]

The Rarity Factor

Demantoid garnet is fairly available in sub-carat sizes. At one carat it becomes scarce. At two carats the gem is quite rare. At three carats the gem begins to increase geometrically in price and at five carats the gem is almost unobtainable.

---

193 As of this writing, most labs will call heat treatment in demantoid garnet and some will also note country of origin. The exception is GIA GTL, which does neither. C.f. http://www.palagems.com/demantoid_disclose.htm.

Tsavorite garnet and diamond earrings. Tsavorite pair shapes from Taita weighing 6.78 carats and trillian cuts (2.13 cts.) from The Scorpion Mine. Photo: Jeff Scovil courtesy Bridges Tsavorite.

Chapter 21
Tsavorite Garnet

"The discovery occurred in 1967 in a small hidden valley in a rugged range of hills about 100 kilometers from Mount Kilimanjaro,"

–Campbell Bridges, 2007

The qualities of tsavorite. The large pale cushion-shaped gem (tonally below 50%) is classified as green garnet. Photo: Jeff Scovil, courtesy Bridges Tsavorite.

Tsavorite garnet was discovered by geologist Campbell Bridges in Tanzania in 1967,[194] which makes it one of the newest of the precious gemstones. Technically a calcium aluminum silicate which crystallizes in the cubic system, it is a green grossular garnet. Grossular is one of seven members of the garnet family. Tsavorite is not the only member of the grossular species; grossulars or grossularites can occur in a range of colors including yellow, orange, pink, and brown. Grossularite, a yellowish orange garnet, was named for the gooseberry, *grossularia*. Green grossular garnet occurs in a formation known as the Mozambique belt, a rich geological formation that works its way down the East African coast from Ethiopia to Mozambique.

Tsavorite Versus Emerald

Tsavorite garnet is often compared to emerald. Given their similar appearance, comparisons are both inevitable and understandable. Emerald, like tsavorite,

---

194 Campbell Bridges states that he found the first traces of green garnet, either tsavorite or demantoid in the African country of Zimbabwe in 1961. Campbell Bridges personal communication, 2005.

is a silicate, containing beryllium rather than calcium. Both are colored green by trace amounts of chromium and vanadium. Here, scientifically, is where the similarity ends. Emerald crystallizes in the hexagonal crystal system. It is doubly refractive: light that enters the gem is divided into two rays. Tsavorite, however, like all garnets, is singly refractive.

The key difference between emerald and tsavorite is refraction and dispersion. The refractive index of emerald is 1.57-1.59; the refractive index of tsavorite is significantly higher at 1.74. This means that light penetrating green garnet is bent at a greater angle than light entering emerald. Even more significant is the difference in dispersion, the way the two gems break up white light into its constituent rainbow colors. The dispersion of tsavorite is 0.28, twice that of emerald at 0.14. This, coupled with the higher refractive index, will disperse the light, adding to the life of the gem. The lower refractive index imparts to emerald what connoisseurs refer to as a sleepy or satiny appearance. Tsavorite will have a crisper appearance when fashioned in most cutting styles. However, when faceted in the traditional step or emerald cut, or when cut *en cabochon*, the visual similarity between the two gems, particularly in their finest qualities, can be uncanny.

Emerald at 7.5 on the Mohs scale is slightly harder than tsavorite, which is about 7.25. Hardness, however, is something of a misnomer when used as a measure of durability. The Mohs scale is really only a measure of scratch-ability. A

Campbell Bridges, discover of tsavorite garnet, with Rebekah Wise at The Scorpion Mine, Tsavo National Park, Voi, Kenya. Photo: R. W. Wise.

stone which is harder is simply one that will scratch one that is softer.

A more important measure of durability is toughness; in practical terms, how well the stone will stand up to setting and the abuse of day-to-day wear. Emerald has a well-deserved reputation of brittleness; it is easily chipped and is problematic, particularly as a ring stone, if worn every day. Tsavorite garnet, on the other hand, is not brittle, is tough enough even for invisible setting, and stands up well to daily wear.

Hue

Tsavorite miner Campbell Bridges has identified three mixtures of hue which, in his opinion, describe the finest colors

Exceptionally fine 1.28 carat forest green tsavorite garnet from Bambali, Tanzania. A slightly bluish green, 75% tone cut by Gene Flanagan, Precision Gem. Photo: Gene Flanagan, courtesy R. W. Wise. Goldsmiths.

I have ever seen was a "Kentucky bluegrass green" from the Scorpion Mine near Voi, Kenya, in Tsavo National Park. The stone had a highly saturated eighty percent green primary hue and a twenty percent secondary blue hue; the tone was eighty percent.

Until recently garnet has been found in every color except blue. Thus any garnet with the slightest pretension toward a blue or bluish hue is regarded with a high degree of reverence.[196] Like emerald, yellow is the bane of this green grossular garnet. A bit of yellow may punch up the saturation, but ten percent or more of it —because of its lighter, brighter tone — begins to dilute the green visually. As a rule of thumb, gems that are visually yellowish should not be considered top color.

Saturation and Crystal

Tsavorite, like many East African stones, is a day stone; that is, it looks its best in natural light. And, although incandescent may actually improve the color, darker stones of optimum tone will often pick up a sooty dark gray to black mask in incandescent light. The stone appears to close up, dramatically reducing transparency and saturation of hue. The darker the tone, the more pronounced the effect. Lighter-toned gems hardly seem affected. In this respect tsavorite differs markedly from emerald, which does not display this

in tsavorite garnet: forest, water, and grass green. Forest green is visually bluish, eighty-five to ninety percent green, ten to fifteen percent blue. Water green appears a bit yellowish: seventy-five percent green, fifteen percent yellow, and ten percent blue. Standing between the two is grass or leaf green, a visually pure green which contains eighty percent green hue with approximately equal amounts of yellow and blue.[195]

In fact, the market tends to favor bluish green stones and I agree. A visually pure green with perhaps fifteen percent secondary blue hue and between seventy-five and eighty percent tone is probably ideal. Unlike African emerald, tsavorite is rarely over-blue. Stones with twenty percent blue secondary, though rare, are extremely beautiful. The most beautiful tsavorite

195. Campbell Bridges, personal communication, 1995.

196 In the early 1990s a blue/red color-change garnet was discovered in Tanzania and more recently in Madagascar. Most of this variety exhibits a strong gray mask. However, at their best, these garnets look very much like the finest Brazilian alexandrite and have a very similar color change.

tendency at all. As with all gemstones, the aficionado is advised to study carefully the effect of shifting light sources, particularly from daylight to incandescent.

Tone

To be considered tsavorite, a gem must have at least medium or sixty percent tone. Stones with tonal values below sixty percent have a watery or pastel appearance and are called green grossular garnet. Seventy-five to eighty percent tone is optimal. Stones with tonal values above eighty percent appear over dark.

The division of color in gemstones into hue, saturation, and tone is useful in discussion. In practice, beauty is a balance between these abstractions in which clarity and cut also play a part. In tsavorite, for example, the addition of yellow to green yields a lighter tone and a more vibrant color. The addition of blue yields a richer hue that is also darker in tone.

Clarity

The market is much more tolerant of inclusions in emerald than in any other gemstone. Tsavorite prices, on the other hand, are markedly affected by the presence of any inclusions which are eye visible. Tsavorite is normally sold flawless; gems with visible inclusions are marked down twenty percent and more, depending on the size and placement of the inclusions. A majority of included stones are cut *en cabochon*. Fine (eye-cleanish) cabochon tsavorite sells for approximately twenty-five percent of the price asked for a faceted stone of the same color. Visible inclusions in a faceted tsavorite garnet can lower the

price as much as forty percent.

Cut

Those who persist in comparing tsavorite to emerald will always favor the classic emerald cut. As mentioned earlier, tsavorite has a higher refractive index and double the dispersion of emerald, giving it potentially much greater brilliance. The emerald cut's fifteen to seventeen relatively large facets restrain the scintillation of the garnet, giving it an appearance which bears an uncanny resemblance to the satiny brilliance of emerald. However, the fifty-plus facets of the brilliant style release the garnet's full fiery potential, allowing it to stand toe to toe with emerald when set with diamond.

The Rarity Factor

Traditionally, tsavorite mining has been clustered in an area approximately one hundred miles northwest of Mombasa, within a forty-mile radius of the Kenya-Tanzania border. In 1990 another source of tsavorite was located in Turkana, in northwestern Kenya near the Ugandan border. Mining commenced in this area in 1994. Gems mined at Turkana are generally larger and darker in tone than those found in the south, perhaps due to a relatively higher content of chromium and vanadium oxide[197] . Towards the end of 1991 a deposit of vanadium-rich tsavorite was discovered in the Gogogogo area of southwestern Madagascar, about forty kilometers north of Ampanihy.

More recently tsavorite has been found

---

197 Campbell Bridges, personal communication, 2003.

at several other locations in Tanzania, including Tunduru and Ruangwa near the Mozambique border. At the turn of the century (2000) a new mining locality was found south of the Taita Hills at Choki Ranch. This site has produced a significant number of larger stones. Given the extent of the Mozambique belt, other deposits of this type of garnet undoubtedly exist.

Tsavorite is decidedly rare in sizes above three carats, and generally unavailable in the market in sizes above five carats. According to the gem's discoverer, the late Campbell Bridges, approximately eighty-five percent of material mined yields stones under one carat, ten percent yields stones above one carat, two and a half percent over two, and one percent between three and five carats. Stones over ten carats are about one tenth of one percent of total production.[198]

Tsavorite's extreme rarity means prices for larger sizes increase geometrically in price. In the early 1970s Tiffany & Co. was offered an exclusive marketing arrangement. Henry Platt, Tiffany's president and the man who gave the new green garnet its name, demurred, because quantities of larger stones sufficient to launch a profitable marketing campaign could not be guaranteed. Prices for fine tsavorite garnet will increase dramatically with size. Stones over twenty carats are priced individually and are true museum pieces.

198 Campbell Bridges, personal communication, 1998.

Two exceptional malaya garnets from Mahenge, Tanzania. Left to right: 6.78 and 6.84 carat. Photo: Jeff Scovil, courtesy Evan Caplan..

Chapter 22
Malaya Garnet

"If malaya is the outcast of the garnet family, it is a splendid outcast indeed."

–R. W. Wise, 1989

The course of the Umba River undulates like an uncoiling snake as it slithers through the parched East African savanna. It was at the river, about four miles west of the Tanzanian town of Mwakaijembe, about a hundred meters out from one of those bends, where the first signs of malaya garnet were found. The time was somewhere in the mid-1960s. Malaya, sometimes spelled malaia, is a Swahili word meaning *out of the family* in the sense of *outcast*. When the rough stones were first found, they were sometimes discarded simply because the miners, looking for sapphire, assumed they were worthless.

"You ask about size and rarity!" Dealer Roland Naftule pauses thoughtfully before responding to the question. "Nowadays anything over five carats is rare. All garnet except rhodolite is generally rare above four to five carats. At six or seven you have another jump. Above twenty, very scarce! Over one hundred carats — it's a museum piece."

A slightly pinkish orange 2.77 carat honey colored malaya garnet of sixty-five to seventy percent tone. Photo: Jeff Scovil, courtesy R. W. Wise, Goldsmiths.

"But, in the early days of the strike, there were rough pieces—water-worn pebbles the size of golf balls." That's how Naftule remembers it. "They cut huge pieces, over a hundred carats; one hundred eighty — that was probably the biggest." Campbell Bridges recalls the early days of the strike: "an abundance" of high-quality rough up to ten grams (forty carats). Figuring in a thirty percent loss in cutting, rough of that size yields cut stones

of almost thirty carats.

The country rock in this area, just south of the Kenya border, is biotite, quartz, and feldspar gneiss. The deposit was alluvial. Stones were found to a depth of about four feet in a long-dead river channel. Traces of malaya garnet have been found elsewhere, in Kenya at Lunga Lunga, and at points along the plain which stretches from the Kenya-Tanzania border to Mgama Ridge in the Taita Hills. More recently this garnet has also been reported in Tanzania at Tunduru, and also in Madagascar near the village of Bekily; as of this writing there is some small-scale production there. Most of this production is of stones under one carat and consists of small parcels brought to local dealers by independent miners.

Technically the mixture of garnet species which makes up malaya is highly variable: 0-83% pyrope, 2-78% almandine, 2-94% spessartite, 0-24% grossular with no more than 4% andradite. As a practical matter the refractive index of approximately 1.765 overlaps grossular but, unlike grossular, which has no distinct ultraviolet spectrum, malaya exhibits distinct bands at 504, 520 and 573Ac. Stones in the following hues with a refractive index above 1.76 and below 1.78 can be safely termed malaya.

Hue

The color range of malaya is broader than spessartite and the orange hue rarely as crisp and pure as the finest of that species. Malaya ranges from a yellowish brown and brownish pink through a cinnamon to a crisp honey brown and reddish brown to a brick or brown orange.

The rarest and most beautiful are the honey peach, cinnamon, tangerine, and pinkish orange hues. Malaya is one of the few gem varieties where the most beautiful stones are not those of one pure spectral hue. Pinkish orange — seventy percent orange with thirty percent pink secondary hue — may be the paradigm. In the very finest stones there is a complete absence of the brown; such stones are very rare. Even pinkish orange stones with a distinct brownish component are desirable.

Some examples, specifically stones from the newer source in Madagascar, are classified as color-change garnets. As the stone is moved from natural to incandescent lighting, the hue will not so much change as intensify, becoming brighter (more saturated) and darker in tone. Thus a stone that is pinkish orange in daylight will often appear reddish under the light bulb.

Saturation

Because of the prevalence of a brown component, malaya rarely exhibits a fabulously vivid hue. Soft, rich, and velvety is a better description. However, due to its high refractive index, what the stone lacks in saturation of hue it will often make up for in brilliance. In our system of evaluation, both gray and brown are thought of as saturation modifiers or masks, as if the stone was being viewed at the bottom of a mud puddle. At times, however, if the brown is itself highly saturated, it becomes a hue, as in beautiful examples of chocolate brown diamonds, tourmalines, and garnets.

The hues between orange and brown are a dominant part of the East African color palette. The total visual effect is what is important. Though orange is preferable, brown-hued stones are not to be dismissed out of hand. That said, the range of hues in malaya are so broad and the brown mask so ubiquitous that a gem of any mixture of chromatic hues with little or no brown should be considered fine.

Tone

Tonally, malaya can run the gamut from forty percent to eighty-five percent or higher. Darker brick orange stones are the norm; a sixty percent tone is probably ideal. In this tonal range the key color can be compared to honey, although stones with a purer orange primary hue can be described as tangerine. Malaya garnets with tonal values under sixty percent begin to appear washed out. Eighty-five percent is overly dark brown and definitely overcolor.

Given its tendency to shift color, malaya garnet should be carefully examined in a variety of lighting environments. This garnet usually darkens appreciably in tone when exposed to incandescent light. In fact, since the hue we call brown is actually a dark-toned orange, darker-toned orange stones will sometimes turn an unattractive brick orange brown when the lighting environment is changed from daylight to incandescent.

Clarity

Malaya garnet is normally eye-clean. Gems with visible inclusions should be avoided.

The Rarity Factor

The days of thirty-carat rough stones are long gone. Since the publication of the first edition of this book (2003) the rarity of this variety of garnet has increased dramatically. Larger examples of the gem have almost disappeared. Malaya has become quite rare in sizes above two carats. Expect big price jumps and two, three, five, seven and ten carats. A twenty-carat gem is a collector's piece.

No longer an outcast, malaya garnet is a modern gemstone success story. Since its introduction in the United States in the mid-1980s malaya has not only knocked the accepted garnet classifications into a cocked hat, it has also established itself surprisingly well among consumers.

An extremely fine 21.50 carat spesssartite garnet from Nigeria. Photo: Jeff Scovil, courtesy R. W. Wise, Goldsmiths..

Chapter 23

Spessartite Garnet

"Everyone knows that yellow, orange, and red inspire and represent ideas of joy, of riches, of glory, and of love."

–Paul Signac, 1905

Spessartite, also called spessartine, is the true orange garnet of the purist hue. The garnet family is scientifically termed a group because its members are related more like cousins than siblings. Members of the garnet group all have an identical atomic structure and crystallize in the cubic system, but have different chemical compositions. Three garnets can occur in a primary orange hue: spessartite, malaya, and hessonite, a type of grossular garnet.

Spessartite has the highest refractive index of the three orange garnets and when well cut has greater potential brilliance. It is "crisper" in appearance than either hessonite or malaya. Due to the variation in refractive index, spessartite measures between 1.78<1.81, malaya 1.765 but less than 1.81, and hessonite between 1.74 and 1.75. Separation of the three varieties is fairly straightforward and easily determined by a qualified gemologist. Spessartite at 7.25 on the Mohs scale is also among the hardest members of the garnet group.

A particularly fine 14.01 carat spessartite garnet from Nigeria. Photo: Jeff Scovil, courtesy R. W. Wise, Goldsmiths.

Spessartite garnet was virtually unknown until the late 19th century. Though rare, spessartite has been found in many places, including Sri Lanka, Madagascar, and Brazil; but historically the finest stones have come from a single source, the Little Three Mine in Ramona, California. This mine, still worked sporadically, has produced perhaps 40,000 carats of facetable spessartite since its discovery in 1903 and was the principal source of this gem garnet from its discovery through most of the 20th

century.[199]. So, until the early nineteen nineties, spessartite was almost too rare to make a market.

A new source was discovered in 1991 at the Kunene River flowing through the Marienfluess Valley in Northwest Namibia, within a stone's throw of the Angolan border. A second mine was discovered two years later about 30 km kilometers south[200]. The newer, low-iron Namibian material was originally marketed under the name "hollandine," but the name that seems to have stuck is "mandarin." Thus, orange garnet labeled mandarin is likely to be Namibian spessartite.

Another new find of spessartite from western Nigeria literally gushed through the pipeline in 1999. This new material was found near the village of Iseyin, about three hours northwest of Ibadan. The best of this new Nigerian spessartite strongly resembled the finest from the Little Three, a saturation that came within a whisker's breadth of the highly saturated gems of Namibia. Suddenly, tangerine orange spessartite became a new force in the gem market.

In 2007 another strike was made in Tanzania at Loliondo, about seven miles from the Kenyan border near Serengeti National Park. Though some are quite exceptional, gems from this location tend to have a strong yellowish secondary hue; the best of the pure orange, however, closely resembles the Nigerian material.

Origin is mentioned as part of the discussion and as a general guide. To paraphrase gemologist G.F. Kunz: gems have no pedigree. Exceptional stones may come from any source, and poor examples can be found at the most celebrated locales. Another early source of very fine spessartite was Amelia Courthouse, Virginia. Unfortunately this source now sits beneath a housing project.

Hue

Spessartite can occur in a range of hues from distinctly yellowish orange through orange to a distinctly brownish orange, orange brown, or reddish brown to brownish red. The primary hue may be orange, brown, or red with any of the others playing the part of a secondary hue. Red spessartite is probably the rarest but the most distinctive stone; those that command the highest prices, however, are a visually pure (ninety percent) primary orange hue with no more than ten percent of brown or yellow as a secondary hue. Dark-toned gems with a fifteen to twenty percent brown secondary hue are called burnt orange and place second in desirability and price. Yellowish orange stones, seventy percent orange with no more than thirty percent yellow secondary hue, can also be strikingly beautiful and place third in both beauty and price. Reddish orange spessartite, which resemble dark-toned rhodolite garnet, comes in last in overall desirability.

Saturation

The finest pure orange spessartite will be exceptionally crisp and brilliant. The

---

199 Brendan M. Laurs & Kimberly Knox, "Spessartine Garnet from Ramona, San Diego County, California," *Gems & Gemology,* The Gemological Institute of America, Winter 2001, p.278, 293.

200 Christopher L. Johnson, personal communication 2015.

best Namibian is without question the most highly saturated. In fact, it will sometimes exhibit visible luminescence, further supercharging the visual effect.

Stones with a yellowish secondary hue will be even more vivid. This is because, as color science teaches, the orange hue achieves its maximum vividness, its gamut limit, at between thirty-forty percent tone, and yellow slightly lighter at about eighteen percent tone. Thus the yellow secondary hue punches up the saturation but also washes out the orange. Brown is the normal saturation modifier or mask found in spessartite garnet. Deep and dark hues of orange naturally appear brownish. So a dark-toned orange garnet is a brownish orange garnet. As the stone becomes tonally darker the hue also becomes less vivid. Stones with the least brown mask will exhibit the most vivid key color. Pure spectral hues are always vivid, so a dull orange hue is also a brownish orange hue.

Tone

Given that deep and dark-hued orange spessartite garnets are by definition brownish, the optimum hue in spessartite tends to be quite light. On a scale where window glass is zero percent tone and coal is one hundred percent tone, the optimum tone for orange is between twenty and forty percent. In darker toned gems, the orange turns to brown.

Clarity

Spessartite garnet is normally sold eye clean. The only exceptions to this rule of thumb are larger stones and the so-called mandarin garnets from Namibia. Spessartite from this source has a slightly different chemical composition, resulting in a particularly vivid saturation. Though some eye clean stones have been found, Namibian spessartite is normally moderately included, containing wispy veils, white puffball and black spicule inclusions visible to the naked eye. Because of their highly saturated hue, even stones with visible inclusions are highly desirable and visibly included Namibian spessartite will sell for higher prices than eye or even loupe flawless gems from other sources.

Crystal

Lighter-toned spessartite garnets, those approaching the optimum tone, will tend to have greater transparency. As noted above, gems from certain localities will tend to have a large number of eye-visible inclusions. These inclusions can sometimes be quite small and give the visual impression of dustiness in the interior of the stone.

Night stone

Like all garnet, spessartite is a night stone. It puts its best foot forward in incandescent lighting. So it is best to examine the gem carefully in daylight before making a purchase.

Cut

Among gemstones, spessartite has the fourth highest refractive index, exceeded only by diamond, zircon, and andradite garnet. When properly cut, spessartite will exhibit exceptional brilliance.

The Rarity Factor

The largest known faceted spessartite from the Little Three Mine weighs 39.65 carats.[201] The largest mandarin garnet, spessartite from Namibia, weighed just over forty carats. One dealer claimed to have seen a Nigerian stone which weighed over seventy carats. Eye flawless Namibian stones of any size are rare and little has been found since the mid 90s. Due to their extraordinary saturation, iron poor Namibian stones will command prices as much as four times that of gems from other sources. Generally speaking, price and rarity increase at one, three, and five carats. Stones above ten carats are considered extremely rare and command prices to match.

---

201 E. Gray, personal communication, 1999.

THE WORLD OF NATURAL PEARLS

Finally, all the Orientals are very much of our taste in matters of whiteness, and I have always remarked that they prefer the whitest pearls, the whitest bread and the whitest women.

–Jean Baptiste Tavernier, 1675

I found Yusuf the diver standing alone, shivering on the foredeck. He wore a threadbare sarong that covered him from his waist to his knees but provided little warmth. His thin arms hugged his undernourished body. He stood looking down at his tender, Nejdi, who squatted by the rail, methodically opening Yusuf's catch with his knife. The diver's shells were always left overnight; they opened easier the next morning. It was a thankless task–perhaps one shell in a thousand held a pearl. Yusuf's teeth chattered in the hot morning sun.

Portuguese slavers kidnapped Yusuf when he was a child and forced him to dive for pearls. He escaped but was still diving. Nearing forty years, he was quite old for a diver, though he looked much older. He spoke good Portuguese, and we became friends.

The first three dives went well. Yusuf came up with his basket brimming with shell. He wrapped his leg around the weighted drop line, adjusted his nose clip, waved to me, and nodded. The tender released the line, and the diver splashed into the water. The rock attached to the end of the line towed him rapidly to the bottom.

A current was running, and Yusuf's lifeline streamed quite far astern. On deck the day was perfect. The sun shone and small puffy clouds scudded across the sky. The sea was soft azure blue. All I could see was his distorted shape in the water. Then a dark shadow moved swiftly over the shoal. Then another, and there was a frantic tugging on the line. This was the danger signal. The tender standing at the stern oar began to point and gesture. "Shark"! he yelled, and began hauling in Yusuf's

Hardhat pearl diving off the Australian coast, circa 1895. In the 1890s some boats working the Australian Fishery still used hand diving crews. Arabs pearlers working in the Persian Gulf used free divers up until 1940. Photo Courtesy The Norman Archive.

lifeline, hand over hand. Crew members on both sides of the boat sprang into action, pulling in their divers.

By the time Yusuf was hauled to the surface, there was little left. The sharks had torn off both his legs and had bitten through half of his upper torso. His face was twisted in a grimace, his dark eyes open and sightless.

That evening we buried Yusuf the diver. He was wrapped in his sarong and gently laid in the center of the tender. He made a pitifully small bundle. We silently rowed the jalboot to the beach and carried him a few yards above the tide line into the sand. The wind had died. All that could be heard was the cry of the sea gulls. Yusuf was buried in a shallow grave just before the evening prayer.

Richard W. Wise, *The French Blue*

The Making of a Pearl

There are two basic types of pearls, natural and cultured. In either case the method of production is essentially the same. An irritant in the form of a tiny marine animal or a spherical shell implant introduced by man is placed or worms its way[202] into the soft mantle tissues of the host mollusk. The oyster in the case of saltwater pearls or clam in freshwater pearls secretes a substance called nacre, a calcium carbonate composed of alternating layers of flattened aragonite crystals and conchiolin, a sort of calcium glue that acts as a binder .

Although a perfectly round pearl has always been the most desired, many of the historically most important natural pearls were not. *La Peregrina* is pear shaped, as was another famous Venezuelan pearl, the *Phillip II*. During the Renaissance, baroque pearls which suggested the body of an animal or person were used extensively in jewelry. The aficionado is advised to view a pearl from all angles much as you would a piece of sculpture. Pearls come in many pleasing shapes. Like a mi a free standing sculpture, shape and proportion play a role. Unlike sculpture, pearls are decorative and are normally set in jewelry and required to enhance the wearer.

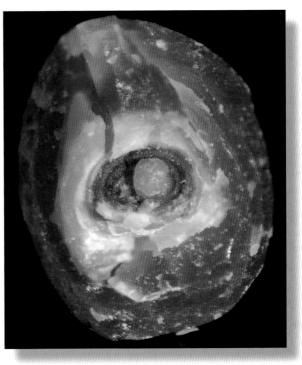

A shard broken off a natural Pteria sterna golden pearl. Image and technical notes, Ana Vasiliu, sample from the research collection of natural pearls at IACT, donated by Perlas del Mar de Cortez.

Natural pearls are first mentioned (2450 BC) by the Sumerians, as coming by ship from the mythical land of *Dilmun*, a country usually identified by scholars as the Persian Gulf island of Bahrain.[203]. Though known for millennia as the *Queen of Gems*, and once ranked above diamond as the most sought after of all gems, natural pearls are rarely seen in even the finest jewelry stores today. The introduction of cultured pearls in the early 1920s eviscerated the natural pearl market and the stock market crash of 1929 completed the gem's precipitous fall. According to author and pearl expert Elisabeth Strack, the price of natural pearls dropped eighty-five percent in one day in 1930 when banks refused to accept

202 According to pearl biologist Douglas McLaurin-Moreno, the generally accepted notion that natural pearls begin with a grain of sand is a myth and that Borer worms, Genus: Polidora, are the usual culprit. http://www.pearl-guide.com/forum/showthread.php?t=4685.

203 Samuel Noah Kramer, *The Sumerians: Their History, Culture, and Character*, (The University of Chicago Press, 1971}, pp. 58, 281-283.

pearls as collateral for loans to pearl dealers.[204]. The gem has never fully recovered.[205].

The market for natural pearls continued at a much reduced pace into the 1950s, after which these ancient symbols of luxury were all but forgotten, and although price lists such as *The Guide* continued to list natural pearl prices, there were few buyers and virtually no market outside the Middle East and India through the end of 20th Century. Few were interested in natural pearls and dealers who understood them had either gone bankrupt, retired or died. This situation is historically unique in the gem market.

The Comeback Kid

Beginning on the cusp of the new Century, the natural pearl commenced a spectacular comeback, fueled, as is often the case, by a highly publicized auction sale. In November 1999, Christie's Geneva sold a single-strand natural pearl necklace, formerly the property of Marie-Antoinette, and later of Barbara Hutton. Comprising forty-one graduated round pearls, measuring approximately 8.50 to 16.35 mm the necklace sold for a world record price of 1.45 million dollars. Five years later, in November 2004, another fabulous natural pearl

Figure 92: The Baroda Pearls. Photo: Courtesy Christies.

necklace, also at one time the property of the same ill-fated French queen, established a new world record selling at Christie's for 3.5 million dollars. In April of 2007 that record too was shattered when the celebrated Baroda Pearls, a double strand of sixty-eight matched "round" natural 9.47 to 16.04 mm pearls sold at Christie's for the extraordinary sum of 7.1 million dollars. Completing the dazzling comeback, in May, 2014, the largest single round natural pearl ever offered at auction, a 17.4mm round weighing 33.15 carats established a world record price for a single pearl when it sold for just under 1.35 million dollars at a British auction house, a hefty five times its presale estimate. The natural pearl had returned with a vengeance.

204 Elisabeth Strack, personal communication 2007. See also Richard W. Wise., GemWise, *The Baroda Pearls, Another Auction Record at Christies*. http://www.thefrenchblue.com/rww_blog/2007/06/09/the-baroda-pearls-another-auction-record-at-christies/

205 Elisabeth Strack, *Pearls*, (Ruhle-Diebner-Verlag, Stuttgart, 2006), p.37.

At the introduction of synthetic ruby in the 1890s, the ruby market teetered, but did not fall and prices soon recovered. In 1941 Linde Air Products began flooding the world market with synthetic star sapphire, in 1949 and the Chatham Company has sold fifty thousand carats of synthetic emerald. None of these events have had any appreciable long-term impact on either market demand or the value of natural gems. Pearls had held their value for millennia. Why then did the advent of the cultured pearl have such a catastrophic impact on the Queen of Gems? Timing? The world was in the midst of the Great Depression and perhaps it may have something to do with the fact that a cultured pearl is not a synthetic; it is a "natural" product albeit one given a hefty leg up by the hand of man.

There are numerous types of natural pearls found in fresh and saltwater in both univalve and bivalve mollusks. Most bivalve mollusks (oysters, clams) produce nacreous pearls. Univalves, (conchs, snails) produce mainly porcelainimous pearls which have a porcelin like luster. Luckily beauty is still the primary criterion and the four Cs of Connoisseurship used throughout this volume can be applied, albeit gingerly, to the connoisseurship of all types of natural pearls. We will focus our attention on pearls produced by saltwater mollusks that are of commercial importance today. As it happens, all of these are from mollusks of the *Genus Pinctada* and *Genus Pteria*, which includes natural pearls from the Persian Gulf, and both natural and cultured pearls from Japan, the South Pacific and Southeast Asia.

Pearls have many sources. Writing of natural pearls, G. F. Kunz famously observed: *Pearls have no provenance.* Several historically famous pearls were from neither Sri Lanka nor from the Persian Gulf, the two most famous sources,[206]. La Peregrina known as "The Incomparable" hailed from the Americas; it was found in Venezuala in the late 16th century. Originally owned by Mary Tudor, this pearl passed through a series of royal hands and bosoms, until it finally came to rest on the most famous décolletage of the last century; it was purchased by Richard Burton and given as a gift to his wife Elizabeth Taylor. Another famous pearl, the Tiffany Queen Pearl, is a freshwater gem (found in 1857) in a stream near Patterson (Notchbrook), New Jersey. This round, highly translucent beauty was once owned by the Empress Eugenie, wife of Napoleon III.

---

206 Historically, sixty percent of the world's natural pearls came from the Persian Gulf and Sri Lanka. See Kunz & Stevenson, *ibid.* p.80.

Hard hat pearl diver surfacing, year 1898.   Photo courtesy: Norman Archives..

Chapter 24
Natural Nacreous Pearls

"This prince possesses the most beautiful pearl in the world, not by reason of its size, for it only weighs 12 1/16 carats, nor on account of its perfect roundness; but because it is so clear and transparent that you can almost see the light through it."

–Jean Baptiste Tavernier, 1675

A particularly fine, rare, natural pearl 8x6.7mm showing strong pinkish overtone and slight translucency. Note the reflection of the light towards the top of the picture. The luster would be best described as misty (Chapter 5). Photo Jeff Scovil. Courtesy of the Edward Arthur Metzger Gem Collection, Bassett, W. A. and Skalwald, E. A.

The factors in the evaluation of natural pearls differ little from those elucidated in the section on the Connoisseurship of Cultured Pearls in Chapter 5.

Symmetry

Acclimated as we are to the ubiquitous cultured Japanese akoya, we expect our pearls to be perfectly round and to match like peas in a pod. This is a main point of distinction between the cultured and the natural product. Very few natural pearls approach perfect roundness. The record-breaking Baroda Pearls are not truly round, but rather *roundish*. In cultured pearls, a spherical bead normally comprises fully ninety percent of the mass of a given cultured pearl. The average thickness of the nacre is usually less than 2mm, so, it is not surprising that a large percentage of cultured pearls available in the market are round. Natural pearls, by contrast, are almost one hundred percent nacre and what is termed round in a natural pearl is barely equivalent to off-round in cultured pearls. Thus, a natural eight-way roller[207] is a rare bird indeed and all things being equal, will command an extraordinarily

207 This refers to one of the traditional dealers' secret tests for roundness. Simply place the pearl on a flat surface and roll it completely around with one finger adjusting the axis by half then by quarters each time the pearl is rolled. Any asymmetry faults will cause an obvious bump in the road. If the pearls are strung, stretch a section of the strand between the index finger and thumb of each hand and twist. Off round pearls will bounce along the strand like the cams on a camshaft of an internal combustion engine.

"The Hope Pearl, one of the largest natural pearls known measures two inches in length by four inches at its broadest and weighs 450 carats. Photo courtesy British Museum

high premium.[208]. The aficionado is advised to adjust expectations accordingly. Other symmetrical shapes, pear, oval and button, will also command a stiff premium though not so much as a perfect round.

Some of the most unusual natural pearls occur in non-symmetrical baroque shapes. The aficionado is advised to consider a pearl much as you would judge a

sculpture. View the gem from all directions in both full and shadowed lighting. Observe the overall shape. Consider the gem's proportions. A pleasing shape is a beautiful natural creation. Baroque natural pearls are highly valued and, historically, have been put to many innovative uses in the creation of jewelry.

Sources

Natural pearls are still fished from the old pearling grounds of Bahrain, the South Pacific, the Gulf of Mexico, the Philippines and other sources though the days of the great pearling fleets off Bahrain, the Gulf of Mannar and the Torres Strait are long past and much of the high-end pearl market centers around old stock, pearls sourced hundreds of years ago.

Matching

The building of a natural pearl strand takes many years. The two strands that comprise the celebrated Baroda pearls (pictured in the Introduction) were distilled from seven strands containing three hundred thirty pearls, owned by the Maharaja of Baroda. According to one eminent connoisseur of pearls, Sir Dorab Tata (1859-1932), prior to the introduction of cultured pearls, it took twelve years to create a sixteen inch single choker strand of exactly <u>similar</u> natural pearls. At that rate a seven-strand pearl necklace would have required eighty-four years to assemble from the pearl markets of the world.[209]

---

208 Surprisingly, true rounds average less than 5% of a farm crop of cultured pearls which may number in the tens of thousands; but with natural pearls it is truly one in a million.

209 Bhat, Dhananjaya, *The Baroda Pearl Necklace-Costliest in History*, The Kashmir Times, July 22, 2012.

Body Color

Natural pearls occur in a variety of body colors. White is generally the most popular hue simply because it is the most flattering to the skin of Caucasian women, and Europe and the United States are the two largest markets. The most desired of all is the white rosé, but rosé refers to the overtone not the body color and will be discussed further. Natural white pearls are often greenish when first removed from the shell. The green may express itself as either body color or overtone. Pearls are normally sun bleached to drive off the green.

Natural pearls are frequently mottled with spots of other colors, uneven distribution of hue and patches of inconsistent luster. The effect on quality and price depends upon the prominence of these defects and their location. If a slight blotch can be hidden in a setting, it has less effect than one which remains prominent.

Simpatico

The pearl has another quality, unique among gems: its beauty can be enlivened or subdued by placing it in contact with a woman's skin. The Spanish word *simpatico*

Natural round black pearl from the Pteria sterna oyster from Mexico. Photo: Douglas McLaren, courtesy Cortez Pearls.

is a term I have chosen to describe this somewhat mysterious quality. A pearl, no matter how beautiful in itself, will seem to come alive if it is placed next to skin of a certain color, texture, and tone with which it is *simpatico*. For example, white pearls, as noted, are normally sympatico with the skin of Caucasian women though less so with women of Asia. Brazilian women, with dusky skin tones, are particularly fond of gray to black pearls.

The simpatico between gem and wearer can be determined with a simple test. The pearl or pearl strand should be placed against the inside wrist of the intended recipient. The inside wrist is a protected area of the skin which rarely tans and is the same color and texture as the delicate areas around the throat and ears, the likely spots where the pearl will be worn. The simpatico between the skin and the pearl becomes readily apparent when several pearls of different hues are compared in this way. Strictly speaking, simpatico is a test of compatibility, not quality although all things being equal, a pearl with the qualities of high luster and prominent overtone, in short a pearl of fine water will enhance the wearer to a far greater degree than one of inferior quality.

Crystal: A Pearl of the Finest Water

Transparency or translucency in the case of pearl is another quality where natural and cultured pearls differ. Pearls are composed of thin layers of translucent

Johannes Vermeer, Girl with a Pearl Earring. Courtesy: Mauritshuis, The Hague.

to transparent prismatic aragonite crystals which resemble the layers of an onion. These layers are mortared together by conchiolin, an opaque form of calcium carbonate. The greater the number and the thinner the aragonite layers and the less conchiolin binder present, the more translucent the pearl. Ideally the upper layers of the pearl should be translucent to transparent. This together with the luster and body color gives depth to the pearl

and defines those gems that have been historically described as being of the finest water.[210] Cultured pearls, due to the large opaque shell bead that forms the nucleus, are never translucent.

White Venezuelan pearls, those produced by Pinctada imbracata and first brought to Europe by Columbus, are reputed to be particularly transparent, almost glassy.[211] Philippine pearls can have thick sections of completely transparent nacre which magnify the inner layers. Pipi pearls from French Polynesia are often semi-translucent.

Luster (Lustre)

Luster is the measure of the reflection of light off a surface. Luster in a pearl is analogous to brilliance in a transparent gemstone.[212] In a fine pearl, the reflection off the pearl's surface is crisp, sharp, and well defined. As quality decreases, the reflection becomes dull and fuzzy. The proper way to evaluate a pearl's luster is to

---

210 Strack, Elisabeth, *ibid.* p. 288.

211 Landman, Neil H., et al., *Pearls, A Natural History*, Harry N. Abrams, NY, 2001, p.54. Two islands off the coast of Venezuela, *Cubaga* and *Margarita*. Discovered by Columbus on his 3rd voyage; these were the source of many of the very large, usually yellowish, pearls purchased by Tavernier and sold to Indian potentates at huge profits. Yellowish pearls are sympatico with the skin tones of East Asian women. Cf: *The French Blue*, Brunswick House Press, 2010, Chap. 49.

212 In her seminal book *Pearls*, Elisabeth Strack states that *luster* has a different meaning when applied to natural pearls. "When natural pearls are graded," she states, "the factor 'lustre' is considered equal to the type of iridescence that is termed orient." The trade is famous for tripping over its own terminology. I know a major dealer who uses the term lustre when he means brilliance. Many natural pearls lack overtone, but it doesn't follow that they are without lustre. *Op cit*, p.375.

Portrait of Hortense Mancini by Jacob Ferdnand Voet, 1675. Pictured wearing her sister Marie's pearl earrings. She had three sisters all believed in sharing.

Matched pear shapes weighing approximately 40 carats each. The Mancini Pearls given by Louis XIV to his mistress Marie Mancini in 1657. The pearl at left is considered to be a perfect drop shape. The pearls sold for $253,000 in 1979. Photo: Christie's images.

position a pearl under an incandescent light bulb. If the pearl is particularly lustrous you will be able to see the shape of the light bulb clearly mirrored in the reflection off the pearl's surface. The more clearly defined the image, the higher the luster of the pearl. If the bulb is clearly visible and well defined, the pearl is of the highest possible luster. Luster can be influenced by a number of factors including the water's nutrient content and temperature. Pearls from warmer sources often exhibit a silky luster with a brushed texture as if the color had been thickly applied with the single stroke of a fine brush. This type of luster is distinctive and quite beautiful.

Surface

Surface is the quality analogous to clarity in gemstones. Skin may be a better

term. The skin of a natural pearl is rarely blemish free, and it is often compared to that of a beautiful woman. Bumps, bruises and dimples, in short, anything that mars the perfect surface is considered a fault or flaw; the more prominent the flaw the more negative. In baroque and grotesque pearls, hills and valleys are less important.[213] To some degree the position of the offending imperfection is important. Those that can be hidden in settings are much less bothersome and will have less impact on the price. Cracks in the nacre render the pearl almost valueless.

Beautiful skin tends to glow. In creating the famous *Girl with A Pearl Earring*, the 17th century painter Johannes Vermeer applied a series of transparent colored

---

213 Pearls with exceptionally tortured symmetry are often referred to as grotesques.

glazes to soften the hard surface of the painted canvas and create the illusion of living flesh, optical depth and luminosity. One critic commented that the pearl itself looked as if it had been painted with an application of a dust of crushed pearls. Similarly, nature adds translucent layers of aragonite crystals to add a soft ethereal component to the skin of a natural pearl.

Orient and Overtone

A fine pearl will exhibit the subtle misty iridescence that connoisseurs call overtone, a glow seeming to emanate from inside the pearl, clinging to its skin like sunlight through an early morning fog, a quality that led the English Renaissance poet Thomas Campion to enthuse: "looked like rosebuds fill'd with snow."[214] The cause of this phenomenon is discussed in Chapter 5. In white pearls, a distinct pinkish overtone called rosé is the most desired and will fetch the highest price though overtones of any number of colors are possible, and overtone contributes measurably to the overall "life" of the pearl. Confused? Submerging the pearl in water is a useful way to separate overtone from body color.

Orient derives from the Latin word *oriens*, meaning "the rising of the sun." Luster is caused by light reflection; overtone and orient are the result of light diffraction. As noted, nacre or pearl essence is composed of translucent layers of aragonite, a calcium carbonate made up of tiny polygonal crystals. In some types of mollusk the structure of nacre includes

a series of tiny grooves, as many as three hundred to the millimeter, which act as a reflection (diffraction) gradient, breaking light up into its constituent colors and creating iridescence or the rainbow effect. The more grooves the more distinct the orient.[215] The number of grooves and the spacing determine the colors we see. Some species and varieties of mollusks have this groove structure; some do not.

The terms orient and overtone are often used interchangeably but describe slightly different effects. Orient is multi-colored and effervescent. It occurs in a number of spectral colors at the same time and is similar in that respect to dispersion in colorless diamonds. It seems to scintillate off the pearl's surface. Orient occurs most often in baroque pearls where the interference is caused by light rays careening off the uneven surface and literally bumping into one another.

Overtone is monochromatic and occurs in large patches that appearing to cling or hover above the pearl's surface. The image of the magnificent round pearl at the head of this chapter exhibits classic overtone.

These two phenomena rarely occur together. It is the combination of orient and/or overtone together with luster and translucency, which create that subtle sense of life and mystery that characterize the finest pearls.

215 Yan Liu, J.E. Shigley, and K.N. Hurwit, "What Causes Nacre Iridescence?" *Pearl World*, vol. 7, no. 1, (December 1999), pp. 1, 4.

214 Richard W. Wise, "A Meditation on Pearls", *Pearl World*, vol. 9, no. 3, July-September 2001, pp 10-

Natural Pearl Pricing

Historically, pearls have been valued based on a number of schemes. Perhaps the most influential is a formula first discussed by goldsmith, Juan de Arfe de Villafañe in his *Quilatador de la Plata, Oro y Pedras* published in 1572. Jean Baptiste Tavernier's bestseller, *Les Six Voyages*, published in 1675 popularized this formula, and it became known as *Tavernier's Law*. Whether de Villafañe developed this system or was simply the first to commit it to paper is a matter of conjecture. De Villafañe began with a carat (four grains). First it was necessary to establish a base price, which was the price of a one-carat pearl. Once that was established, the formula used was to square the weight of a given pearl and then to multiply the product times the base rate. De Villafañe increased the base rate by a third for pearls weighing above eleven carats (44 grains) or 11.5mm[216]. Some dealers still use this method. The trick, of course, is to get two dealers to agree on the base rate.

Compared to cultured pearls, natural pearls are in another price league entirely. According to a 2006 pricelist[217] issued by *The Guide*, an industry pricing publication, a fine 5mm round natural pearl will cost two thousand percent more than an equivalent cultured akoya pearl. Using this same pricelist, a 9mm natural pearl will cost nine hundred ninety percent more than its cultured counterpart. Why the drop in percentage at 9mm? At that size, the increased rarity of the cultured product dictates that the cultured pearl prices begin to increase geometrically.[218]

"Worked" Pearls

From earliest times man has attempted to improve upon nature. The physical structure of a pearl is analogous to the layers of an onion. It is sometimes possible to improve the luster and/or the symmetry of a pearl by a process known as peeling. If the outer layers of nacre are opaque and dull, specialists, known as *pearl doctors*, will sometimes physically peel or grind through the upper layers hoping to find finer, more translucent layers within. Also, older pearls with a damaged outer skin can sometimes be improved by peeling the damaged layers.

The thin layers of nacre in cultured pearls preclude peeling, though surface bumps are sometimes ground down and most pearls are polished. Natural pearls subjected to peeling are known in the trade as *worked* pearls. Peeling normally leaves physical traces detectable by experts under magnification. All quality factors being equal, a worked pearl will be subject to discounts of twenty-fifty percent and sometimes more.[219] The discount depends upon the extent and visibility of the working. Pearl dealer Monte Stern suggests that up to ninety percent of natural pearls are worked to some degree.

---

216 A 4-grain pearl (1 carat) is approximately 5.23mm in diameter. A 1-grain pearl is 3.32mm. See Kunz & Stevenson, *The Book of the Pearl*, The Century Co., N.Y. 1908. pp. 329-331. Kunz speculates that de Boodt was chronicling an ancient system imported from Asia.

217 *The Guide* no longer publishes price information for natural pearls.

218 Geoffrey A. Dominy and Tino Hammid, photos. *The Handbook of Gemmology*, (Amazonas Gem Publications, 2014), p. 756.

219 Thomas Ortega "Monte" Stern, Ashok Sancheti and Sunil Jain, personal communications, 2015.

An 8.96 carat fine conch pearl. Photo: Jeff Scovil, The Edward Arthur Metzger Gem Collection, Bassett, W. A. and Skalwold, E.A.

Chapter 25
Conch Pearls

"Although conch pearls occur in a range of colors, pink are usually the most desirable. "Pearls over ten carats are rare, but pearls have been observed as large as twenty-five carats."

–Emmanuel Fritch, 1987

Unlike pearls produced from bivalves, pearls produced by univalves are non-nacreous. The Gemological Institute of America describes these "pearls" as calcareous concretions and distinguishes them from true pearls. Nonetheless, conch pearls have become increasingly popular in jewelry, and some knowledge of these lovely concretions may be of interest to the eclectic collector.

Visitors to the Caribbean and Central America have become familiar with pearls produced by the queen conch (Strombus gigas). Another pearl of this type is the melo pearl from Vietnam. Conch pearls are beautiful and have become increasingly popular in jewelry. These pearls have a similar composition but a different structure, and they are often described as having porcelain like luster.

Like their natural nacreous cousins, conch pearls have begun to be taken seriously. In November 2012, a "baby pink" conch pearl bracelet (pictured) designed by Cartier in the 1920s for Queen Victoria Eugene of Spain fetched a hefty $3.5 million at auction.

This lustrous example of a conch pearl is a medium dark pink hue with a distinct flame structure and desirable symmetrical oval or egg shape with the characteristic flame pattern. Photo: Blaire Beavers, courtesy Monili Fine Jewelers.

Shape/Matching

Round conch pearls hardly exist. The preferred shape is a symmetrical oval, preferably with tapered ends. Egg and pear shapes rank second and third in desirability and value. As with all natural pearls, the standard differs greatly from their cultured brethren. Matching is a matter of just getting close.

Hue

Like the conch shell, pearls can vary from pale rose to light through dark pink to dark orangy red. Conch pearls are frequently mottled in appearance, and mottling, if present, is a definite negative. An orangy secondary hue is often present and is the usual secondary hue found in conch pearls. Purer hues are generally more sought after by connoisseurs. However, a highly saturated pink with a violetish secondary hue is quite rare and the most sought after in the marketplace.

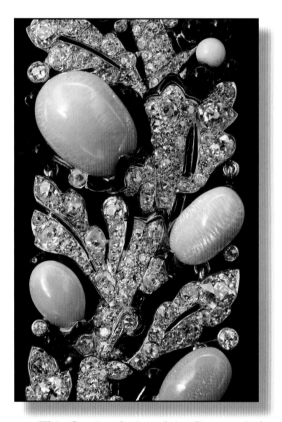

This Cartier designed Art Deco period platinum brooch with "baby pink" conch pearls sold for $3.5 million in 2012. Note the overall lack of symmetry common in natural pearls, but unacceptable in cultured.

Surface

The skin of a conch pearl is best described as porcelain like or silky and sometimes velvety. *Wet silk*, essentially a hard sub-vitreous luster, is preferred. As with other pearls, any blemish or mottling is a negative.

Phenomenon

The upper layers of the conch pearl are composed of intertwined aragonite crystals which often produce a lovely flame-like pattern distinctly visible to the unaided eye and it is an important grading criterion. The effect is sometimes described as similar to a cat's-eye, but that is really not a helpful comparison. When turned in the light, the flames appear to dance in a fiery diorama as the pearl is rotated. The more distinct the flame structure, the better. Pearls from other mollusks, including bivalves such as the giant clam, may also exhibit flame patterns.

Fading

The vivid pink to red hues of conch pearls are subject to fading if the pearl is left exposed to the sun for extended periods.

THE CULTURED PEARL

The romance of the South Seas! The exotic mystery of the black pearl! It is this combination that has lured me to travel over ten thousand miles to Manihi Island, a remote speck of land in the Tuamotu island group, three hundred fifty miles northeast of Tahiti. It is only in these far-off islands that the natural conditions exist for the culture of this unique gem.[220]

The Manihi atoll is a doughnut-shaped flat ribbon of coral surrounding a central lagoon more than twelve miles across. Several pearl farms are located in the lagoon. This sheltered body of water provides an ideal habitat for the pearl oysters as well as sufficient protection for the pearl farmers against the windswept dangers of the open Pacific.

Our expedition gets underway shortly after dawn and consists of a French interpreter, our Tahitian boatman and guide Toputu, and me. Our destination: the pearl farm Kata-Kata.

With an intuitive skill passed down from a thousand generations of seafaring ancestors, Toputu pilots our small wooden runabout through a maze of coral heads lurk just beneath the surface, waiting to tear the bottom out of our fragile craft.

Crossing the lagoon, we pass several pearl farms. The layout of each is similar. Several small wooden huts face the lagoon. A neatly constructed wooden wharf juts from the beach approximately two hundred yards out into the lagoon. Perched at the end of the wharf is a small wooden building that serves as the farm's workhouse. Pairs of wooden piers spaced about twenty feet apart extend from the workhouse; connecting the piers is a horizontal gridwork of steel pipes. Suspended by ropes from the pipes are slender cages of wire mesh that both house and protect the pearl oysters.

The trip takes a bit over two hours. Upon our arrival at the Kata-Kata farm we are heartily greeted by the farm's manager, Momo Paia, and his wife, Tipati. Both are native islanders sporting the natural bronze-tan complexions that Western women would kill for. Both are dressed casually, Western style, but with the relaxed friendly manner that is one hundred percent Polynesian.

---

220. The waters of Tahiti itself are too cold for this oyster to grow. Black pearls can also be found farther north of Tahiti in the Gambier and Cook islands. In earlier times, natural black pearls occurred in both Panama and Baja California. These pearls, however, were produced by a different species of oyster, *Pinctada mazatlanica.*

Aerial view of Manihi Atoll, three hundred fifty miles north of Tahiti, French Polynesia. A small section of the lagoon can be seen in the foreground. Protected lagoons like this serve as a perfect incubator for the black pearl oyster (Pinctada margaritifera). Photo: R.W. Wise.

Responding to my curiosity about the meaning of the farm's name, Momo laughs. From Tahitian, he explains, "Kata-Kata" translates as "lots of laughs."

Momo leads the way out along the wharf to the headquarters shack. Inside the one-room unpainted structure, blackboards chart the progress of the cultivation. In one corner is a small kitchen presided over by Tipati. Diagonally across the room is a glistening white enameled lab table where the implanting operation is performed.

Pearl making is an irritating proposition for the pearl oyster. In nature, the pearl nucleus— a foreign substance, often a worm—works its way into the soft mantle tissues of the oyster, causing irritation. To combat the discomfort, the oyster secretes nacre. More commonly called mother of pearl, nacre is a two-part calcium carbonate. The nacre covers the nucleus, enlarging it and making the oyster even more uncomfortable. The mollusk responds by secreting more nacre until, layer upon layer, the pearl takes form.

Mr. Sadao Ishi Bashi works at the Kata-Kata pearl farm. He is a slender Japanese man of medium height with a bright smile and an excellent command of English. Mr. Ishi Bashi's job is to implant the nucleus into each pearl oyster. Sadao, as he likes to be called, has been performing this operation for over two decades. A marine biologist by training, he learned his trade at a Japanese-owned company in Australia. Sadao gave me a step-by-step explanation of pearl farming techniques as I watched him perform the delicate operation.

The Kata-Kata Pearl Farm, Manihi Island, French Polynesia, is protected by the encircling lagoon. Oysters are implanted inside the building; pearl oyster cages can be seen hanging from the metal grid just visible in the foreground. Photo: R. W. Wise.

Between October and February, when the waters of the South Pacific are at their warmest, the black lip oyster (*Pinctada margaritifera*) begins its reproductive cycle. The young oysters begin as "spat," microscopic offspring given off by the parent mollusk in the tens of millions. After several days of free floating in the lagoon, the spat not consumed by predators begin to look for something to attach themselves to. To gather the spat, the pearl farmer simply takes a bundle of miki-miki branches, a scrub bush native to the Tuamotu islands, wraps the bundle in chicken wire, attaches it to a buoy, and drops it into the lagoon in the area where the oysters are breeding. The woody branches of the miki-miki provide an ideal resting place for the spat. The bundles are then gathered and placed in a protected area of the lagoon. After two years, the spat have matured into full-grown oysters, six to eight inches in diameter.

The mature oyster is removed from the water. A worker drills a small hole in the heel of the shell. Using a special tool, another worker gently pries the shell open, sliding a small bamboo wedge between the valves to keep the shell from closing. The oyster is then ready for Mr. Ishi Bashi and the grafting operation.

I watch closely as Sadao slides the oyster into a stainless steel clamp mounted about ten inches above the work surface. The clamp holds the animal in place while the implant is performed. Sadao carefully selects a seven-millimeter mother-of-pearl sphere with a long-handled stainless steel spoon, similar to a dentist's instrument, and slides the spoon between the partially opened valves of the oyster. He then places the sphere, along with a tiny square of mantle tissue taken from another live oyster, in the area surrounding the gonad sac, the oyster's reproductive organ. The addition of this mantle tissue helps stimulate the formation of nacre. He then removes the wedge, allowing the valves to close.

The pearl nucleus, Sadao explains, is cut from the shell of the pig-toe, a variety of Mississippi River mussel. The size of the implant depends upon the size of the oyster receiving the graft. The black tip can reach a diameter of twelve inches and a weight of eleven pounds. Normally a sphere with a minimum diameter of seven millimeters is used but a very large oyster may take a sphere of up to twelve millimeters.

Once the operation is completed, the oysters are returned to the lagoon. Suspended by a length of monofilament passed through the drill hole, the mature implanted oysters are attached inside a

Tending the crop. Pearl farming, Atlas Pearl Farm, Penyabangan off the Indonesian island of Bali. Photo: Robert Verspui.

long narrow wire basket and hung between the piers until harvest time.

Kata-Kata will farm an average of sixty thousand oysters. Forty to sixty percent of the oysters will fail during the two-year cultivation period. On average, approximately six percent of the total crop will be finer quality pearls.

Back in Papeete, I ask Ronald Sage, a well-known dealer, about the black pearl's unique color. "It is completely natural," he says. Although it is called black, the color of the pearl varies from a cream color, called poe nono, through gray, poe motea, to a black, called poe rava. Sage is referring to the body color of the pearl.

The basic or body color is only one criterion in the connoisseurship of pearls, and it is of secondary importance. Poe rava or poe nono, neither is inherently more beautiful that the other; symmetry, texture, size, luster, and orient are more important.

The Cultured Standard

The first successful pearl culturing was developed in China in the fourteenth century. In the early years of the last century, two Japanese researchers, Tokshi Nishikawa and K. Mikimoto,

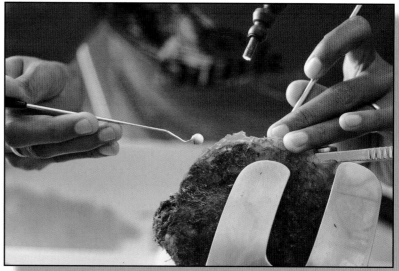

Implanting an oyster. The implant is normally cut from the shell of a freshwater mussel. Photo: Robert Verspui.

working independently, developed a process to artificially introduce a nucleus into the oyster. Patented in 1907,[221] this technique made it possible to produce this extremely rare pearl in much larger quantities. In 1908, Mikimoto established the first commercial pearl farm in Japan.

Cultured pearls began to appear in force in the market in the late 1920s, causing a dramatic fall in the price of natural pearls. It has taken forty years for the aftershock to subside. It was not until the 1960s that cultured pearls were completely accepted.

The cultured pearl is a gem market anomaly. Natural pearls are more expensive than cultured pearls but, outside the Middle East, until recently there was essentially no market for natural pearls.

The three types of pearls that will be discussed in the following chapters

Assortment of cultured pearls from various sources. Photo: Chin Cheng Choa, courtesy Otimo International.

— black South Sea, white South Sea, and Chinese freshwater — are all cultured. The Japanese akoya pearl, produced by the *Pinctada martensii* oyster, has been omitted from this discussion. This is not because the akoya is an inferior pearl. In fact, if left to its own devices, Pinctada martensii produces a highly lustrous and superior pearl. However, assembly line methods currently practiced by the Japanese produce a pearl with such thin layers of nacre, and, so highly processed (bleached and dyed), that it more closely resembles a manufactured product than it does a true pearl.

---

221 It has been suggested that an English marine biologist William Saville-Kent established the first pearl farm on Albany Island in 1906 and that both Nishikawa's stepfather visited the island and told his stepson about Saville-Kent's techniques. Cf: George, C. Denis, "Debunking a Widely Held Japanese Myth," *Pearl World, The International Pearling Journal,* pp. 10-23

Tahitian black pearls with various overtones. Photo: Tibor Ardai, courtesy Assael International.

Chapter 26

The Tahitian Black Cultured Pearl

"Greenish black pearls are perhaps valued higher than any other colored pearls, if they have the proper orient; this is probably partly owing to their rarity."

–G. F. Kunz, 1908

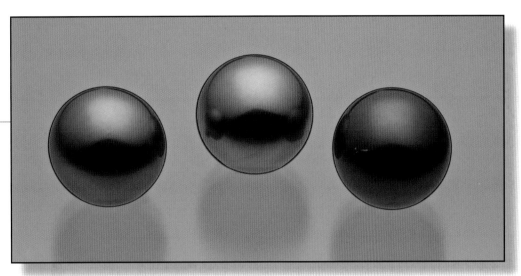

Three exceptional 11-11.5mm round Tahitian black pearls photographed in diffused lighting. Reading left to right; the first two have green/pink overtones, the last, reddish with a hint of blue. The diffused lighting makes it difficult to judge the quality of the pearl's luster. Photo: Jeff Scovil, courtesy R. W. Wise, Goldsmiths.

History

The black pearl was not known at all in the West in ancient times. Black pearls first came on the market in 1845.[222] Originally they were not highly esteemed, selling for a fraction of the price brought by white pearls from the traditional sources in the Persian Gulf and the Gulf of Mannar. The Empress Eugenie, wife of Napoleon III, was the person most responsible for bringing the black pearl into fashion. After the fall of Napoleon III, a fine necklace owned by the former Empress was auctioned at Christie's for twenty thousand dollars, the equivalent of several hundred thousand of today's dollars.

222. Kunz and Stevenson, *The Book of The Pearl*, pp. 29-30.

Body Color and Overtone

The finer examples of black pearl will normally appear greenish, bluish, pinkish, or violetish in tones of gray through black. This appearance is a combination of two factors: body color and orient.

Body color can be thought of as a tonal continuum from gray to black. Overtone is the result of light entering and refracting through the alternating semi-translucent layers of aragonite (crystalline calcium) and conchiolin, a type of calcium that acts as a binder. The result is similar to the effect of oil on water. Orient is the ephemeral fuzzy glow, like sunlight through an early

morning fog that appears to emanate from and cling to the surface of the pearl. Overtone can be separated visually from body color by observing the pearl under a light bulb. The overtone will be seen in the direct reflection of the bulb. The area surrounding the reflection will exhibit the body color. In fine black pearls this test is rarely necessary, as the orient, if present, is normally quite distinct.

It is this distinctive overtone that gives the black pearl its life and its air of mystery, and characterizes the finest of these pearls; it is the defining quality of the Tahitian black pearl.

Specific overtones are given fanciful names. Deep green is called *flywing*. The combination of green and pink is termed *peacock*; a dark-toned body color combined with pink is called *eggplant*. Occasionally pearls have a pure purple or pure blue overtone. Tiffany's gemologist G.F. Kunz (see quotation beginning this chapter) held that the green orient, or flywing, was the rarest and most valuable. Writing in 1908, he was speaking of natural black pearls. In cultured blacks, green is the most common orient color. Pink and blue come next in rarity, followed by peacock. Purple is by far the rarest and in my opinion the most beautiful of all.

Luster

Luster, the reflection of light off the surface of the pearl, is essentially the brightness of the pearl. The brighter, the better! In a fine pearl, this reflection is crisp and sharp. As quality decreases, the reflection becomes less distinct. Use direct lighting. Hold the pearl under a lamp and examine the bright reflection of the lamp on the surface of the pearl. The more distinct the reflection of the light source, the higher the luster.

The luster of the black pearl is somewhat softer than the luster of the akoya or the Chinese freshwater pearl. In the relatively warm waters of the South Pacific, nacre accumulates at a much faster rate than it does in the colder waters off the coast of Japan. However, nacre that accumulates in colder water tends to be closer grained, producing a potentially higher luster than is possible farther south. The luster of both the black and white South Sea pearls is softer than that of cold water pearls.

Symmetry

The more perfect the shape, the greater the value of the pearl. Round is the most favored shape, followed by pear, oval, and button. Asymmetrical shapes are classified as baroque. Prices of baroque pearls can vary widely. All nuclei implanted in cultured pearls begin as perfectly round and blemish-free; however, nature plays some interesting tricks during the pearl's growth. As a result, the average harvest yields only a very small percentage of perfectly round pearls. Pearls can be found in many unusual shapes.

Texture

Bumps, blemishes, tiny pits, or anything intruding on the pearl's surface is considered a negative. A silky flawless skin is the most desirable. Most pearls have slight imperfections; the issue is how

much they disturb the eye. Dealers are most concerned with how the pearl will "face up," that is, whether imperfections will still be visible when the pearl is set. Pearls with surface cracks are almost worthless.

The Rarity Factor

Black and white South Sea pearls are, speaking generally, the largest of all pearls. This is a result of two factors: the size of the mollusk itself and the rate of nacre accumulation. The shellfish, due to its large size, can accept large nuclear implants and this, coupled with relatively rapid nacre accumulation, results in very large round pearls. The larger the pearl the rarer it is and the higher the price that will be asked for it. Large mollusks can accept either a large number of small spherical implants or a very small number of larger diameters. It takes at least twenty-four months to lay down the two millimeters of nacre required to produce a fine black pearl. Thin-skinned pearls lack sufficient nacre to produce a fine luster and a distinct orient. Thin-skinned black pearls often exhibit a muddy brownish secondary hue or mask.

Since larger pearls command much higher prices, the usual grower's strategy is to produce a smaller quantity of larger pearls. With the increasing production of South Sea pearls, both black and white, this situation has begun to change. Growers are beginning to address market demand for smaller pearls. Prices for round pearls begin at 8mm. Black South Sea pearls under 10mm were difficult to find for. Prices increase at a reasonable percentage rate to 12mm. Prices for pearls over 12mm increase at somewhat larger percentages. Above 16mm, prices become negotiable because, at these sizes, rarity becomes an increasingly important factor in the value equation.

The spread of pearl culturing throughout its growing region is bound to affect prices. Although Tahiti has dominated the market for several decades, nascent industries in the Cook Islands and off the coast of Baja California are just beginning to challenge French Polynesia's market dominance. The spread of the pearl culturing industry is likely to continue. The prognosis for the short term is, therefore, a continued softening of prices.

Cultured natural color Chinese tissue nucleated pearls. Photo: Jeff Seovil courtesy, R. W. Wise, Goldsmiths.

Chapter 27
Freshwater Cultured Pearls from China

"The finest quality [pearl] that comes from China compares favorably with the finest from Japan."

–John Latendresse, 1992

China was one of the first to master the art of culturing. The Middle Kingdom began producing cultured pearls in the fourteenth century. In the 1960s a state-controlled industry introduced freshwater cultured pearls to the world market. Originally cultured using the wild *Cristaria plicata* or cockscomb mussel, initial production was of low quality baroque pearls, disparagingly known as *rice crispies*.

In the early 1980s, Chinese farmers abandoned the cockscomb in favor of the thicker shelled *Hyriopsis cumingii* or "three-cornered shell" mussel. Implantations of this new mollusk resulted in a breathtakingly superior freshwater pearl. By the early 1990s these pearls began to appear in force on the world market. Pearl culturing in China had come of age.[223]

Today Chinese pearls are filling a niche hitherto occupied by Japanese Biwa and akoya pearls. Ironically, just as demand for fine freshwater pearls has increased, and the term "Biwa pearl" has come to connote the very finest in freshwater pearls, actual production at Lake Biwa has declined to the point of nonexistence. Pollution is the culprit! Industrialization surrounding Lake Biwa has sounded the death knell for Japanese freshwater pearls.

The Chinese are producing two types of freshwater pearls: tissue nucleated and bead nucleated. Tissue nucleation uses only a thin segment of living tissue from a donor mollusk to stimulate the development of the pearl.[224] Bead nucleation (explained in detail in Chapter 15) is a relatively newer technique in China.

Tissue-nucleated pearls, which are by far the largest group produced, are mainly baroque pearls and are available in many bizarre and amazing shapes. Bead nucleation, as of this writing, has produced round pearls as large as fourteen millimeters.

Color

223 Richard W. Wise, "The New Face of Chinese Freshwater Pearls," *Colored Stone Magazine*, January 1992, p. 22.

224 The term tissue nucleation is a bit of a misnomer. The tissue implant grows into a sac and does not become the nucleus of the pearl. This type of pearl has, in effect, no nucleus at all. Fuji Voll, personal communication, 2003.

Chinese pearls come in a variety of hues including pink, apricot (yellowish orange), peach (pinkish orange), champagne (slightly pinkish yellow), plum (reddish violet), bronze (reddish brown), and every shade in between. Unlike black pearls, these pearls can be bleached white by prolonged exposure to the sun or by soaking the pearl in a bleaching agent for several hours. Natural color Chinese freshwater pearls should be stored in a darkened environment in order to preserve the natural pastel color, since they may fade with long exposure to sunlight. Color in pearls is not a part of the quality equation. Apricot is not more beautiful than champagne. That is a question of preference and simpatico (see Chapter 5) and is

beginning to replace tisssue nucleation as the dominent technique..

Overtone

Chinese pearls appear to have relatively opaque nacre. In smooth, relatively round pearls, the orient will exhibit itself as a slight darkening of the body color, plus perhaps a bit of pink.

Overtone can best be judged if the pearl is viewed against a color which is close in hue and therefore neutralizes the body color of the pearl. Diffused daylight is the best viewing environment. In this lighting the overtone is seen in the actual reflection of the light source; the other color seen is the body color. Incandescent light can sometimes produce the opposite effect,

Multi-color cultured Chinese freshwater pearl strands of exceptional luster and orient. The over-tone wreaths the body color on each individual pearl, like a halo. The bright spot at the center of each pearl bespeaks exceptional luster. Photo: Jeff Scovil; courtesy of M. Freeman & Co.

bleaching out the color toward the center of the round pearl. In such cases, the orient may be found in the surrounding halo.

Orient; the Rainbow Effect

Baroque China pearls tend to exhibit a rainbow effect, a quality unique to this type of pearl. This is particularly true of the more baroque tissue-nucleated variety. The more texture, the more pronounced is this effect. Rainbow iridescence must be distinguished from true orient. Though beautiful and desirable, it appears to be a surface effect, not the result of refraction. Rainbow iridescence is probably caused by light reflecting in different directions off the pearl's surface. This phenomenon is called interference; light rays literally bump into one another, resulting in the breakup of white light into various spectral colors.

Symmetry

Symmetry is the least important factor in evaluating a pearl. A majority of Chinese pearls are baroque, occurring in many strange and fanciful shapes. Baroque pearls, like fine abstract sculpture, may assume shapes which in no way detract from and in fact contribute to the beauty of the pearl. While it is true that curved surfaces bring out the beauty of the orient in a pearl, and that perfection of form does carry a substantial premium in the marketplace, the requirement that a pearl be perfectly round seems, at least to me, somewhat arbitrary.

Baroque pearls present the flexible aficionado with an opportunity to acquire a beautiful gem at a price which is dramatically less than the price a pearl with comparable luster, orient, etc. would bring in a pearl with perfect symmetry.

Fine white cultured South Sea pearls from Australia. Photo: courtesy Paspaley Pearls

Chapter 28
South Sea Cultured Pearls

"Not a chap among you knows the value of a given piece of pearl. That's how I can help you. I'll undertake to value without any fee for any of you chaps and my valuation stands as an offer too. If none of the big bosses will give you more bring it back to me and I'll pay the fee all right."

–Louis Kornitzer, 1900

In addition to black pearls, the sultry waters of the South Pacific also produce white pearls. Three species of pearl oyster are found in southern Pacific waters. The most important of these,

A large, fine round white South Sea cultured pearl from the Atlas Pearl Farm, Penyabangan, Northern Bali. Photo: Robert Verspui, Tears of the Moon.

the *Pinctada maxima*, is the largest of the pearl oysters. Three subspecies or varieties of Pinctada maxima are the white lip, silver lip, and the yellow or golden lip oysters.[225] These three siblings produce pearls which range from a white to silver white through yellow to a distinct golden hue.

South Sea pearls should not be confused with the Japanese akoya pearls, which are a product of the *Pinctada fucata* (*martensii* oyster called *akoya-gai* in Japanese. Both have a white body color. The Japanese oyster rarely produces pearls over 9mm. In fact, 9mm is a very large akoya pearl. The *Pinctada maxima* regularly grows pearls over 15mm and, if left to its own devices, will produce pearls as large as 20mm.

Pinctada maxima is found throughout the southwest Pacific from the Ryukyu Islands to the Arafura Sea, along a band running within twenty degrees latitude north and south of the equator. Until the current decade, the vast majority of white cultured South Sea pearls were produced off Australia's northern coast. Recently farms in the Philippines and Indonesia have begun to produce pearls from *Pinctada maxima*, more specifically the golden lip variety. These

225 The largest known pearl produced by the *Pinctada maxima* oyster is 24mm in diameter.

pearls occur in various tones of yellow through golden.

Body Color

South Sea pearls come in several hues. but white is predominent in the market.. Pearls, like faceted stones, will sometimes show a secondary or modifying hue. Some of these secondary hues are more desirable than others. Green, for example, is something of a negative factor; even a slight tint of greenish hue to the otherwise white body color will lower the value of the pearl. Gray is a double-edged hue. If it is dull, it is a negative. If it is bright it is called silver and adds measurably to the pearl's appeal.

Simpatico

South Sea pearls, as we have seen, are not all white, though white is the most sought after and thus the most expensive. This is because the white body color is simpatico (see Chapter 5) with skin tones typical of northern Europeans. People of the developed world simply have more money to spend on pearls; hence demand is greatest for these colors. The pearl itself occurs in white to golden, passing through various tones of yellow along the way. The actual body color has little to do with quality; simpatico is really a measure of compatibility. All other factors being equal, the subtle nuances of hue found in the South Sea pearl are of equal beauty.

Luster and Overtone: White South Sea Pearls

Luster and overtone are two of the most important criteria to be used when evaluating the beauty of a pearl. The white variety of South Sea pearl rarely exhibits more than a bit of overtone; that is, a contrasting chromatic color such as that displayed by the finer black South Sea pearls. Occasionally a fine white pearl displays a bit of pink.

The overall luster of these pearls is somewhat softer than that of pearls produced in colder waters. Soft luster coupled with translucency can result in a pearl with a soft misty surface suggesting a lake-bred early morning fog. Thus if one were to apply the luster test described in the discussion on pearl connoisseurship in Chapter 5, the most one could say is that the luster of the South Sea pearl is, at best, a good luster. The misty surface does impart a sense of life to the pearl's skin, which lends the South Sea pearl a unique, ephemeral sort of beauty.

Luster and Overtone: Golden South Sea Pearls

The golden South Sea produced by the golden lip variety of *Pinctada maxima* shares the softer luster of the South Sea varieties generally, but will exhibit a distinct overtone. In fact, in finer pearls, the yellow to golden color is part overtone and part body color (see the section on connoisseurship in pearls, Chapter 5). It can best be described as a golden hologram which seems to cling somewhat tenuously to the pearl's skin and to follow it when the pearl is rotated. In the finest gems, it emanates from the skin and appears to hover over the surface of the pearl.

Size

South Sea pearls are, speaking generally,

Natural colors of cultured South Sea pearls from the Pinctada maxima oyster. The gem in the foreground is a South Sea golden exhibiting the darker halo-like overtone. Photo: Robert Verspui, Tears of The Moon.

a fine South Sea pearl. Since larger pearls command much higher prices, the grower's usual strategy is to produce a smaller quantity of larger pearls. With the increasing production of South Sea pearls, both black and white, this situation has begun to change. Growers are beginning to address market demand for smaller pearls. Prices for South Sea round pearls begin at 10mm. South Sea pearls under 10mm are very difficult to find and pearls under 9mm hardly exist at all. Prices increase at a reasonable percentage rate to 14mm. Pearls over 14mm increase at somewhat larger percentages. At 16mm prices become negotiable because, at these sizes, rarity becomes an increasingly important factor in the value equation.

The spread of pearl culturing throughout its growing region is bound to affect prices. Although Australia has dominated the market for several decades, emerging industries in Indonesia, Burma, Vietnam, and the Philippines are beginning to erode Australia's market dominance. The prognosis for the short term is, therefore, a continued softening of prices for these fine southern beauties.

the largest of all pearls. This is a result of two factors: the size of the mollusk itself and the rate of nacre accumulation. Nacre accumulates up to ten times faster in the warm waters of the southern Pacific than it does off the coast of Japan. The shellfish, due to its large size, can accept large nuclear implants; coupled with nacre accumulation, this results in very large round pearls. The larger the pearl the rarer it is and the higher the price that will be asked for it.

Large mollusks can accept either a large number of small spherical implants or a very small number of larger diameters. It takes at least thirty months to produce

31.29 carat peridot necklace. Photo: Jeff Scovil, The Edward Arthur Metzger Gem Collection, Bassett, W. A. and Skalwold, E.A.

Chapter 29
Peridot

Originally called *topazios* and considered an undercooked emerald, peridot has been known since ancient times. The Biblical source of peridot was the island of Zabargad located in the Red Sea, which, according to Pliny, was first worked in the 2nd Century BC. Peridot is widely distributed. Sources that currently produce include Burma, Arizona and Pakistan as well as Norway, Brazil, Australia, Tanzania, Madagascar and China (Xinjiang and Hubei Provinces). Peridot is also the only known extraterrestrial gem. Peridot crystals are often found in meteorites.[227]

In the mid 1990s a large strike of high quality gem rough sourced from the Sapat Valley, Northwestern Frontier Province of Pakistan first appeared on the market. As of this writing (2016) this source is still being produced though the material which yielded stones up to one hundred carats is mostly mined out.

Extremely fine example of peridot from Pakistan. Stone has a pure green hue with no trace of yellow. Photo: Jeff Scovil, courtesy R. W. Wise, Goldsmiths.

Pakistan is the major supplier of larger fine peridot. Most other sources do not currently produce much of note. Currently there is some peridot coming out of Burma and production from San Carlos Apache Reservation in East Central Arizona is mostly in gems less than two carats.

---

227 Peridot crystals are a major constituent of a type of meteorite known as a pallasite. Despite their celestial origin, gems cut from pallasites tend to be pale and grayish and foggy, see Sinkankas, John, et al., "Peridot As An Interplanetary Gemstone", *Gems & Gemology*, The Gemological Institute of American, Spring, 1992), pp.43-51.

Hue/Saturation/Tone

The hue of peridot is normally described as olive green, meaning a yellowish-grayish/brownish green. However, the finest peridot is a rich medium dark toned (60-70%) green. A majority of gems and almost all smaller sizes show a distinct yellowish secondary hue. This is one of the few green gems that does not include blue as a secondary hue. It is the elimination of the yellow secondary hue defines the best of the best in peridot. Stones with five percent (almost imperceptible) or less of a visible yellowish hue should be considered the finest of the fine.

Gray is the usual saturation modifier or mask found in peridot, but brown is also common. The gray is rarely prominent, but it does impart a distinct cool dullness to the color. The best way to separate it is to compare a number of stones side by side. Green is a cool color, but gems lacking the gray mask will sport a visibly sunnier hue.

Clarity

Peridot is normally found eye-flawless. Gems with visible flaws are of little interest to the connoisseur.

Crystal

As noted above, peridot is capable of exceptional transparency. Gems from Pakistan exhibit a crisp transparency, particularly when compared to Burmese peridot, which appears slightly sleepy in comparison. Fine stones will exhibit fine crystal.

The Rarity Factor

Peridot can be found in very large sizes. In the early years of the Pakistani production, stones between twenty-five to one hundred carats were not uncommon according to one of the partners involved in the mine.[227] The largest known faceted stone weighs in excess of three hundred carats, and five to ten carat stones, mostly from the Pakistani source, are fairly abundant in the marketplace. The per carat price normally increases to twenty carats and declines thereafter.

227 Wayne Thompson, personal communication 2014

Chapter 30

Topaz

"A topaz presented by Lady Hildegard, wife of Theoderic, Count of Holland, to a monastery in her native town, emitted a light so brilliant that prayers could be read witout aid of a light. A statement that may be true if the monks knew the prayers by heart."

–Oliver Cummings Farrington, 1903

The 2nd century BC merchant's guide, *The Periplus of the Erythraean Sea*, referred not to the gem we know today as topaz but to other species of gems. The appellation *topazion* referred to peridot, citrine or both depending upon the writer. True topaz remained unknown or at least unrecognized until the 18th century. The Konigskone Topaz Mine in Saxony was the first commercially exploited deposit of true topaz. The mine, discovered in 1734 and worked until 1800, produced some bright, gemmy, mostly yellow crystals. The confusion persists. *Smoky topaz*, *Madeira* and *Brazilian topaz* are all varieties of quartz.

Commencing in the early 18th century, more than half of the world's then known supply of gold was removed from the verdant hillsides surrounding the quaint Brazilian town of Ouro Preto. Today, these same hills hold almost all of the entire world's known commercial reserves of imperial and precious topaz.

Small deposits of topaz have been found in the northern Brazilian state of Para, and in Mexico, Sri Lanka, Burma, Pakistan, in Russia's Ural Mountains and in Colorado

Dark peach (reddish orange) "sherry" topaz, gem and crystal from outside Ouro Preto, Minas Gerais, Brazil. This sixty-seventy percent toned orangy-red, is among the rarest hues of topaz. Photo: Harold and Erica Van Pelt, courtesy of Kalil Elawar.

and Utah. *Ouro Preto* — the name translates as black gold — is the only location which is currently producing commercial quantities of natural gem-quality topaz.

Here some clarification about natural

color topaz as distinguished from blue topaz is in order. Natural blue — in fact, any topaz in the blue/green color range — is an extreme rarity in nature. The ubiquitous blue topaz that is seemingly everywhere in the market today is common colorless topaz which has been color-enhanced through a combination of irradiation and heat treatment.

Imperial Versus Precious

Natural color topaz is usually divided into two types: *precious* and *imperial*. There is some confusion as to the distinction: some experts consider precious topaz to be

34.52 carat topaz with strong multicolor effect. The hue grades from yellow to orangy red. Photo: Mikola Kakharuk, courtesy Nomads.

any topaz in a certain color range. Others, such as *The Guide*, a standard dealer's reference, limits the term precious topaz to stones which do not exhibit multicolor effect! These are terms best avoided.

Faceted topaz can be strongly dichroic because the C axis of the crystal is often a darker, richer hue than the AB axis. When topaz is cut with the AB axis face up, particularly in the long pear, oval, and marquise shapes which insure the best yield from the rough, the darker hue of the C axis bleeds into each end of the gemstone, showing a richer, more saturated hue at each tip of the finished gem. An emerald cut topaz may not exhibit multi-colors, but an oval almost certainly will. The only color that is consistently monochromatic is yellow topaz.

Hue

Natural color topaz occurs on a color/rarity continuum from yellow through orange, cinnamon pink (peach), orange pink (ripe peach), pink, light violet, dark orangy red (hyacinth), violetish red, pink, pinkish red and red. Prices follow this same line with yellow hues priced lowest and violetish red ones fetching the highest prices. Generally speaking, the redder the better. The peach and cinnamon colors are the most characteristic. Topaz possesses a liquid or soft limpid brilliance, the result of characteristic refraction coupled with a high degree of transparency.

Pink or Purple

Topaz often occurs in a light-toned violet to purple hue. At lighter tones, these hues are difficult to separate from pink and are often lumped under the term pink topaz. Purple, like pink, is a modified spectral hue. Purple occupies a space on the color wheel halfway between blue and red; pink is a lighter shade of red. The distinction is important because true pink topaz, even purplish pink topaz, is much rarer than light purple. The purer the pink hue, the rarer the stone. Thus, the astute collector, who has trained his eye to discriminate between pink and purple, may find a buying opportunity in a parcel of pink topaz.

Saturation

Brown is the dominant saturation modifier or mask found in topaz, although gray is also found, particularly in pinker stones.

Tone

Using a tonal scale where window glass is zero percent and coal is one hundred percent tone, the optimum tonal range for topaz is fifty to eighty percent. Below fifty percent, the color begins to pale and wash out; at eighty percent, which is rare, topaz begins to lose the liquid effect and appear overcolor. The optimum tone increases as the hue changes from yellow to peach, to pink and finally to red. Twenty percent is optimum tone for yellow, forty to sixty percent for the peach range, and fifty to seventy-five percent for the pink to red range. Because of the liquid aspect of the key color, gems of lighter than optimum tone may be preferred if the connoisseur favors a more delicate hue.

Nightstone

Topaz is a true lady of the evening, a nightstone, one of the few gems that looks its best in incandescent light or candle light, and holds up well in low-light environments. Topaz tends to bleed a bit; that is, lose both color and tone in fluorescent light and in daylight.

Mechanized mining at the Capão Mine outside the village of Rodrigo Silva, Minas Gerais, Brazil. Photo: R. W. Wise.

Clarity

Topaz is normally eye-flawless. Stones with visible inclusions are sold at a deep discount.

Treatments

The traditional color enhancement for topaz is heat treatment. This technique, known as pinking, is performed under relatively low temperatures, at times over the open flame of a miner's campfire.

Figure 110: A rare 5.88-carat example of fine red topaz. The consistency of color in the face up mosaic leads some dealers to term this "Imperial" topaz. Photo: courtesy Meyer & Watt

Stones with some pink or bluish pink can be turned a purer hue using this technique. Although experienced dealers claim to be able to separate heated from unheated stones by eye, pink topaz also occurs naturally and, at the time of this writing, there is no gemological test that can separate natural color from pinked topaz.

Colorless topaz is run through a linear accelerator which alters the atomic structure of the material, then heat treats it to turn it blue. Connoisseurs do not take blue topaz seriously as a gemstone. Recently a new process was announced which turns colorless topaz pink through a combination of heat and high pressure. This process is entirely different from the gentle heating that will "pink" a light pink or peach topaz.

Limited Production

Most of the topaz currently on the market can be traced to a single mine, Capão, about five kilometers from the small village of Rodrigo Silva. This village is almost dead center of the two-hundred-ninety-square-kilometer topaz belt running in an east-west direction, west of the city of Ouro Preto. *Capão*, "big lid" in Portuguese, is one of two large-scale mechanized mines currently operating. The second, Vermelhão, is situated about twelve kilometers to the east.

Capão is an open pit operation with two large pits, each carved over two hundred feet down into the verdant hillside. The mine currently employs forty-seven workers and is highly mechanized. Bulldozers are used to open the pits, and most initial sorting uses German-made hydraulic sluices and sieves, with some final sorting done by hand. The huge amount of water necessary for mining operations is drawn from a nearby lake.

There are a number of smaller mines in the area, including Don Bosco and Garimpo which are strictly hand operations worked intermittently by independent miners called *gariempieros*.

The Rarity Factor

Exceptional topaz is rare in every color and size. Generally stones above twenty carats will decrease in price on a per carat basis.

Oxcart, Mogok, Burma. Photo: R. W. Wise

RUBY & SAPPHIRE

The road is narrow and winding, the sky a gunmetal blue and cloudless. The day is hot! A cloud of brown dust hovers over the single lane road paved only in places. The monsoon rains that will soon turn it all into a muddy quagmire have not yet arrived. The road leads to Mogok and the ancient ruby and sapphire mines of upper Burma.

Riding through the countryside, we pass through villages, seeing houses of plaited bamboo, creaking bullock carts, and peasants, reed-thin and brown as dirt, plodding along the road. The green fields of rice stretch along both sides of the road, like a well-tended lawn, on toward the horizon. Farmers in conical straw hats bend over rice paddies in a tableau from centuries past.

Mogok is a provincial town one hundred seventy miles west of Mandalay, Burma's second largest city. Since antiquity the valley in which the town is situated has been famous as the legendary Valley of the Serpents. According to the ancient tale, somewhere in the mystic East was a nearly bottomless valley carpeted with glittering gems. Poisonous serpents stood guard over the gems. Merchants seeking the stones tossed the sticky carcasses of skinned sheep into the valley. The gems stuck to the meat and the great eagles circling the valley floor would swoop down, grasp the meat in their talons, and fly it back to their nests to feed their young in the high rocky crags surrounding the valley. Then while the great birds were off hunting, men would climb up to the nests and retrieve the precious stones. Among the stones found in or near this valley were ruby, sapphire, peridot, and tourmaline.

Deep, verdant, enveloped in mist and surrounded by snaggletoothed peaks, the valley today remains the stuff of legend. The Burmese government kept it closed to foreigners for over thirty years, and in that time the modern world all but passed it by. Beginning in the early 1990s some westerners were allowed to visit, but special government permits were still required. Traditionally stones have been found on the valley floor, in the streambeds and catch basins, and in the limestone caves that honeycomb the mountainsides surrounding the town.

It was on the sultry afternoon of our second day in Mogok town that I got my first glimpse of a legend. I was with my friend Joe Belmont, one of Asia's premier gem dealers. Thus far we had seen

Kanase women work the tailings in a stream issuing from a large mechanized mining operation in Mogok, Burma. They sieve through the gravel looking for ruby, hoping to find small gems overlooked in the washing process. This privilege, the Burmese version of social security, is restricted to the widows and orphans of miners. Photo: R.W. Wise

mostly sapphires, some of the limpid, rich blue beauties for which Burma is famous. We had just finished lunch at an outdoor restaurant specializing in soup. The customer selects from an array of condiments; pork, chicken and some not so easily identified. I pointed and smiled. The partially cooked noodles were plunged into a huge cauldron of boiling stock. My finger did the talking and chicken, pork and beansprouts were added and ladled into a bowl. A dark-haired Burmese approached our table. He was thin, in his forties, dressed in a Western shirt and the traditional cotton skirt, or lungyi. His face was dark and angular with high prominent cheekbones. He was a ruby dealer and apparently knew our agent. They greeted each other with extravagant courtesy. The dealer had heard we were in town and had something special to show us.

After the introductions and ritual pleasantries, the dealer escorted us up two blocks along the dusty main street. We passed through an open-fronted food market. At the back of the store, we ascended a staircase of dark teak. The workroom of a bakery was visible below the stairwell and the rich scent of baking bread accompanied us up to the second floor. The walls and floors were richly paneled in polished teakwood. He smiled and motioned us to a set of comfortable chairs clustered around a coffee table. Market sounds and the warm rays of the westering sun filtered through open casement windows overlooking the street.

We were offered tea and delicacies still warm from the ovens below. Two parcels were placed before us. After a sip of tea, Joe opened the first paper and after a brief look handed it to me without comment. Inside was a two-carat cushion-shaped ruby of the finest color I had ever seen. I looked at

Joe, our eyes locked briefly; his glance confirmed the insight that had tripped off like a flashbulb in my mind. In my twenty years' experience this was my first glimpse, the Pigeon's Blood!

> *Asking to see the pigeon's blood is like asking to see the face of God.*

–Khun Cha, 1982

The color was new to me yet somehow I knew it for what it was, perhaps because I had seen everything but! The stone was a primary red, the hue and tone like a rich tomato gravy that has simmered for hours on the stove: a rich, deep-toned pure red. I took a deep breath in an effort to control my hammering heart. My friend reflected for a moment, then asked the price. The response was, of course, astonishing! Negotiations continued for about an hour, but in the end our offer was not accepted. The next day proved luckier.

Sapphire comes in many colors. It is one of two varieties of the gem species corundum. If it is red, it is ruby. If it is any other color, it is called sapphire.

10.08 carat blue Ceylon sapphire in antique Buccellati setting. Cora N. Miller Collection, Peabody Museum, Yale. Photo: Jeff Scovil, courtesy R. W. Wise, Goldsmiths

Chapter 31
Blue Sapphire

"The characteristic color of the Sapphire is a clear blue, very like to that of the little weed called the "corn flower," and the more velvety its appearance, the greater the value of the gem."

–E.W. Streeter, 1879

They say you never forget your first love. Gemologically speaking, blue sapphire was mine. I remember my first date with a Kashmir . . . ah! That limpid velvety blue, that sleepy bedroom glow. It was back in the 80s in Bangkok that I met my first Burmese sapphire, a saucy royal blue, deep hued with just a touch of purple. Its vibrant saturation gave me a thrill. I didn't realize just how lucky I was; it took me ten years to find another as fine.

From the early 1980s sapphires were mostly from Australia, Thailand, and Sri Lanka. Kashmir stones were a fading legend and, except for a trickle across the Thai border, Burma blues were just a

pleasant memory. Thai stones were available but were dark or black, often opaque. Australians were greenish; Sri Lankans were the best—many were heat-treated, it's true; but the finest were just a step in saturation below Burmese and occasionally displayed a sleepiness similar to Kashmir.

Record breaking 35.09 carat Kashmir sapphire captured in this remarkable photograph. Variously described as "cornflower" and "royal blue," on May 13th, 2015, this fabulous top color, velvety purplish blue gem of an ideal 80% tone sold at Christie's for 7.45 million dollars, besting the 2013 record for a Burma sapphire by $155,000 per carat. Photo: Alistair Grant, Corbis.

Beautiful sapphires are still to be found, but the cast has changed. Sri Lankan stones are still in reasonably good supply. Australia

is reportedly producing better blues. Thai production has declined significantly. Increasingly we hear about new finds in Africa, from places unknown just a few years ago. In Madagascar, the big name in the early days of this century, production has slackened. As of this writing (2016) Madagascar and Sri Lanka are the most productive.

Color: the Two Standards

Historically, blue sapphire has been judged based on two paradigms, the best from Burma and the finest of Kashmir. The best of Burmese sapphire is a vivid blue to vivid purplish blue, a pure medium dark toned blue primary hue with a secondary hue of five to ten percent purple. This is usually described as "royal blue." Burma sapphire is set apart by its transparency (crystal) and the vivid crispness of its hue. Kashmirs, by contrast, are a vivid purplish blue hue with ten to fifteen percent purple, a hue often described as

The 62.02 carat Rockefeller Sapphire. Considered to be one of the world's finest Burmese gems, was purchased by John D. Rockefeller from the collection of the Nizam of Hyderabad in 1934. The gem subsequently sold at Christie's in 2001 for $3.01 million dollars or almost $49,000 per carat. Its value today is a matter of speculation.   Courtesy, Christie's Images.

"cornflower."[228]

Saturation in the best Kashmir stones surpasses Burmese. The crystal is not so crisp and the color will often have what is described as a sleepy quality, a result of myriad numbers of tiny inclusions, sugar-like grains which are difficult to resolve under the highest magnification. Light refracting through this microscopic Milky Way is diffused and this gives the stone an overall sleepy or velvety appearance. These inclusions also reflect light, dispersing it throughout the gem and reducing extinction. Kashmir sleepiness contrasts with the robust brilliance and transparency of a Burma or Sri Lankan stone. Sri Lankan stones are sometimes found with a similar sleepy quality caused by similar if somewhat larger grained sugary inclusions. One famous dealer once told me in strictest confidence,

228 The use of terms such as "cornflower" illustrates the problem of comparing the color of a gemstone to another natural substance. Cornflowers themselves have a violet component yet, in the gem trade, cornflower is usually used to describe a pure blue hue. Powder blue is probably the more precise term.

strictest confidence, "the finest Kashmir I have ever seen was from Sri Lanka."

Kashmir sapphire was found on one side of one hill in the Indian state it is named for, and was effectively mined out by the 1930s. Burmese sapphire has also been in short supply since the 1930s. The new Burma ruby diggings at Mong Hsu produce almost no sapphire. In and around Mogok the ratio of ruby to sapphire is 100:1. With the resumption of hard rock mining in the late 1990s, a few stones per year find their way from the old mine areas of Mogok into the Bangkok market. Which is the best? Connoisseurs disagree but, historically, fine Burmese stones cost at least fifty percent more than Ceylon sapphires, and Kashmir stones more than twice the price of Burmese.

That ratio has recently been scrapped as Kashmir sapphire prices began to escalate after 2005. The SSEF lab described it as *royal blue*, the American Gemological Lab labeled it *cornflower*. In May of 2015 a magnificent 35.09-carat top color Kashmir sapphire set a new world record price at auction when it sold for $7,450,000 or just over $200,000 per carat. In October of 2015 that record was bested when the 27.68-carat Jewel of Kashmir sold at Christie's for $245,000 per carat, and Kashmir sapphire became the new darling of the auction markets. Can Burma blues be far behind?

Ceylon/Sri Lankan Sapphire

In terms of availability, Ceylon or Sri Lankan sapphire has been the quality standard bearer for the past half century. The Ceylon gem may look just like its Kashmir and Burmese brethren, appearing either cornflower or royal blue; with, as mentioned above, a Kashmir-like sleepy

An extremely fine 2.05-carat unenhanced Ceylon sapphire. A rich slightly purplish blue of 75% tone exhibiting exceptional crystal. Photo: Jeff Scovil. Courtesy R. W. Wise, Goldsmiths.

appearance. However, the relatively larger size of the inclusions that cause the effect in Ceylon stones translates into a crisper sort of sleepiness which is arguably less subtle than the same characteristic in stones from Kashmir. Ceylon sapphires may also achieve a vivid royal blue like a fine Burma blue and may also exhibit a beautiful clear, crystalline quality.

When seeking a fine sapphire, the collector-connoisseur is advised to avoid labels and look at the gem on offer. Ceylon sapphire has its own distinct and beautiful look; the natural color stones often have a bit more of a purple secondary hue and may be crisply transparent.

New Contenders from Africa

The Tunduru deposit was once described as the most important discovery in fifty years. Tunduru is in southeastern Tanzania hard against the Mozambique border. The best of the Tunduru stones have a deep royal blue color similar to Ceylon, in the words of Joseph Belmont, a dealer noted for his fine eye, "Tunduru stones have a much better crystal."[229] The best Tunduru stones were a step up from Ceylon sapphire and only a half step down from fine Burmese sapphire. Unfortunately the Tunduru stones were quickly mined out and little has been heard from this source since the mid 1990s.

At this writing blue sapphire from Madagascar is a smaller force in the market. The best of Madagascar comes from an environment geologically similar to Ceylon and closely resemble sapphire from that source with perhaps a bit more consistent purple secondary hue.

Australia, Steady Supplier of Blue Sapphire

Australia remains a steady supplier of blue sapphire to the world market. Sapphire is found at a number of sources in Queensland and New South Wales.

For many years Australian stones had the reputation of being greenish and over dark with tonal values of ninety percent or more. Sapphire, like ruby, emerald, and tsavorite garnet, is judged by the purity of its primary hue. A little violet is desirable, but green is the bane of blue sapphire. Thai dealers often bought the best of the Australian production, heated them and sold them labeled "Ceylon" in the Bangkok market — and still do. However, with new sources and advanced heating technology, many heat treated stones, of finer color, are available today.

America the Beautiful

Alluvial sapphire deposits were first discovered in Montana's Missouri River in 1865. Three other sites — Dry Cottonwood Creek, Rock Creek, and Yogo Gulch — were added before the turn of the century. Yogo Gulch, the only hard-rock deposit, was mined steadily until the late 1920s, when it was abandoned; production resumed in the 1980s and has sputtered along in fits and starts.

Yogo sapphire is often described as cornflower blue, a rich purplish blue hue that has been erroneously compared to Kashmir. Generally, the finest of the Yogo stones have a distinctive "steely" appearance, the result of a slight gray mask. These Montana beauties are of uniform color, relatively free of inclusions, and resist heat treatment. Unfortunately, rough Yogo sapphire occurs in flat tabular crystals and rarely yields faceted stones in sizes above one carat.

Sapphire from the three other sources

229 Joseph Belmont, personal communication, 1995.

mentioned have been of little commercial importance until recently. Although huge quantities of sapphire have been taken from the Missouri River and both Rock and Dry Cottonwood creeks, these areas produced mostly colorless to pale-toned (twenty to thirty percent) mixed hued stones of little beauty. Advanced heat-treating technology has significantly altered the situation. These techniques have raised rough yields from Rock Creek from as little as eight percent to as much as eighty percent facet-grade gem material.

The best of these blue sapphires display a rich (eighty to eighty-five percent) blue primary hue, with a pinch (five percent) of violet and a slight (ten to fifteen percent) gray to gray-green modifier. Due to the apparent green secondary hue, Montana stones from these sources never approach the finest sapphire qualities. They have been given fanciful names like *wintergreen* and excite little more than local interest. Blue Rock Creek sapphire most resembles high-grade commercial quality stones from Australia.

The Best of the Blues: Hue and Tone

Blue sapphire, like ruby, is a primary color gemstone. The purer the primary hue the better. In practice this means that a dark-toned eighty percent primary blue hue with no more than a ten to fifteen percent secondary purplish hue is most desirable.[230] Some connoisseurs prefer a distinct purplish secondary hue because it adds a velvety richness to the blue; others prefer a purer, more open blue of slightly lighter (seventy-five percent) tone. This range of hues is considered the finest color in sapphire.

A comparison between the top Kashmir and the top Ceylon sapphire illustrated above is a case study in the interaction between saturation and tone. The color blue reaches its maximum saturation, its gamut limit, at eighty percent tone (Chapter three). The Kashmir record breaker pictured at the head of this chapter is precisely the ideal tone. Now consider the Ceylon gem. It is approximately seventy-five percent tone, five percent below the gamut limit for optimum saturation. The other name for saturation is the quantity of color.

Visualize a syringe filled with black or a dark purplish blue and mentally inject a bit of the liquid and picture it dissolving into the gem. You can see how the addition of just the smallest quantity of black would add depth and richness to bring the hue to almost the same level as the Kashmir gem, though not quite. The Ceylon blue appears steely, slightly grayish, when compared to the Kashmir, which sits at the very pinnacle of color. There is something in the chemistry of Kashmir gems that seems to

230 C.R. Beesley, personal communication, 1990 and 1998. According to Beesley, the best sapphire in the world would have no more than a seventy percent pure blue hue. The other thirty percent would be a combination of all secondary hues, including some green. Beesley describes this color as "vivid purplish blue." Minor hues are not necessarily visible to the eye. See Beesley, *Colored Stone Training Manual*, p. 15.

allow these gems to achieve a richness and purity of hue that gems from other sources rarely, if ever, achieve.

Green is the bane of blue sapphire. Any visible hint of green brings a stone's value crashing down into the commercial range. The problem is that all blue sapphire has a greenish component when viewed at certain angles to the C axis. It is the cutter's job to cut the stone so that this green is not part of the face-up appearance.

Saturation

Gray is the normal saturation modifier in blue sapphire. Often it will be found mixed with green in lower-quality stones. A slight gray mask will introduce a cool or slightly steely quality in the normally warm hue of a sapphire. All pure chromatic hues are vivid. If the key color appears dull and/or cool, a gray mask is the probable culprit.

Multicolor Effect

Since blue sapphire is a primary color gem, the closer it comes to exhibiting a uniformly pure primary blue hue the better and more desirable it is. The face-up mosaic of a gemstone, however, is far from uniform; each facet may exhibit variations in the gem's key color. Some facets may appear bright, some dull; some may display a dark tone, others a medium- or light-toned blue.

Multicolor effect has several causes (see Chapter 4). Sapphire is dichroic, i.e., light entering the gem divides into two rays, one violetish blue, one greenish blue. In addition, the stone may be zoned: colorless

zones are juxtaposed against zones of color. Light rays passing through colorless zones lose color. Also, the pavilion facets of the gem cause a light ray entering the stone to reflect at least twice within the stone, absorbing color as it goes giving you an idea why the face-up scene in blue sapphire may be less than uniform.

In ruby and sapphire a negative type of multicolor effect is traditionally called *bleeding*, manifested as a lightening of tone and a loss of saturation when the stone is shifted from natural to incandescent lighting. At lighter tones, blue becomes pastel, less saturated, and washed out. Bleeding is a good analogy: the color is drained from the stone just as blood is drained from the body. The effect is similar: the stone becomes pallid, and the life, figuratively, is drawn out of it. Bleeding in sapphire may be described as weak, moderate, or strong. The more apparent or stronger it is, the greater the fault.

Kashmir sapphire contains little or no chromium, which appears to be one cause of the bleeding. What this means is that Kashmir sapphire, unlike Burma sapphire, will not intensify the purplish secondary hue, but will maintain its hue as the viewing environment is shifted from daylight to incandescent.

Crystal

With gemstones there is one truth: no matter how fine the stone, somewhere there is a better one. Sapphire at its zenith — that is, a stone seeming to have everything: color (hue, saturation, and tone), clarity, and marvelous make — still requires a velvet

transparency to reach the very pinnacle of quality.[231] Heat treated blue sapphire often will close up in incandescent lighting. This may have little effect on color, so it can't be described as bleeding; the crystal simply becomes turbid, dark, and murky as the lighting environment is shifted from natural light to the light of the bulb.

Some stones will have the three Cs (color, clarity, and cut), but very few also have the diaphaneity, the good crystal — the fourth C, imparting a quality that is rich, crisp, and velvety all at the same moment. It is this quality which finally separates the very finest from the rest of the herd. Fine stones fitting this description may come from any source. Beauty is its own best pedigree.

Texture

Color in sapphire often occurs in zones which follow the hexagonal outline of the sapphire crystal. Zones of rich color will alternate with colorless areas. This is particularly characteristic of gems from Sri Lanka. Due to zoning, sapphire will often show what experts call texture. This means that the zones are sometimes visible face up, causing the color to appear uneven. Even color is very important in sapphire; it is the cutter's job to integrate the zones so that the face-up color appears to be even. What is seen through the side or back of the

stone is of little importance. Sometimes the lapidary's attempt to even out the texture by eliminating the effect of the natural zoning in the crystal leads to poor symmetry.

However, stones which appear lopsided below the girdle or those that have off-center culets are more tolerated in sapphire than in most other gem species. Symmetry faults which would be considered major flaws in diamond are understood and accepted as necessary in sapphire so long as they occur below the girdle and do not create a lopsided girdle outline.

Heat Enhancement

No one is quite sure how long heat-enhanced sapphire has been in the market for hundreds of years, certainly.[232] Heat treatment, known as burning, was reported in India as early as 2000 BC.[233] However, it was not until the 1970s that the technology to achieve very high temperatures became available, and heat treating began to be practiced on a grand scale. Some lighter Ceylon stones (thirty to fifty percent tones) are unheated. But most of the finer Ceylon stones, as well as a good portion of the Madagascar stones currently in the market, are heat enhanced.

Heating has a negative effect in blue sapphire. Heat-treated blues have generally poorer crystal than unheated stones. The heating process tends to reduce transparency or muddy the crystal. [234] If all other factors are equal, the very best natural

---

231 Kashmir sapphire is something of an exception. The diffused sleepy effect, together with the tiny inclusions that produce it, reduce transparency in Kashmir and Kashmir-type sapphire. In such cases, the beauty of this unique phenomenon makes up for some loss of crystal; although Kashmir sapphires with this attribute can hardly be described as limpid, the best still retain a high degree of transparency.

232 Tagore, *Mani Mala*, vol. 1, pp. 243, 455.

233 Nassau, *Gemstone Enhancement*, p. 25.

234 Joseph Belmont, personal communication, 1997.

color sapphire will be more beautiful than the very best burned sapphire. For example, of the top ten blue sapphires in the world, two through nine may be heat enhanced, but number one will be natural color. The exact opposite occurs when ruby is subjected to heat treatment. Heated or unheated, which should the aficionado consider? It may come down to budget. In general, natural color sapphire will sell at a minimum of fifty percent above the price of a comparable heat-enhanced stone.

The Rarity Factor

Exceptional blue sapphire is rare in any size. Stones over twenty carats are available. In most gem species and varieties, stones larger than those readily usable in jewelry tend to decrease in price (per carat) above twenty carats. However, as noted, the record prices currently being paid, by collectors, for large very fine gems, means this is not the case with blue sapphire.

Chapter 32
Padparadscha Sapphire

"It is GIA's opinion that this color range should be limited to light to medium tones of pinkish orange to orange-pink hues. Lacking delicacy, the dark brownish orange or even medium brownish orange tones of corundum from East Africa would not qualify under this definition. Deep orangy red sapphires likewise would not qualify as fitting the term Padparadscha."

–Robert Crowningshield, 1982

adparadscha is a corruption of the Singhalese word *padmaragaya*, which is composed of two words, *padma*, lotus; and *raga*, color.[235] Thus a padparadscha color sapphire is the color of the lotus, in this case the oriental lotus (*Nelumbo nucifera* 'Speciosa'). As with other cases discussed in this book, such descriptions often create more problems than they solve. What color is the lotus? As a bud it is a beautiful, delicate shade of reddish pink, but as it opens the pink shades into yellow.[236]

As a further source of confusion, some experts believe that only sapphire from Sri Lanka, the original source, may be properly called *padparadscha*. This is another manifestation of the innate conservatism of the gem trade. The experts might be forgiven if the only other possible contenders were the brownish orange pink African padparadscha sapphires found in the gem gravels of Tanzania's Umba

A 20.84-carat "padparadscha" sapphire sold by Christie's Hong Kong in 2005. The vivid saturation pushes the stone beyond the pastel hues as defined by the LMHC. Photo courtesy Christies, Hong Kong.

River.[237] However, in recent years, fine padparadscha stones have been found in Vietnam's Quy Chau mines and, more recently still, in newly discovered gem-

---

235 R. Crowningshield, "Padparadscha: What's In a Name?" *Gems & Gemology*, Spring 1983, p. 31. Dealers from Sri Lanka invariably call yellowish orangy pink sapphire padparadscha.

236 Ibid.

237 Some excellent quality stones have been found at Umba River, comparable to the finest from Sri Lanka. Excellent quality stones with a high degree of saturation have also been found in Vietnam. See Hughes, *Ruby & Sapphire*, pp. 398, 414. The finest examples that this writer has ever seen were from Songea, Tanzania.

producing areas of Songea, Tanzania and southern Madagascar. [238]

This natural 3.09-carat pinkish orange sapphire embodies the essence of padparadscha. It is similar in saturation/tone to the gem pictured above. The color is pinkish orange, fifty percent tone. Note the delicate blending of hues and the exceptionally limpid crystal. GIA-GTL "padparadscha" grading report issued in 2013. Photo: Jeff Scovil; courtesy of R.W. Wise, Goldsmiths, Inc.

In 2005 the Laboratory Manual Harmonization Committee (LMHC), a committee representing seven of the most important gemological laboratories, issued LMHC Information Sheet #4 with a definition of how the member laboratories defined padparadscha sapphire. The definition closely followed Crowningshield's definition quoted at the head of the chapter. In limiting the color as light to medium pastel tones, it did effectively preclude more highly saturated pink/orange sapphire and eliminate gems with uneven color distribution as well as those with secondary hues additional to pink and orange or those with any sign of a brown mask. According to sapphire expert Richard W. Hughes, one of the finest

padparadscha sapphires, the 20.84-carat oval sold at Christie's in 2005 at a then auction record price of $18,000 per carat, exhibits a vivid saturation which falls outside the LMHC's definition.[239] This is an interesting development. If Hughes is correct, this would be the first time a gem of more vivid saturation of any species or variety of a more vivid would be classed as less desirable. A dark toned pink is red and a dark toned orange is brown, but a medium toned pinkish-orange to orangy pink? This is an excellent example of the consequences of a universal grading system. Uniformity comes at a price. In this case it also may present an interesting buying opportunity for the astute aficionado.

Hue and Saturation

Today it is generally accepted that the term padparadscha may be applied to delicately colored "light to medium tones of pinkish-orange, orangy pink to orange-pink hues."[240] The actual percentages of pink and orange hues cannot be defined generally. It is a question of the individual gem. Additional secondary hues—or, rather, tertiary hues such as yellow or violet —push the definition. Any combination may be acceptable. The key word is delicate. Padparadscha sapphire may exhibit either a brown or gray mask. In darker-toned gems with an orange primary hue, the darker-toned orange hue may shade into brown. This is characteristic of Umba River stones.

238 Hughes, p. 221.

239 Richard W. Hughes, *The Meaning of Words*, http://www.ruby-sapphire.com/padparadscha-sapphire.htm. Hughes is the president of Lotus Gemology Laboratory in Bangkok.

240 Ibid. Crowningshield, "Padparadscha," p. 35.

The Morgan padparadscha sapphire, currently in the American Museum of Natural History in New York, is considered a paradigm of padparadscha gems. Note the distinct yellowish secondary hue (multicolor effect) toward the center of the stone. Photo: Tino Hammid, © Gemological Institute of America.

A gray mask is more prevalent in gems with a primary pink hue.

Tone and Crystal

Light to medium tones coupled with a high degree of transparency, or good crystal, translates into a delicate effect in padparadscha sapphire. The tonal range is from thirty to sixty-five percent. Stones with tones above sixty-five percent are too robust to qualify as padparadscha. Stones with less than thirty percent tone may fit the definition, but are too pale of hue to be of interest.

Multicolor Effect

Sapphire is a dichroic stone. Since padparadscha is by definition a mixture of hues, multicolor effect is very much present in gems of this type. In many cases, multicolor effect in padparadscha sapphire may include some element of yellow. The most famous gem of this type, the one-hundred-carat Morgan padparadscha, in the collection of the American Museum of Natural History, exhibits strong multicolor effect, including a definite yellowish secondary hue. Ideally the hues should be well mixed. Sometimes an elongated oval or pear shaped stone will show a divided color that is a pinkish orange hue at the center with a darker, more visually pure orange or orangy yellow toward either end. This might be described as parti-colored or a sort of topaz effect. While this might be a plus in imperial topaz it is not in a padparadscha sapphire. A uniform pinkish orange to orangy pink is preferable to a stone that is particolored.

African Bird of Paradise

Pink-orange to red-orange stones from the gravel of Tanzania's Umba River have been marketed for years as African padparadscha sapphire. A majority, though by no means all, of the gems from this source have a marked brownish mask. Umba River stones have poorer crystal and are darker (sixty-five to eighty-five percent) in tone. The combination of hue, saturation, and tone is best described as robust rather than delicate. The true padparadscha expresses itself in a petal-soft visual vibrato. If padparadscha is the color of the lotus, it can be said that most Umba River stones find their floral soul mate in the blossom called bird of paradise (*Strelitzia*

reginae).[241]

Rocky Mountain Padparadscha

Perhaps the finest and most interesting sapphires produced from the Rock Creek and Gem Mountain deposits in Montana are the medium-toned orangy pink violet to violetish orange sapphires. A majority of these stones are tricolor with zones of orange and pink and with a secondary hue of lavender (light violet). When faceted in the brilliant style, the three colors tend to mix into a unique padparadscha-type stone which is violetish pink orange. The finest of these Rocky Mountain padparadscha are a light to medium tone (forty to sixty percent) with almost no brown mask, allowing an exceptional saturation.

All sapphire from this source is routinely heat-treated. Before treatment, Montana stones are extremely light in tone. Sixty percent of the stones heat treat to various tones of bluish green to greenish blue. In 1997 I completed a quality analysis of sapphire from Gem Mountain (part of the Rock Creek deposit outside of Phillipsburg, Montana) commissioned by the American Gem Corporation. At that time approximately two million carats of sapphire rough had been dug, heat-treated, cut, and sorted by the company. Of this entire production, less than one hundred stones were graded padparadscha; fewer than ten of these stones weighed over one carat. Heat treated or not, padparadscha is a very rare sapphire.

241 Richard W. Wise, "The Colors of Africa," *Jewelers Quarterly Magazine*, 1989, Designer Color Pages, pp. 7.

Caveat Emptor

Padparadscha sapphire is not, strictly speaking, a variety of sapphire. That is to say, there is no scientific gemological test to establish that a particular sapphire is worthy of the name. Since the first publication of this book, a committee of gemological laboratories has standardized the definition and a number of gemological laboratories have begun to use padparadscha as a descriptive term on their grading reports. Due to the gem's rarity most labs use a printed chart which defines the parameters of hue/saturation and tone they will term padparadscha. Printed media never approaches the saturation of hue possible in transparent media so the use of the term on lab reports is likely to be construed quite conservatively.

Padparadscha sapphire is also routinely heat-treated. Members of the LMHC will not use the term for heat treated or irradiated gems.

Chapter 33

Ruby

"At a carat there is a price, at two carats that price doubles, at three carats the price triples . . . at six carats there is no price"

–Jean Baptiste Tavernier, 1676

erhaps the oldest mining area on earth, the Valley of the Serpents, is about two miles wide and twenty miles long, and includes the town of Mogok. This valley is part of the Shan States, an area controlled by the Shan tribe for hundreds of years. In 1886 the British annexed the Shan States, and in 1888 mining rights were leased to a British company, Burma Ruby Mines Ltd., which worked the mines with sporadic profitability until final bankruptcy in 1922. Following World War II the Shan States were grafted onto Burma in a poorly conceived, untidy little state named the Burmese Federation.[242] In 1962, General Ne Win took power in a military coup, moved the Burmese army into the Shan States, and banned independent mining. Thus began a thirty-year period of isolation which ended only in 1990, with a partial economic reform that allowed the reopening of mining in the Mogok stone tract.

A new strike of Burma-type ruby was found in Mong Hsu, about halfway between Mogok and the Thai border, in

The 25.60 carat Sunrise Ruby. Due to size and quality, perhaps the rarest ruby on earth sold in May 2015 at Sotheby's for 30.3 million dollars. Graded by the Gubelin Lab as of Burmese origin (Mogok) and pigeon blood color. Photo: courtesy Sotheby's.

1991. Though a few fine untreated stones are found, much of the ruby from this source has a definite violet to purplish cast which can be removed by heat treatment.[243] Through the first decade of the new century, most of the available Burma-type

242 Richard W. Wise, "In Search of the Burma Stone," *Jewelers Quarterly Magazine*, 1988, Designer Color Pages, pp. 8-11.

243. Adolf Peretti et al., "Rubies From Mong Hsu," *Gems & Gemology*, Spring 1995, p. 4.

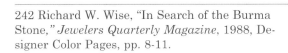

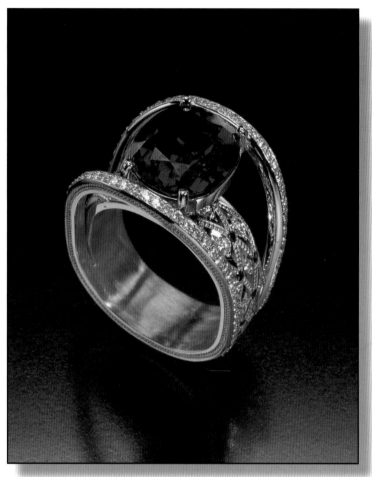

Fine ruby is where you find it. An exceptional 7.23-carat Thai-type ruby from Cambodia. Due to pronounced iron content, the stone is a slightly orangy (20%) red, but exhibits crisp diaphaneity (crystal). Some connoisseurs believe that an orangy secondary hue "frames" the red primary hue. Ring design by Zoltan David. Photo: Jeff Scovil, courtesy R. W. Wise, Goldsmiths

ruby under two carats was from this source. As of this writing (2016), the Mong Hsu deposit has slowed to a trickle and production from Mogok is sporadic at best.

Burma; The Standard

What makes Burmese ruby so important is that the Burma stone, clarity and cut being equal, is far superior to the Thai ruby, the standard bearer in the market between 1962-1989 after which Burmese goods returned to the market. Why is this so? In

a word: iron! Thailand's Chantaburi-Trat mining district is iron rich. During the ruby's formation, more than one hundred fifty million years ago, trace amounts of iron became part of the gem's chemical composition. Iron lends the Thai ruby its characteristically brownish cast and quenches the natural ultraviolet fluorescence of the ruby crystal. This means that Thai ruby is brownish and barely fluorescent under ultraviolet light.

By contrast, the pure white metamorphic marbles of the ruby mining districts of Burma are iron poor. Conditions in Burma are ideal for the formation of ruby crystals (aluminum oxide with trace amounts of chromium) which are an exceptionally vivid red. These crystals fluoresce strongly under ultraviolet light. As any diamond lover knows, ultraviolet fluorescence, while technically invisible to the naked eye, supercharges the saturation of the gem. This causes the color to radiate like the juxtaposed hues of an Op-Art painting or glow like a blue-white diamond. Ruby from Burma and adjacent areas will often fluoresce slightly in visible light. The absence of the diluting effect of iron, coupled with florescence, gives the Burma ruby a supercharged saturation.

Ruby of similar appearance has also been found elsewhere, including Pakistan, Afghanistan, Vietnam and Mozambique. However, *Burma-type* is the more appropriate term to use when discussing ruby from these sources. Stones from these areas have

small internal differences of little real concern to the collector. Vietnamese stones, for example, may be even more fluorescent than Burmese, and they characteristically have a pinker secondary hue. As always, it is the beauty, not the pedigree, of the gem that is the key issue.

This does not mean that all consumers prefer Burma-type to Thai ruby. Thai stones generally have a purer red hue. Thai ruby was virtually the only option from the early 1960s through 1989, when Burma reopened and the new mining area was discovered at Mong Hsu.[244] Some ruby lovers who have grown up with the Thai stone prefer it to Burma ruby. It is interesting to note, however, that since the re-emergence of the Burma stone, Thai rubies have almost completely disappeared from the market.

New Contenders From Africa

So long as the issue was between Thailand and Burma there was no real contest. However, since the first edition of this book appeared in 2003, several new ruby sources have been discovered on the continent of Africa. Small deposits of ruby from East Africa began titillating

Extraordinary 6.03 carat natural Winza ruby. Photo: Jeff Scovil. Cora N. Miller Collection, Peabody Museum, Yale. Courtesy R. W. Wise, Goldsmiths.

the gem world as far back as 1973. I recall seeing a large parcel of vividly hued, cabochon-cut ruby from Kenya's John Saul Mine displayed at the Tucson Gem Show in the late 1980s. These strongly fluorescent gems boasted a vivid red primary with a vibrant pink secondary hue. Unfortunately despite the vivid color, the gems were highly fractured and the strike was too small to make much of an impact on the market.

Since the first discovery of large deposits of sapphire at Ilakaka, Madagascar, the world's largest island has become a major producer of both ruby and sapphire. In 2005 the first significant deposit of ruby was found east of the capital city of Antananarivo, near Andilamena outside the village of Moromanga. A year later a more important find was made twenty-eight miles

244 In the mid- to late 1980s, I made several pilgrimages to Bangkok and along the Burma border to Mae Sot and Mae Sai in search of Burmese gems. Diligence might yield, at most, five or six fine rubies normally in sizes of less than one carat. No Burmese sapphires were available..

away at Vatomandry.[245] Moramanga has produced some large exceptionally fine stones. The latest and possibly greatest find was made in the spring of 2012, just outside the small town of Didy, about one hundred twenty miles northeast of Antananarivo. Frequently sporting a distinct orangy secondary hue somewhat similar to Winza stones, Didy rubies will hold their vivid color in all lighting environments.[246]

In 2008, a modest deposit of high quality ruby was located in central Tanzania near the town of Winza, followed by two strikes in Mozambique; the first in Niassa Provence near the village of M'awize in 2009, the second near the town of Montequez in Cabo Delgado Province. The best material comes from Namahumbire, near the town of Montepuez. In the trade it's known as Montepuez ruby.[247] As of this writing, a majority of stones in the marketplace are from a four hundred square kilometer mining concession at Montequez, a partnership between the Mozambique government and Gemfield's, a publicly traded company based in London.

Like the rubies of Thailand, gems from

A very fine 3.33-carat color ruby Montepuez ruby from Mozambique. The gem exhibited very strong ultraviolet fluorescence. Photo: Jeff Scovil. Courtesy R. W. Wise, Goldsmiths.

these East African sources all contain traces of iron. However, in most cases the percentages are significantly less than that found in the classic Thai material resulting in gems exhibiting variable weak to moderate to a high degree of fluorescence under short and long wave ultraviolet.[248] Some of the finer stones from these newer sources personally tested by the author do exhibit moderately strong to very strong ultraviolet fluorescence which compare

245 Pardieu, V., and Wise, R. W., "Ruby Boom Town, Mines of Madagascar Part 1", *Colored Stone Magazine*, 2005. http://www.rwwise.com/madagascar1.html.

246 Hughes, R. W., *Ruby & Sapphire, A Collector's Guide*, Gem & Jewelry Association of Thailand, Bangkok, 2014, p.96.

247 Richard W. Hughes, personal communication 2012.

248 Schwartz, Dietmar & Pardieu, Vincent et al, "Ruby and Sapphires from Winza Central Tanzania". *Gems & Gemology*, Winter 2008, p.332.

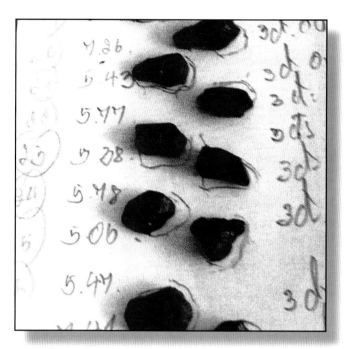

Doing the homework. Dealer's notes on a multimillion-dollar lot of Mozambique ruby rough offered for sale at a Gemfield's auction. Photo: R. W. Wise.

favorably to the best of Burma.[249]

The Winza gems were quickly mined out (2008) and, in general, exhibited hues unique and distinctly different from the classic Burmese pigeon blood color.[250] The reds are pinker with often a trace of orange and lack the purplish (blue) secondary hue characteristic of the

classic source. The best of these stones exhibit a fresh, exuberant, *primavera* hue resulting from the combination of a high degree of transparency (crystal) coupled with a relative freedom from inclusions, specifically the dense cottony concentrations of rutile crystals often found in gems from the classic source in upper Burma. As a result, the finest offerings from Winza have attracted a cult following among ruby aficionados.

The best of the Mozambique gems are of varying fluorescence and exhibit hues very much in the classic vein and, as stated, the author has seen several stones that rival, in hue, saturation the best Burma has or had to offer.[251] As of this writing (2014), production is strong and estimated to comprise eighty-five percent of all gem business being transacted in the gem capital of Bangkok.[252] Gemfield's holds auctions twice a year. At a recent auction (2015), a single parcel yielded cut gems of fine color up to twenty-five carats. Mozambique gems are offered at lower, but constantly increasing prices. A new strike near Zahanema National Park in northeastern Madagascar has come on line. According to GIA's peripatetic gemologist Vincent Pardieu, ruby from this source

---

249 The willingness of gem laboratories such as GRS, Bangkok and Gubelin, Lucerne to use the term *pigeon's blood* on grading reports hinges on the degree of ultraviolet fluorescence. As of this writing (2015) Gubelin has yet to see a stone from any source other than Burma with sufficient levels of fluorescence. GRS, the leading ruby testing lab, has used the term to describe some of the best gems from Mozambique. The author tested two very fine gems from Mozambique, a 3.00 carat oval and a 3.33 carat cushion (2013), both exhibited strong fluorescence. *Gubelin Newsletter #41*, December 17, 2013. Dr. Adolf Peretti, personal communication, 2013

250 The lack of blue or purplish secondary hue is odd given the distinctive blue zones found in many stones from this source. C.f. Peretti, A., *Contributions to Gemology No. 7, Winza Rubies Identified*, GRS, Gem Research, Swiss Lab., 2008, pps. 58-60

251 Ruby expert Richard Hughes agrees but notes that due to varying iron content Mozambique ruby can resemble fine rubies from Thailand and the best of Burma and that the presence of iron tends to shift the hue of Mozambique rubies away from purplish toward pure red. C.f. Richard W. Hughes, *Red Rain, Mozambique Ruby Pours into the Market*, 2015. http://www.lotusgemology.com/index.php/library/articles/316-red-rain-mozambique-ruby-pours-into-the-market.

252 Joseph Belmont, personal communication 2014.

has characteristics similar to ruby from Mozambique. With the decline of ruby production from Burma, the gap in price between African and Burmese gems will continue to narrow though Burmese ruby will always command a premium.

Hue or What Color is the Pigeon's Blood?

The Traditional View

Pigeon's Blood is one of those terms that has been tossed around for years, but rarely defined. The term first appeared in English in the year 1839.[253] Over the years, this writer has heard at least four definitions, none of which agreed. In 2012 an unpublished manuscript written in 1934 by the noted gemologist J. F. Halford-Watkins containing detailed essays on quality assessment in ruby and sapphire was edited and published in 2012.[254] One of the world's great experts, Halford-Watkins was a director of Burma Ruby Mines. Ltd. He lived in the mining districts working as a company valuer of ruby and sapphire for several decades until his death in 1937.

According to this author the ruby dealers of Mogok neither used nor understood the term pigeon's blood.

Halford-Watkins maintains that the term is of great antiquity and that it is of Chinese not Burmese origin. Pigeon's blood or number one color is, according to Halford-Watkins, a pure red with no trace of any modifying, specifically blue, secondary hue. Under the dichroscope, a pigeon's blood stone will exhibit two identical pure red rays.[255] A tiny bit of blue qualifies the stone as number two color and a bit more as number three.

Hue

Although a visually pure red is the most desirable, pure hues are seldom encountered in nature; thus when describing gemstones in this volume we speak of primary and secondary and occasionally tertiary hues. Obviously, with ruby, red is the primary hue; pink, orange, purple and violet are the possible secondary hues. Burma-type ruby tends toward the pink. Experts disagree on which of the secondary colors is preferable. Most, of course, prefer a true-red red, others believe a bit of orange frames and pumps up the saturation of the red hue. This is further complicated by the fact that ruby may sometimes exhibit more than one secondary hue. A pinkish-orangy-red is not beyond possibility.

After some thought, this writer agrees with the late Mr. Halford-Watkins. The blue secondary hue normally reads as purple or purplish (red + blue = purple) in finer, medium-dark toned gems. Purple is a modified spectral hue located in the visible

253 Bram Hertz, *A Catalogue of the Collection of Pearls and Precious Stones Formed by Henry Philip Hope, Esq.* (1839) Primary Source, Rare Books Club, POD. 2015, p. Oriental Rubies Drawer – 4a, p.16. Hertz also uses the term *couleur de sang de boeuf* (beef blood) to describe color in ruby.

254 J. F. Halford-Watkins., Hughes, R. W., editor. *The Book of Ruby and Sapphire*, RWH Publishing, 2012, p.56. http://www.ruby-sapphire.com/halford-watkins-book-of-ruby-and-sapphire.htm

255 The dichroscope is a gemological instrument which visually separates the two rays in a dichroic gemstone. For an explanation of dichroism, see glossary.

spectrum half way between red and blue. Purple reaches its maximum saturation, its gamut limit, at about sixty percent tone. That makes it potentially the darkest secondary hue. Orange achieves maximum saturation at a light thirty percent tone and pink is, by definition, a light toned red.

Red reaches its optimum saturation at eighty percent tone; thus, a vivid orange or pink secondary hue may be said to visually dilute the primary red, perhaps reducing the tone below its potential gamut limit. At darker tones, pink becomes red and orange morphs into brown. Therefore, a slightly purplish red is preferable particularly when it is to be set in yellow gold. Given the plethora of new sources, however, the aficionado should cultivate a flexible attitude toward ruby color. There are some particularly fine Winza and Mozambique gems available in the market.[256] Other factors being equal, a medium dark stone with fifteen percent or less of any combination of secondary hues is of fine color and should be treated with great respect.

The aficionado must be cautious when viewing rubies, particularly when the stones are presented in parcels of the traditional saffron-yellow paper or as in Mogok on polished copper plates. Many years ago I was told that this was because saffron is considered sacred by Buddhists and is the proper color on which to display the King of Gems. While the statement is partly true, Buddhist monk's robes are indeed

5.04 carat ruby from Didy, Madagascar. Photo: Wimon Manorotkul, courtesy Lotus Gemology and Crown Color.

yellow; the real reason is that yellow is the compliment of blue and complimentary colors cancel each other. Yellow will mask a bluish secondary hue, making the gems so displayed appear to be a purer red hue.

Saturation

Ruby has been described as "a gem of barbaric splendor". This is largely due to the strong ultraviolet fluorescence typical of the gem. Although it is true that the human eye cannot see light in the ultraviolet range, the fluorescence at times laps over into the visual spectrum, in effect supercharging the saturation of the color. This is particularly true of Burma-type ruby from Vietnam, which often glows in natural daylight. Other than the finest rubies from Mozambique, few other gem species achieve the level of saturation of the Burma-type stone.[257] In ruby the normal mask or

---

256 As of this writing, there is little of note from any of these sources being sold in the Bangkok market. Joseph Belmont, personal communication, 2015.

257 The only other gems with comparable saturation are the tourmalines of Paraíba, Brazil, spessartite garnet from Namibia, and most recently the hot pink spinels from Mahenge, Tanzania, and the so-called Jedi spinels of upper Burma.

saturation modifiers are gray and brown with brown predominant. Due to iron content, most ruby from Madagascar and some gems from Mozambique are distinctly brownish. If the hue of a Burma-type ruby appears dull, a gray secondary (mask) is the most likely culprit.

Tone

Ruby exists in a tonal range between sixty and eighty-five percent tone. Since pink is actually a red of pale saturation and light tone, gems of less than fifty percent tone are by definition pink sapphire. The color red reaches its optimum saturation—what color scientists describe as its gamut limit—between seventy-five and eighty percent tone, the ideal tone in ruby. Gems with tonal values below sixty percent appear washed out; those with tones in excess of eighty percent appear over dark (overcolor) and lose transparency.

Crystal

As stated, heat treatment tends to clarify and improve the transparency of ruby. Burma ruby will often appear dense due to high concentrations of inclusions and relatively weak transparency. This effect is often mitigated by the visual impact of fluorescence, which imparts an almost searing brilliance.

Multicolor Effect

Like sapphire, ruby is a primary color gem. The closer ruby comes to exhibiting a visually pure red, the better and more desirable it is. However, like all gems, ruby also can display a multicolor effect. Think of the face-up gem as a patchwork quilt:

each facet is a distinct swatch of color and may exhibit variations in the key color.

Multicolor effect has several causes (see the discussion of the multicolor effect in Chapter 4) Ruby is dichroic; for example., light entering the gem divides into two rays, one purplish red, the other orangy red. Also, ruby is often zoned, with colorless zones juxtaposed against zones of color. Light rays passing through colorless zones lose color. In addition, the pavilion facets of the gem cause a light ray entering the stone to reflect at least twice within the stone, absorbing color as it goes. Add to this the fact that some facets may appear darker in tone, some lighter, the effect of dichroism, and you have an idea why the face-up scene may be less than uniform.

In ruby and sapphire, a negative type of multicolor effect has been traditionally called *bleeding*, a loss of saturation and tone when the stone is shifted from natural to incandescent lighting. At lighter tones, red becomes paler (less saturated) and pinkish. Bleeding in ruby may be weak, moderate, or strong. The more apparent or stronger it is, the greater the fault. With the rise of the LED and the outlawing of incandescent bulbs in the United States and Europe effectively altering the night-time viewing environment, this is no longer of much consequence. Unless you are planning to live by candlelight, aficionados should concentrate on the daylight-viewing environment.

Ruby Versus Pink Sapphire

Is it red or is it pink? This is a difficult question, since pink is not a distinct hue;

Is it pink or red? This 5.88 carat pink sapphire exhibits dramatic multicolor effect. The upper half reads as red, the lower pink. Photo: Jeff Scovil, courtesy of Manavi International.

The connoisseur should pay attention to the primary hue. Is it pink or is it red? Since pink is a lighter-toned red, the assumption might be that a pink stone will be light-toned and pale in saturation. This is not always the case because the normal secondary hues, purple and orange, achieve their gamut limits, their maximum saturations, at darker tones than does pink. The situation is particularly difficult if the secondary hue is purple. Purple achieves its gamut limit at about sixty percent tone and, of course, may be even darker. A pink gem with a strong secondary of purple in tones above sixty percent may appear quite dark in tone. Stones like this are often described as magenta or fuchsia pink. Purple pink gems in darker tones will often read as red to the inexperienced eye.

it is a light-toned pale red. In fact, the distinction between ruby and pink sapphire is of modern vintage, and has sparked a lot of debate within dealer circles. Some experts want to call all red corundum ruby, whatever the tone; others prefer to maintain the distinction.

The issue has meaning because the market recognizes a distinct difference between the prices of ruby and pink sapphire. The latter has a price structure close to that of blue sapphire. The former brings a substantially higher price which escalates towards the stratosphere as size tops two carats.

Hue: Pink Sapphire

A very fine 11.91-carat pink sapphire with a distinct purplish secondary hue. Photo: Allen Kleiman, courtesy, A. Kleiman & Co.

The distinction is real; the term has meaning that we all understand. But, where does pink end and red begin? This is a question that can only be answered by comparing one stone to another, the laborious path that all true aficionados who hope to achieve mastery must tread. Compare, compare, compare!

Clarity

Eye-flawless ruby, particularly unheated ruby, is extraordinarily rare and stones of this description will command high prices. In some cases, small concentrations of broken straw-like rutile inclusions, even when visible, may be beneficial, as they act to break up and scatter the light throughout the stone, reducing extinction. The question is how prominent or disturbing the inclusions are, and how much they detract from the beauty of the stone. This, finally, is a judgment call. Because of the gem's extreme rarity, the collector-connoisseur may find that otherwise beautiful stones with a few eye-visible inclusions are acceptable particularly in larger stones.

Texture

Color zoning or texture is another possible fault in ruby. Due to characteristic zoning, ruby may exhibit a brushed effect face up. A fine gem should exhibit a uniform smooth texture-less appearance when viewed in the face-up position.

Heat Treatment

No one is quite sure how long heat-enhanced ruby and pink sapphires have been in the market — perhaps a thousand years, probably much longer.[258] Paleolithic man heat treated flint. Low temperature heat treatment was reported in India as early as 2000 BC[259] and the 11th century polymath Al Beruni describes a technology in use that differs little from techniques in use today.[260] Beginning in the 1970s precision technology became available allowing treaters to heat stones to precise temperatures in carefully controlled environments, and heat treatment began to be practiced almost universally.

Over ninety-five percent of rubies currently on the market are heat enhanced. Almost one hundred percent of those from the newer Burmese source at Mong Hsu were heat-treated and a good proportion of gems from Mozambique and Madagascar are treated as well. Generally speaking, a heat-enhanced ruby will have a better visual appearance than an unheated stone. Heat tends to enhance the color, clarity, and crystal of ruby. Some of the stones from Mong Hsu, when heated, will assume a particularly pure red hue with very little or no secondary hue.[261] Pink sapphire is also routinely heat treated. Natural unenhanced stones will command premiums of fifty to one hundred percent and more over comparable heated stones, and that percentage increases with size.

The Rarity Factor

258 Tagore, *Mani Mala*, vol. 1, pp. 243, 455.

259 Nassau, *Gemstone Enhancement*, p. 25.

260 Al-Beruni (973-1052) *The Book Most Comprehensive In Knowledge on Precious Stones*, 1040-1048 AD, Adam Publishers and Distributors, New Delhi, 2006, p. 37.

261 C.R. Beasley, personal communication, 1998.

As the epigraph at the head of this chapter suggests, fine ruby is rare in any size, and prices begin to increase geometrically at two carats. At five carats rarity and price accelerates into the stratosphere, and at ten carats a fine gem is almost priceless. The new source in Mozambique does produce gems in a full range of sizes up to twenty-five carats and more. However, much of the Mozambique rough is quite flat and tabular, and gem dealers unwilling to sacrifice weight, often cut large stones which are often flat, windowed and lacking in brilliance.

All areas producing ruby also produce pink sapphire. Pink Burma-type sapphire will command a distinct premium and is similarly rare in larger sizes. However, larger pink sapphire, particularly from Sri Lanka, which produces almost no ruby, make larger pink corundum far more available than ruby. Prices for Sri Lankan and other non-Burma pinks increase at five, seven and ten carats but far less precipitously than ruby.

Jade dealers, Kowloon. Photo: R. W. Wise

JADE

Chu Shen Hi squatted stiffly in the bare red dirt in the back yard of his thatched hut. Chu was a short old man, thin and wiry. His only garment a pair of ragged shorts. He had the look of a scrawny chicken that had been left too long in the oven, with skin like wrinkled parchment stretched over his bony frame. Chu was completely bald, his face shrunken against sharp cheekbones. His deepset eyes, dark and lively, stood in stark contrast to a complexion the color of bisqued clay.

Overhead the dense jungle growth all but blotted out the sun. Only a few errant rays of diffused light managed to slip through the tangled green canopy. It was hot as only the jungle can be, the air steamy, the smell dank. It was near the end of the monsoon and despite the heat, Chu felt each one of his eighty plus years.

Slowly and methodically, Chu dropped the boulder into the battered iron washtub. Brows knit in concentration; he watched the rock's descent through the water and listened to the dull clunk as it hit bottom. "My knees hurt and this proves nothing," he thought. Frowning he lifted the boulder from the tub and hefting it up on his shoulder like a shot-putter he rested it against his right cheek. Pausing a moment to savor the rock's cool surface, Chu shifted position a quarter turn right, reached down and plucked a rusted steel bolt from the packed earth. Working methodically and breathing through his nose deep into his abdomen to maintain his mental equilibrium, Chu struck the bolt against the surface of the rock. Rotating and striking he listened for any discrepancy or changes in tone, which would suggest a crack or an impurity within the boulder. He could hear his father's words; "Listen with the ear of the heart." Chu rotated the boulder, tapped and listened until his shoulder began to numb and he could feel a dull ache radiating down his upper arm. To his practiced ear the tone was as clean and resonant as a temple bell. Finally he finished the examination and placing the boulder on the ground he absently rubbed his aching shoulder. Chu's reputation was something of a legend along the border for he was believed to have a golden hand. Those possessing the golden hand were believed to have the intuition, the ability to somehow see past the skin into the interior of a jade boulder. This amused Chu himself for he knew that his so-called intuition was sometimes a matter of technique but more often a question of luck.

At one end of the egg-shaped boulder a small section, about two inches square, of the outer skin

that covered the rock like the rind of an orange had been ground away and polished. This section called a "maw" or eye had created a window revealing a small portion of the interior of the rock, showing the color of the jade. Chu spat and rubbed saliva into the maw. The color was a uniform vivid emerald green. A thin layer of yellow rind, oxidation, that the Chinese call "mist" surrounded the green, identifying the boulder as river jade. Lifting the boulder to

Rebekah and Richard Wise pricing jade at the jade market, Quangzhou, China. Photo: Richard W. Hughes

take advantage of the light, Chu studied the color.

It is well known among gem dealers that the mind has a poor memory for color. Many years ago Chu's father taught him this fact and taught him also that color memory could be reinforced by reference to the sense of taste. Each time Chu examined a piece of jade he tried to relate the color to a certain taste or food. Thus, he both saw and tasted each color. In this case his eye and tongue agreed, this jade was what the old men called khem-khem, a Thai word literally translated as "salty," what the Burmese called yay kyauk, the finest imperial green.

Unfortunately Jade is a dense mineral and, in its finest qualities only semi-transparent. The color revealed by the maw might be only surface or it might run through the entire boulder. Translucency, or shui fen – literally "water content"– the degree to which light penetrates the stone, together with texture – zhi di – are two of the most important criteria in evaluating jade. The skin of the boulder made it impossible to see anything other than two square inches of polished stone. The trick was to judge the depth, which the Chinese measured in fen, about 3.6mm each. Chu had examined the boulder with minute attention, attempting to measure the fen and looking for show points, places where the interior of the rock showed though the skin. There were none and judging translucency with such a small section of stone was of very limited value.

Purchasing such a boulder was a game of chance, one played for very high stakes in which all the cards but one were dealt face down. Custom forbade Chu from cutting another maw or altering the rock in any way. To do so would oblige him to pay the full asking price. So far only one cut had been made and the boulder might change hands several times before another cut was made.

Why didn't the owner simply cut up the stone? That would ruin the game. There was only one card sitting face up on the table but that card was the ace of spades. The asking price was, of course, astronomical but if the color ran true, it was nothing in comparison to the value of the jade.

Imperial jadeite pendant. Photo: Sky Hall, courtesy Mason Kay.

Chapter 34

Jadeite, the Enigmatic Gem

"Green jade is rare and deep green jade even rarer. It is a treasure among treasures. The archer's ring carved from this jade is more lovely than the dark-green leak, its color is as bright as bamboo shoots."

–Emperor Qianlong,1785

5.34 carat imperial jadeite cabochon set in a man's ring. Photo: Jeff Scovil. Courtesy: The Edward Arthur Metzger Gem Collection, Bassett, W. A. and Skalwold, E. A.

The term jade refers to two distinct minerals, jadeite and nephrite. Jadeite is a member of the pyroxene mineralogical group, and nephrite is a member of the amphibole group. Nephrite, also known as Chinese jade, was the only known type of jade in Asia prior to the discovery of jadeite, in the late 18th century, in North Central Burma between the Uru and Chindwin Rivers.[262] Nephrite has been known and appreciated by the Chinese principally as a carving material for several thousand years. Though there exists a hierarchy of quality in nephrite, the artistic quality of the carving itself is central to the appreciation of a nephrite object. Only jadeite fits the definition of a gemstone, a material valued principally for its visual appearance[263].

Originally the two, alike enough in visual appearance and working

---

262 Howard, Kim Be, *Jadeite*, Canadian Institute of Gemology online pdf. pps 1-4.

263 Jadeite is often carved into traditional shapes. Its value, however, is based primarily on the quality of the material.

characteristics, were considered to be the same mineral until 1863 when French mineralogist A. A. Damour (1808-1902) determined their differing chemical composition and proposed the name jadeite.[264]

It is fair to say that although Western science failed to make the distinction until the late 19th century, jade carvers in China undoubtedly understood the difference. As American jade expert Richard Gump noted, the vivid saturation of jadeite contrasts with "the soapy almost aged-looking hues of nephrite." Jadeite is translucent, in gemological terms, semi-transparent. "In general," Gump suggests, "...the colors of jadeite tend toward vividness...while the colors of nephrite are greasier, denser and heavier. Nephrite looks soft, jadeite hard."[265]

Jadeite was also known to the Mesoamerican cultures of the New World, the Olmec, Toltec, Aztec and Maya from the 11th century BC. These cultures used jadeite much as the Chinese used nephrite, principally to carve aesthetic and votive objects, but they also recognized the extreme toughness as utilitarian objects as well.

The English word comes from the Spanish: *piedra de ijada*, or colic

stone. When, in an attempt to dissuade the conquistador from a trek inland, Montezuma sent Cortez a sumptuous gift including large pieces of jade, he did not realize that he was simply whetting the Spaniard's appetite for conquest. Bernardino de Sahagún (1499-1590), a Franciscan missionary working in Mexico, compared jadeite to the highly saturated emerald green of the Quetzal feather ("Piedra trasparente...muy verde"). The good padre further describes the stone as having the transparency and density of obsidian, a semi-translucent form of volcanic glass. Given the time of his writing (1575), he had probably not seen the true emeralds of Columbia, which had only just been discovered. He compared four gems in total to emerald. He was perhaps thinking of gray-green sapphire, which prior to the Colombian discovery, was known as oriental emerald.[266] The fact is, Mesoamerican emerald green jadeite from Central America is, as of this writing, more hyperbole than fact.

The source of Mesoamerican jadeite was lost until its recent rediscovery in Guatemala's Motagua River Valley.[267] Although there is some attractive semi-translucent blue jadeite found in Guatemala, Mesoamerican jadeite lacks the purity, saturation of hue and transparency of the Burmese. From the perspective of beauty, it simply does not compare. The

264 Damour, A. Alexis. *Analayse du jade Oriental réunion de cette substance à la Tremolite, Annales de Chimie et de Physique*, 3rd ser., Vol. 17, 1846, pp.469-474. Prior to the advent of modern gemology, there were many materials that were considered jade by the Chinese.

265 Gump, Richard, *Jade, Stone of Heaven*, Doubleday, N. Y., 1962, p.187.

266 Tavernier, Jean Baptiste, *Travels In India* (1689), Vol II, p.363.

267 Ward, Fred, "World Jade Resources, "Arts of Asia, vol. 29, no. 1, 1999, pp. 68-71.

discussion here will focus entirely on Burmese jadeite[268].

Jadeite, a silicate of sodium and aluminum, is, like nephrite, a rock. This means that the mineral composition of jadeite will vary. Unlike a majority of the gem varieties discussed in this book, which are mono-crystalline, jadeite gems are not cut from a single crystal; jadeite is a polycrystalline aggregate, an interlocking mosaic of tiny granular to fibrous crystals. Due to its internal structure, jadeite can never be completely transparent; the best jadeite can only be described as translucent or semi-transparent.

Jadeite, called *fei cui* or "Plumage of the Kingfisher" by the Chinese, is, aesthetically speaking, little understood in the West. Jadeite connoisseurship has been, almost exclusively, the province of the Chinese. Few Western writers have tackled the subject and those who have, have mostly attempted to translate and interpret the evaluation criteria used by the Chinese.[269] After some years of study, this writer has concluded that this approach is needlessly

complex and confusing.

The main thesis of this book is that beauty is the ultimate criterion in the evaluation of gemstones and that the basic criteria of judgment, the Four Cs of connoisseurship are universal. If this is so, it should be possible to evaluate jade within this context. Why should jade be any less (sic) scrutable than gems such as ruby and sapphire, the appreciation of which can be traced to Asian cultures no less venerable than the Chinese?

There is no question that the subject is fairly dripping with cultural associations. For the Chinese, jade is the *Stone of Heaven*, the intermediary between heaven and earth. Nevertheless if jade is to be appreciated in the West, it must be comprehensible within a Western cultural context. With the addition of a single criterion, texture, jadeite fits quite neatly into the author's Four Cs of connoisseurship (color, cut, clarity and crystal), the system used to evaluate all other gem varieties discussed in this book. It is, in fact, one of the few gems where transparency is always discussed as an overt grading criterion.

Jadeite occurs in many colors and as with all gemstones, color is king. From the point of view of desirability and price, there is green jadeite and then there is everything

268 Miller, Anna M. "Mesoamerican Jade", *Lapidary Journal*, vol. 54, No. 11, 2001, pp.29-31.See also Morley, Sylvanus G., *The Ancient Maya*, Stanford University Press, 1956.

269 Christie's Hong Kong, A *Chinese Approach To Jade,* Christie's Magnificent Jadeite Jewelry Auction Catalog, October 30, 1995, pps. 18-23.

Mdivani-Hutton Necklace, twenty-seven perfectly matched beads graduating from 15.4-19.20mm: Holder of the current record for highest price paid at auction April 7, 2014, 27.44 million dollars. Close-up view of center beads. Note the purity of hue, translucency and the wet-waxy look across the surface. Photo courtesy Sotheby's Hong Kong.

else. Speaking of value, though something of a simplification, the hues of jadeite, generally speaking, break down into the following hierarchy of value with Imperial green at its apex:

The Hues of Jadeite in Order of Value:

1. Green	5. Yellow
2. Lavender	6. Black
3. Ice jade (colorless)	7. White

Jadeite: Imperial Jade (*Lâo Keng Zhong*)

The color of jadeite, like other green gems, can be graded by evaluating the three elements of color: hue, saturation and tone. Like emerald, green jadeite derives its color from Chromium so it is not surprising that green jade is often compared to emerald and, like emerald, the finest color, known as *imperial* jade or sometimes *old mine jade*, can be described as a medium-dark toned, vivid visually pure green hue similar to that of fresh chives,[270] or the finest emerald. Secondary hues in jadeite incline towards yellow and percentages of secondary hue which are less than ten percent are almost impossible to discern. Unlike emerald, the market prefers a slightly darker toned pure green balanced between yellow and blue.[271] If one secondary hue is dominant, it is fair to say that a smidgen bluer is

270 Mok, Dominic., *Fei Cui Jadeite Jade Smart Grading System*, Asian Gemological Institute and Laboratory Ltd.

271 In most gemstones considered in this book, a blue secondary hue is preferred to yellow. U Nyan Thin writing in his (sic) *Myanma Jade* mentions a slightly yellowish green as the most desirable texture. At another point in the text, he discusses transparency and suggests that a fine grained texture with a hint of blue is the most desirable. Add the blue to the slightly yellowish hue and you have a balanced green. See, Thin, U Nyan, *Myanma Jade*, Mandalay Gem Association, 2002, pp. 71-72.

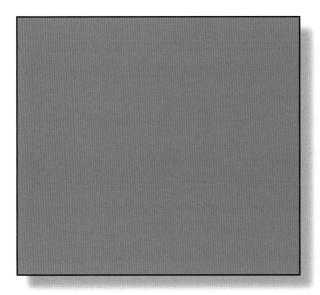

Comparison: Extreme close-up of the center bead of the Mdivani-Hutton Necklace (above) shows a slightly yellowish green hue when contrasted with a close-up of the key color of a fine Colombian emerald.

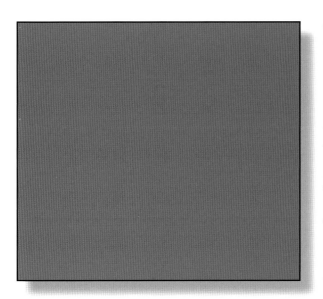

Comparison: Extreme close-up of the key color of a very fine Colombian emerald. Note the distinctive blue-ish secondary hue that contrasts sharply the yellowish secondary hue of the jade.

preferred over a bit more yellow. However, as the comparison shows, the term emerald green should be used quite gingerly when applied to jadeite.

Saturation and Tone

Saturation refers to the vividness of the hue. The Chinese term is *cui*. Tone refers to the darkness or lightness of the hue and the Chinese word is *nong*. [272] These are both intellectual abstractions that may be separately discussed but are inter-related in the physical object. The hue green achieves its optimum saturation, its gamut limit at approximately seventy-five percent tone (see Chapter 3).

In Chinese connoisseurship, highly saturated green imperial quality gems are termed *sharp*, *bright* or *hot* in the same sense as we use the English term; hot pink. This is also true of other colors. As with almost all gemstones, the brighter, more saturated the hue, the better the color is judged to be. The Chinese use taste to reinforce sight as an aid to color memory. (see Chapter 4). The color of a highly saturated example of green jadeite is referred to by the Chinese as *laijao or chili pepper*.[273]

Color Distribution

The color of a given gem may be quite unevenly distributed with vivid hued swathes of color alternating with

---

272 Hughes, Richard, et al., "Burmese Jade The Inscrutable Gem, Part II, Jadeite Trading, Grading and Identification," *Gems & Gemology*, The Gemological Institute of America, Spring 2000. Hughes translates *nong,* I think incorrectly, as saturation and describes the best color as medium rather than medium-dark in tone.

273 Christie's, ibid, p. 21.

lighter toned sections or white or colorless sections. This may be thought of as analogous to multi-color effect in the face up mosaic of a faceted gem. In either case, the more even the distribution the finer the gem.

Unevenly colored gems featuring more than one chromatic color such as green and orange (red) can also be interesting and valuable. Gems with patches of three or more chromatic hues are quite rare and all other factors being equal, will command a price relative to the saturation and tone of the juxtaposed hues together with the pattern.

Crystal

Crystal (transparency) is used overtly as a grading criterion in all Chinese systems of jadeite evaluation.[274] This is unusual as most modern systems of gemstone evaluation ignore this obvious criterion. The Chinese word is *zhong*, a term which includes texture.[275] In the old days it was said that crystal without color was of little value, but with the discovery some twenty years past of quantities of colorless ice jade sufficient to make a market, crystal has become important an increasingly criterion in the evaluation of colored jadeite. In the evaluation of colorless ice jade, crystal is, in

fact, the preeminent determinant of value. Some systems of jade evaluation refer to jades with high translucency as "glass jade" or just "glassy."

It is interesting to note that jade expert Eric J. Hoffman describes the transparency of the material as follows: "The most highly sought jadeite is the famous "Imperial Green" whose clear, flawless transparency and pure emerald green color give it the appearance of a drop of green oil."[276] This description precisely mirrors the terminology emerald experts use to describe the finest color/crystal in what is called *old Mine emerald* (chapter 11)."

Internal Glow

Some cabochons appear to glow from within. Jadeite is an aggregate. If the crystals are prismatic and line up perfectly, the gem will glow specifically around the edges of the cabochon when it is viewed perpendicular to a light source.[277] Translucency, fineness of grain and the fashioning of the piece into the preferred double convex cabochon may also play a part in producing or enhancing this characteristic the Chinese refer to as *Ying*. Ying is a subtle effect often visible at the edges of the cabochon and its presence will materially increase the beauty and value of a given gem.

Due to its dark tone, the crystal of gems of the finest color can also be described as slightly blackish, ashy or sooty, like

274 Ng and Mok, as well as the anonymous author of A Chinese Approach to Jade (op. cit. Christies, 1995).. See also Richard W Hughes, et al, Burmese Jadeite http://www.palagems.com/burma_jade_pt2.htm. All cite transparency as a specific grading criterion.

275 Christies, *ibid.* pps.18-19.*Zhong* sub-divides into two qualities; *shui fen* or water content and *zhudi* which is texture as we understand the term. *Zhudi* can be either fine or rough. In the final analysis the closer the grain, the greater the degree of transparency (crystal).

276 Hoffman, Eric J., *Jade, Symbol of Beauty, Nobility and Perfection*, http://hoffmanjade.com/Adornment_Jade.pdf

277 Dominic Mok, personal communication, 2012.

Hawking "jade" bangle at the Jade Market in Guangzhou, China? Maybe! If natural, the high degree of translucency would mark them as of fairly high quality. More likely these bracelets are dyed chalcedony. Photo: R. W. Wise.

the black soot seen on the chimney of a kerosene lamp.

Texture

Due to its fibrous nature, jadeite's crystal is never as pure or transparent as that of emerald. Jadeite is composed of prismatic to feathery crystals between ten microns and one centimeter in size. Thus, jadeite will always exhibit some texture and is, at best, only semi-transparent. The extent of the texture is dependent upon the size and compactness of the tiny crystals of which the gem is composed.

The texture may be even or uneven, fine or coarse depending upon the tightness and consistency of the grain. The finer and more even (less apparent) the texture, and the higher the degree of transparency, the finer the gem. The fineness and evenness of the texture make the difference between jade that is water clear and jade that is honey clear. The Chinese consider these two sub-categories transparency (*sui fen*) and

texture (*zhudi*) separately but they can be considered attributes of crystal in the evaluation of the finished gem.

Clarity

Like all gem varieties, jadeite may have inclusions. Jade, like all colored gemstones, is evaluated by eye using the visual standard. Any eye visible inclusions are a negative. Fractures, even completely internal fractures are particularly undesirable. The extent to which inclusions mitigate value is dependent on the effect each has on the beauty of the given gem. The highest standard is perfection, a fine-grained semi-transparent visually flawless gem. White cottony inclusions which inhibit transparency are a negative as is any sort of opaque spotting. Long translucent silky inclusions are less problematic as they may be beautiful in themselves. Inclusions of vivid color against a lighter toned body color may actually enhance the gem's beauty and therefore value.[278].

Cut

The Chinese prefer oval cabochons with a moderate dome and a convex base known as double cabochons. The perfect ratio is 1x2x3 (depth, width, length). Though traditional carved jadeites such as *pi's*, a roughly donut shape with a central hole, and gourds command a lower price. Carved surfaces usually indicate that inferior material has been removed and smooth

---

278 Jeff Mason, personal communication, 2015.

surfaces are preferred.[279] There are some exceptions, but they are rare.[280]

The Wet Look

A few decades ago, a leading hair product manufacture declared: "The wet-head is dead." In jade connoisseurship, however, the wet look lives on. Connoisseurs refer to a fine piece of jade as looking wet. This has to do with the crystal and the quality of the polish, which in turn has to do with the quality of the material. Finer jade is closer grained and takes a higher, more uniform polish and will look as if it is full of water. The Chinese often use taste to reinforce color memory. Visualize biting into a slice of cool honeydew melon. Poorer quality material may show what is described as an orange peel like texture under 10x magnification. Luster can be judged by viewing the sharpness of the reflection of the light source; the more distinctly it mirrors the light source the higher the luster.

Prior to the introduction of diamond grits, jade was polished using its own dust, and jade cabochons had a distinctly dry waxy luster. A poorly finished cabochon will have a dull lifeless surface with little surface reflection (luster); however, absent quality, a well-polished stone may still appear dry and lifeless.

Apple Green Jade
(Xim Keng Zhong)

The second color of jade is a lighter, brighter hue, a fifty to seventy percent toned vivid, very slightly yellowish green. The secondary yellow hue is less than ten percent and barely discernible to the practiced eye. The tone is lighter than that of the imperial hue. The Chinese frequently use the color of indigenous fruits and vegetables to discuss and describe the color, so searching for a comparison easily understood in the West, the hue/tone is analogous to that of a Granny Smith apple without the ashy look of the darker toned Imperial jade.

Flower Green Jadeite
(Huã Qing Zhong)

Flower green is an uneven mottled gem with patches of both medium dark-toned vivid and light green hues with perhaps some white. The Chinese consider jades of this description to be number three in quality.

Jade is a Nightstone:

Jadeite like most gemstones will change in differing light sources. Incandescent light brings out the saturation and the blue secondary hue. Fluorescent light suppresses the saturation and daylight increases the translucency. Old time dealers prefer a sixty-watt standard incandescent light bulb,[281] though these days finding one may be a challenge. As with all gemstones discussed in this volume, jadeite should be viewed and evaluated in all possible light sources.

The Other Colors of Jadeite

Lavender Jade

279 *ibid.* 2012.

280 Richard W. Hughes, personal communication, 2012.

281 Christies, *ibid.*, p.22.

Exceptional lavender jade cabochon with purple primary and little or no secondary hue next to an exsceptional imperial jade cabochon Photo: courtesy Mason Kay, Inc.

Lavender jadeite has a purple primary hue. The possible secondary hues are red, pink and blue. Pink is the preferred secondary hue because a little pink adds piquancy to the hue. The normal saturation modifier or mask found in lavender jade will be gray. The finest lavender will usually be courser grained and less transparent than green jadeite. The finest lavender is a medium rather than the medium-dark tone preferred in imperial green. Transparency tends to reduce the saturation in pastel hued gems.[282]

Blue Jade

Subtract the pink component and add a gray mask and the result is what is known as blue jade.[283]

Ice & Melon Jade

Both jadeite and nephrite are colorless in their pure form. In ancient times,

colorless nephrite jade was highly valued. Colorless jadeite was not well known outside China, but the discovery in the late 20th century of significant quantities of pure ice jadeite has stimulated the market's interest. The finest ice jade is totally colorless. Ice Jade is valued chiefly

Ice jade Pi photographed through a jewelry store window. Note the icy wet look of the pendant. Jade market, Guangzhou, China. Photo: Richard W. Hughes.

---

282 Jeff Mason, personal communication, 2012.

283 *ibid,* 2015.

for its transparency and freedom from internal inclusions. Place a jade piece on a newspaper. Allowing for distortion due to the shape of the stone, the text should be easily readable beneath a piece of ice jade. As translucency decreases, so does value.

Much of what is considered ice jade is slightly milky, therefore only semi-transparent and more or less the (sic) color of non-fat milk. The best ice jade will also exhibit the transparent, wet internal glow discussed above.

Occasionally the aficionado will see a piece of jade which is almost colorless, but retains just the slightest bit of green, lavender or another color. This is known as melon jade. A bite of cool honeydew or muskmelon will aid and reinforce the understanding. Melon jade will command a premium above ice jade.

Red Jade

Hue

Red jade is colored by percolating ground water rich in iron oxide and is limited to the first half inch of the skin of a jade boulder. Red jade can be, but is rarely red; it is orange to reddish orange to orangy red.

The confusion in nomenclature may be the result of the fact that in ancient times, red nephrite was described as being the color of a cock'scomb, an appendage which is considerably more to the red than red

jadeite depending, of course, upon which cock is being discussed.

Saturation/Tone

Orange is a big color (hue), bold and brassy. Thus, red jade can be highly saturated. Orange reaches its optimum tone at between thirty and forty percent tone. Fine red jade will be light and bright.

Crystal/Texture

Red jade may be every bit as finely grained as green jade and every bit as transparent, and as with green jade, its desirability increases as the gem approaches semi-transparency. [Most red jade has poor texture.]

Yellow Jade

Hue

Yellow jade like red jade has a strong orange component. It probably should be referred to as yellowish orange.

Black Jade

Hue

What we call black jade is, generally speaking, an extremely dark (95%) toned green. Most black jade jewelry found in the market is nephrite.

Saturation/Tone

Tonally black jade is very dark, 90% plus green. It is fair to say that the higher the tonal value, the closer the hue gets to 100% tone (pure black), the better. At this

tonal level the green hue is over-color and dull. Stones at this tonal level are essentially opaque and the beauty of the gem is more a function of its surface luster. The surface of the stone should appear wet rather than dry and lifeless.

Mutton fat jade carvings. Photo: Jeff Scovil. Courtesy: The Edward Arthur Metzger Gem Collection, Bassett, W. A. and Skalwold, E.A.

White Jade: Differing Standards

"On examining jades, one should regard those white in colour as the best, though yellow and blue jades[284] *are not without value…Warm and mellow, one can feel when one fingers them, as if some unearthly stream is flowing into one's hand."*

—Wang Tso, 1388

The preceeding quote, taken from the *Ko Ku Yao Lun*, the oldest known treatise on Chinese connoisseurship, refers specifically to nephrite jade. It was written in the late 14th century, four hundred years prior to the discovery of Burmese jadeite.

The author goes on to state that *mutton fat*, slightly yellowish lard-like white hue,[285] is of the highest value, and nephrite the color of ice was considered of second quality.[286] The aficionado may hear terms like mutton fat bandied about. It is useful to understand that there are two standards of quality relative to the two types of jade and that the standards differ markedly. White nephrite stands at the apex of quality while white jadeite lies close to the bottom.

In modern connoisseurship, a jadeite jade cabochon of a pure white hue with a fine texture and a moderate translucency would be considered the finest of its type. A semi-transparent specimen would be milky and would be poor quality ice jade. White is not a chromatic hue; it is the neutral combination of all hues.

284 What was called Blue jade in the 14th century was probably Gem Chrysocolla. This gem material (see Chapter 12) brings high prices in Taiwan where it is called blue jade. cf. David, Percival (Sir), ed & trans., *Chinese Connoisseurship, The Ko Ku Yao Lun*, Praeger Publishers, N. Y., 1971, p.120.

285 Gump, *ibid*, p.68. See Taylor, I Grant, *Chinese Jade*, Grants Ltd., London, 1932, p.12.

286 David, Percival (Sir), ed. & trans., *Chinese Connoisseurship, The Ko Ku Yao Lun*, Praeger Publishers, N. Y., 1971, p.120.

The jade market, Guangjhou, China. Photo: R. W. Wise

discontinuities, all other treatments are definite negatives and should be disclosed to the buyer. Prices for fine quality jadeite have increased as much as four hundred percent in the past decade.[287] It is best to assume that anything that can be done will be done and insist on a grading report (certificate) from a recognized gemological laboratory before considering an important purchase.

287 A pair of 26x22mm Imperial green cabochons sold for 6.6 million dollars at Christie's Hong Kong in November, 2014.

The Rarity Factor

The Chinese value jadeite by size rather than weight. Gems less than 10x8mm (approximately 3 carats) are considered small. Medium stones are sizes up to 16x12mm (4-10 carats). Large stones are those over 16x12 (10-12 carats and up). Rarity normally increases with size. Nephrite, by contrast, is available in huge boulders that are used for carving.

Treatments: Buyer Beware

Jadeite is routinely waxed, heated, bleached and impregnated with polymer and dyed. Other than surface waxing which is done ubiquitously to fill in surface

OPAL

Life in the Outback is Hard . . .

We wake before dawn, our bodies glazed in dried sweat. The heat drives us, half asleep, stumbling out the door of the iron-roofed shack. The full moon casts a dim ghostly film, turning the landscape the color of gray flannel like the interior of some forgotten crypt. The desert winds assault us. The bushflies will arrive with the sun to suck the moisture from any exposed body parts; they will be our companions until sunset.

Thirsty is not the word – the inside of our mouths are dry as parchment. Water in this part of the outback is drawn from artesian wells drilled three thousand feet into the sandstone. The brownish liquid comes up hot and stinking of sulfur. At this remote location it must be trucked in, a trip that takes ten hours over dirt track. By ten o'clock the temperature passes the hundred-degree mark; by noon it stands at one hundred twenty.

The mining is mechanized at the Cragg Mine. Test holes are dug with a Caldwell drill, a giant, truck-mounted auger, which in minutes can drill a thirty-inch hole thirty feet into the red-clay sandstone. This is a job that took days with pick and shovel. The dirt the auger brings up is a dark brick red caused by high concentrations of iron. If traces of opal are found, James Evert, Vince's son and

Life in the Outback. Miner's shack, Mainside, Queensland, Australia. Photo: R. W. Wise.

The long road to the opal fields, five hundred miles into the Queensland outback. One hundred fifty million years ago this whole area, called the Great Artesian Basin, was part of a vast inland sea, an area larger than the state of Texas. Photo: R. W. Wise.

the camp foreman, will lower himself into the hole with a flashlight for a closer look.

We are standing in an area called Mainside, five hundred miles into the Queensland outback. One hundred million years ago, this area, which geologists call the Winton Formation, was an inland sea larger than the state of Texas.

The vista is broken by a series of tabletop buttes, island sentinels in the midst of a rolling plain dotted with trees and bushes. "Trees can be good indicators," says mine owner Vince Evert. "Some blokes dig near gidgee trees ~ but gidgee grows in the white sandstone. The opal lies under the red." With a single motion, Evert sweeps off his bush hat and wipes his brow with the back of his forearm. "Mallee bush grows in the red sandstone, so it can be a good indicator. Lapunyah, too; its roots grow deep into the faults."

Faults – cracks between strata that run up to the surface – expedite the formation of the gem. Opal begins as water-borne silica which percolates down along the fault lines and is caught between layers of semipermeable sandstone. Over the millennia the water leaches away, leaving a layer of concentrated silica. In some cases the silica hardens into a single layer; in others, it fills fractures in previously formed ironstone nodules.

The Cragg Mine, where the Everts are currently digging, was originally worked in the 1890s. Old timers with colorful names like "Texas Jack" and "Silk Shirt Joe" pioneered these fields with little

more than their swag and tucker (bedroll and provisions), a pick, and a shovel. "Nowadays it's D-8 Cats, augers, and backhoes." Vince shakes his head, reflecting on the changing times.

This year they "need a win," as Evert puts it. Last year they found little opal and the cost of the diesel fuel is driving them rapidly towards bankruptcy. A week later the monsoon comes unexpectedly with its driving rains. The rains fill the pits with water and the clay-rich soil turns to glue. The mine area becomes a series of little islands. Mining stops and many of the miners must be evacuated by air, setting back mining for the rest of the season. Life in the outback is hard. . . .

The Discovery of Opal

Did the Romans know opal? Stones may have been traded down the Danube from the single known ancient source in the northern heart of the Slanskevrchy Mountains of the old Kingdom of Hungary outside the villages of Opálbánya (Dubnik) and Červenica in what is now Slovenia. Pliny mentions opal though some say he was talking about iris agate. The first written mention of the gem mining in the area is dated 1568. The Emperor Rudolf II issued the first mining lease in 1597.[288]

Johann Menge first discovered Australian opal in 1849. The first commercial mine was established in Queensland at Listoral Downs in 1872. Later came the discoveries at White Cliffs in 1890, Lightning Ridge in 1902, Coober Pedy in 1915, and Mintabie in 1931. Due to the relative ease of extraction – Australian opal is sedimentary while Hungarian is volcanic and can be extracted only by mining the hard rock – Hungary's mines closed in the late 19th century and Australia came to dominate the world market and until recently was responsible for ninety-five percent of world production of gem opal.

Australia's dominance came under challenge in 2008 when large deposits of volcanic opal were discovered at Wegel Tena in the province of Wollo, Ethiopia. Ethiopian opal first entered the gem market in 1993 when a few stones from an unknown source, some with an unusual chocolate brown body color, were offered to Nairobi dealer Dr. N. R. Barot. This deposit was located in 1995 in the Menz Gishe district of Shewa Province at Yita Ridge. While this latter deposit has produced some opal, large quantities of what became known as Wello opal began flooding the gem market in 2010.

Despite some political issues over aboriginal rights, Australia continues to produce high quality opal. There is new mining activity around Winton in central Queensland. At Lightning Ridge, source of the crème de la crème of black opal, there is reasonable production. In the north, The *Allawah* and *Mehi* fields continue to produce. To the south, *We Warra* has been a good source of seam opal. In the west, *Bitter Sweet* is yielding some spectacular nobbies. In the venerable Coober Pedy district, there have been new finds of black and crystal opal at *Allan's Rise, Olympic* and *11 Mile* fields.[289]

288 Peter Semrád. *The Story of European Precious Opal from Dubnik*. Granit Ltd, Czech Republic 2011. P 90.

289 Damien Cody, personal communication 2016.

Ethiopia Versus Australia

An overwhelming percentage of Ethiopian, specifically Wello opal, has either a white or gray semi-crystalline body color. The play of color is remarkably consistent both in color and pattern like multicolored rags on a clothesline flapping in the breeze. Wello gems are easily paired and made into suites. They compare favorably to the old Hungarian stones and to the better grades of white-based opal from Coober Pedy. In some smaller gems, orange and green predominate with pink and occasionally some red seeming to float in a colorless crystalline body. These can be striking. In medium to larger gems the orange to grayish tinged body color often mutes the color display.

Wello opal rarely exhibits the pure vivid blues, reds and greens or the variety of patterns that characterize the best of the Australian black or boulder opal. You rarely find large juxtaposed blocks of color but, when you do, they can be beautiful and relatively less expensive than comparable gems from Australia.

Most Ethiopian opals are the hydrophane type. This means the gem is porous and will absorb liquid. This is a durability issue. Care must be observed in cleaning. Hydrophane opals will also absorb inks and dyes.

Chapter 35
Opal

Opal Types

Traditionally, gem opals have been classified by background or body color into seven types: *light*, *dark*, and *black* opal, plus *white crystal*, *dark crystal*, and *black crystal*. Recently the Australian Opal Dealer's Association has revamped and simplified the classification system. In the new nomenclature, opal body color is divided into four types: *light*, *dark*, *boulder*, and *black*. Stones that are transparent to semi-translucent are termed crystal. However, the older categories are more descriptive so I have used them below, with the new classifications given in parenthesis.

White (light) Opal (N7-N9)

White opal, the most familiar, has a translucent milky to opaque white background. Most of the current production comes from the fields in south central Australia at Coober Pedy and at Wegel Tena in Ethiopia. White opal is the opal normally seen in commercial jewelry .

Borealis, a rare Lightning Ridge black opal. This is a red multicolor opal of the rare harlequin pattern (N1-N2). Photo: Jeff Scovil. The Cora N. Miller Collection, Yale Peabody Museum, courtesy R. W. Wise, Goldsmiths.

A fine Australian white opal (N7). Photo: Rudy Weber, Courtesy Cody Opal.

Crystal Opal

Translucent to transparent opal is called crystal opal. Crystal opal may be light, dark, black, or have no discernable body color at all. Crystal opal is found at many sites in Australia and at Wegel Tena in Ethiopia. Under the new system of classification, any opal that is transparent to translucent when held to the light is called crystal.

Gray (dark) Opal (N5-N6)

Translucent to opaque opal with a gray body color is termed gray opal. The Mintabie Field in central Australia is the primary source for opal of this type, although much of what is found at Lightning Ridge is properly termed gray opal.

Semi-black (dark) Opal (N4)

Semi-black opal has a body color darker than gray, but not quite black. Semi-black will be semi-translucent when held to the light. Semi-black can be found at Mintabie and in central Australia at Lightning Ridge.

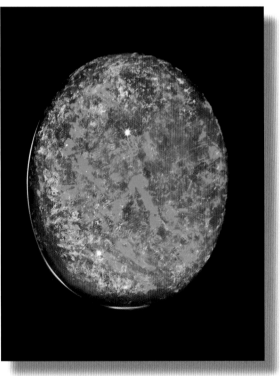

Australian black crystal opal on black background. The gem in semi-transparent when held to the light. Photo: Jeff Scovil, courtesy R. W. Wise, Goldsmiths.

| N1 | N2 | N3 | N4 | N5 | N6 | N7 | N8 | N9 |

The Australian Opal Dealers Association has adapted a new classification system which breaks down body color into nine categories from dark to light (N1-N9). Photo: R. W. Wise.

Black Opal (N1-N3)

Opal that is very dark in body color is called black. According to the older classification system, opacity is the key which divided black from semi-black. That is somewhat at odds with the more widely held view that the stone should appear opaque but may show some translucency when held directly to the light. The most highly valued blacks have what miners call "a bit of the blue", that is a blueish coal-black background. Lightning Ridge is famous for its black opal. A small amount has also been found in the western part of the Australian state of Queensland and at Virgin Valley in the state of Nevada.

visible face up, called opal with matrix, and gems with no visible inclusions called opal in matrix. Boulder opal is found at a number of fields in western Queensland. Opal in matrix is also mined at Andamooka and has been found in the Brazilian state of Piauí.

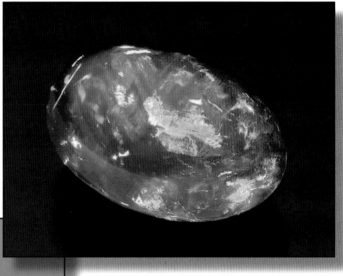

8.23 carat Fire opal from Mexico. Fire opal is a type of crystal opal. Much of this material has a vivid orange body color and little if any play of color. Photo: Jeff Scovil, The Edward Arthur Metzger Gem Collection, Bassett, W. A. and Skalwold, E. A.

Black boulder opal with matrix from Queensland (N1). The ironstone matrix is visible on the back but not on the face of the gem. Photo: Rudy Weber, Courtesy Cody Opal.

Boulder Opal

Opal with sandstone or ironstone as part of the cut gem is called boulder opal. The Gemological Institute of America classifies two types: gems with ironstone

Fire Opal

Fire opal can be easily recognized by its reddish orange to red body color. This type of opal is normally translucent to semi-translucent with little play of color. Opal of this type is found in Mexico, and Ethiopia. It is valued chiefly for its highly saturated translucent body color. Fire opal falls outside the Australian classification system. Generally speaking, the finest fire opal will have a distinct play of color coupled with a rich orangy to orangy-red body color.

A Hierarchy of Values

There is some difference of opinion among experts about the relative value of the different opal types. Other quality factors being equal, all agree that black opal is the most valuable. This is because the gem's black body color sets off the play of color, resulting in a breathtakingly vivid display. Boulder opal follows in potential beauty. Boulder-blacks — boulder stones with a black background or body color — can have exceptionally vivid saturation, sometimes surpassing black opal. Boulder, however, occurs in thin layers over ironstone, so this type of opal will rarely have the depth (transparency) of a black, which is solid opal. Depth allows for a more complex visual scene. However, as I have suggested throughout this volume, ignore pedigree and consider each stone individually.

Grading Opal

Opal is unique. However, some analogies can be made between opal and faceted gems. Opal does exhibit color, which can be categorized on the basis of hue, saturation, and tone. The percentage

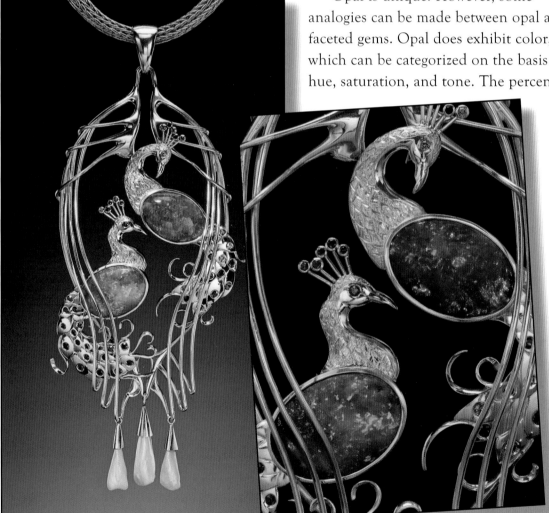

18k and 22k opal pendant in the Art Nouveau style. Photo: Jeff Scovil. Courtesy of Cora N. Miller Collection, Peabody Museum, Yale and R. W. Wise, Goldsmiths.

Closeup of Art Nouveau pendant at left. Photo Jeff Scovil, Courtesy of Cora N. Miller Collection, Peabody Museum, Yale and R. W. Wise, Goldsmiths.

of the face-up gem that exhibits play of color can be judged in the same way that brilliance or the quality of the cut is evaluated in a faceted stone. The degree of transparency (crystal) is important in the connoisseurship equation, particularly in what is known as crystal opal.

Play of Color

Play of color may occur in any or all of the spectral hues: red, orange, yellow, green, blue, and violet.[290] This phenomenon is the result of light refraction, similar to the effect of oil on water. The internal structure of opal is composed of microscopic silica spheres which are packed together in layers like racks of billiard balls. Opal contains about six percent water, which collects in the spaces where the spheres meet. Light reflecting though the spaces causes the phenomenon called play of color.

Gem of Many Hues

There is a rough hierarchy of opal colors based on rarity. In descending order the list is red, orange, yellow, violet, green, and finally blue.[291] In the marketplace, there is red — and then there are all the other colors. All would agree that the presence of red, particularly a dark-toned (seventy-five to eighty percent) visually pure scarlet hue adds dramatically to the rarity and price of an opal.

Opals are named by their dominant hue. A gem with fifty-one percent or more

red play of color is called a red stone. However, a stone exhibiting a single color is not valued as highly as one which shows several. An opal that exhibits three or more distinct hues is termed a multicolor. Multicolor gems are the most valuable. *The Guide* reserves its highest rating for an opal with a dominant red (over seventy-five percent of the surface) plus two additional hues. Paul Downing, a respected opal expert and connoisseur, prefers a red-blue multicolor with one or more additional hues.[292]

Saturation Comes First

"Opal is like a light bulb: the brighter it is, the better it is." This is how one well-known expert summed up the central truth of opal appreciation. In other words, saturation — not the hue, or the lightness or darkness of the hue, but the vividness of the hue — is the central criterion in opal grading.[293] A simple test: the farther away you can hold the stone and still see the play of color, the better the stone. On the lighter side of connoisseurship you might say that there are four-footers, five-footers, and then there are ten-footers!

Color — hue, saturation, and tone — should be observed in all possible lighting environments. Indirect sunlight (skylight), as always, is the standard, but the gem should be observed in incandescent and fluorescent lighting as well. Standard fluorescent lighting brings out the blues and greens but will often flatten the reds,

290 Damien Cody, personal communication 2002.

291 Purple is the rarest hue. Purple occurs in opal only when translucent red and blue patches are layered one on top of the other. Light passing through both patches may on rare occasions produce a purple play of color.

292 Paul Downing, *Opal Identification and Value* (Tallahassee, Florida: Majestic Press, 1972), p. 116

293 Andrew Cody, *Australian Precious Opal* (Melbourne: Andrew Cody Pty. Ltd., 1991) p. 53.

A 28.43 carat Lightning Ridge Black opal (N7) set in 22k w/diamonds. Photo: Jeff Scovil, The Cora N. Miller Collection, Peabody Museum, Yale, Courtesy R. W. Wise, Goldsmiths.

while incandescent lighting shows off the reds to best advantage.[294] Purple is not part of the opal spectrum. I have seen only one opal with purple play of color and that stone had a red color bar underlying a blue bar in a semi-transparent black opal.

Daystones, Nightstones

294 Richard W. Wise, "Australia, Thy Name is Opal," *Colored Stone Magazine*, March/April 1991, p. 7.

The term nightstone was originally coined to describe opal which held up well in indirect or shadowed lighting. I like the term so much that I have adapted it to classify any gemstone which looks its best in incandescent lighting. In opal, the criterion is more specific. Opals which look good in intense direct light but dull out in indirect or shadowed lighting are not as desirable as those that hold their beauty. As always, the aficionado should observe the stone in all indirect lighting environments. Opals that deaden in indirect lighting are called *daystones*. Stones that qualify as *nightstones* are beautiful in daylight as well.

Crystal

Generally speaking, in opals with dark or black body color, opacity is preferred and will command a premium. Conversely, in lighter-toned white (light) and colorless, crystal gems, the higher the degree of transparency the more vivid the colors and the more the stone will fetch in the marketplace. [295] Black opals tend to have a transparent layer over an opaque layer.

Performing the Test

If the opal is placed directly under an overhead light source and rotated three hundred sixty degrees, there is a direct analogy with brilliance, the quantity of light reflected from the crown of a faceted gemstone. As with faceted gems, the whole face of the stone is rated one hundred percent. With opal, the percentage of nonchromatic or dead areas is deducted from this figure to arrive at the total percentage of brilliance or play of color. If

295 Damien Cody, personal communication, 2002.

half of the stone is dead, it can be said to have fifty percent play of color. The greater the percentage of the face that shows play of color the better the stone. This seems simple, but it is just the beginning.

Some opal is highly directional, that is, the play of color is visible only when the stone is tilted steeply away from the perpendicular under overhead illumination. This is a fault. It is true that a little directionality can be a plus, as in a rolling flash pattern where the color darts like a prairie fire across the face of the stone as the stone is tilted side to side. However, if the stone shows large areas of extinction when viewed on axis and/or requires more than a twenty-degree tilt from the perpendicular to show its colors, it should be discounted substantially.

Opals are full of surprises! If the stone is to be worn, it should be oriented and viewed as it is to be worn. For example, a pin or pendant is worn on the body more or less vertically. This means that the stone will normally be oriented almost perpendicular to an overhead light fixture. The stone should be held at the appropriate angle to the light source and observed in a variety of lighting environments. In some cases the opal may look better than it looked in the grading position. This doesn't improve the stone's theoretical grade, but it does add measurably to its desirability.

The presence of matrix — that is, ironstone in boulder opal or any other inclusion — is also deducted from the theoretical one hundred percent. The presence of matrix, particularly in the face-up gemstone, is something of a flaw. Its effect can be mitigated, however, by the composition. If the juxtaposition of pattern and play of color against visible matrix is generally pleasing it is less of a negative.

Shape

The shape of the stone is a factor in opal evaluation and has a definite effect on price. Regular shapes — specifically, symmetrical ovals — command a higher price than other shapes and free forms. A cabochon with a high dome is also preferred to a low dome because it increases the angle of visibility of the play of color. Free forms may present a real buying opportunity for the collector who is attracted to unusually shaped stones.

The Opal as Art

8.45 caratBlack opal (N7) with the extremely rare peacock's tail pattern set in a custom designed, handmade butterfly brooch with accompanying sketch by R. W. Wise, Goldsmiths. Note that although the background color is dark gray rather than coal black it is dark enough to be classified black opal. Photo: Jeff Scovil, courtesy R. W. Wise, Goldsmiths.

A work of art gives us a glimpse into the mind of the artist. A good work of art may give us insight into our own mind as well. In this respect, opal is a work of art. Some might argue that this couldn't possibly be true, that man's only contribution is to shape and to coax the beauty from the rock. But the cutter does more than that. Opal colors occur in juxtaposed layers — like a pile of wet leaves. The lapidary is an important part of the equation: he has to decide when to stop cutting away the layers. No lesser artist than Michelangelo held that the finished sculpture was already contained within the material (marble) and his job was merely to liberate it.

Opals relate more closely to nonrepresentational paintings than they do to faceted gemstones. Terms such as composition, balance, and depth, terms borrowed from the fine arts, become relevant in the contemplation of a fine Opal.[296]

Cubism is an artistic style which views the two-dimensional canvas as an art object rather than simply as a vehicle to represent a bowl of flowers or a mountain range.[297] As such, cubism was the most radical departure in art making since the Renaissance. Qualities such as depth are achieved not through the illusion of perspective, but by the juxtaposition of colors, textures, and shapes. Cubism led directly to abstract art, whose practitioners used color (and to a lesser extent texture and shape) to evoke emotion directly without the aid of any type of representation. In this sense, an opal is very much like an abstract (nonobjective) painting.

Grading Patterns

It gives "atmosphere to the stone," to borrow a painter's phrase, and, like a fine impressionist picture, suggests more than it definitely expresses.
–Sydney B.J. Skertchly, 1908

If this description is correct, we seem to be left hopelessly mired in subjectivity. Not quite! It is still possible to talk about opal as one would a painting, to consider the stone as a total composition.

Composition

A pleasing composition is largely a question of harmony and balance. In any given stone, how well do the various colors and shapes harmonize and balance one another visually? How well do the areas of color play off against the background and areas of visible matrix? Since the composition of an opal is dynamic — that is, it changes depending on its orientation to the light — what is the effect of that movement? Do the overlay and juxtaposition of shape and color give a feeling of depth? Finally, how do all of these factors, combined with the three Cs—color (hue, saturation, tone), clarity, and cut — affect the viewer?[298]

Traditional Patterns

Connoisseurs particularly prize certain

296 Richard W. Wise, "Queensland Boulder Opal," *Gems & Gemology*, Spring 1993, pp. 12-13.

297 Douglas Cooper, *The Cubist Epoch* (London: Phaidon Press, 1970), p. 263.

298 Wise, "Queensland Boulder Opal."

traditional patterns. Pinfire, a pattern of densely grouped points of light something akin to neon Milky Way, is the most common and the least desirable. Harlequin, a multicolor pattern of overlapping angular blocks or diamond shapes (resembling the costume of the clown figures made famous in the paintings of Pablo Picasso) is the most sought after.

There are so many variations that the whole question can become confusing and seem needlessly complex. As a rule of thumb, patterns which contain large distinct segments or blocks of color are more desirable than those made up of small dots or ragged patches.[299] Thus, pinfire is the least expensive; harlequin will command the highest premium.[300] Rare traditional patterns such as mackerel sky, rolling flash, peacock, flagstone, and Chinese writing, the more literal the better, are also very collectible and will command a substantial premium among connoisseurs.

Color, texture, proportion, balance are important criteria in evaluating a work of art or a fine opal! "Dream in Lavender" 9x6", gouache on paper, by Rosanna Bruno 2014.

The Psychology of Opal

Psychologists have for years used a technique called the inkblot or Rorschach test to plumb their patient's subconscious. A series of inkblots are shown to the patient, who tells the doctor what he sees. This technique is used because psychologists have long recognized that these seemingly nonrepresentational shapes evoke certain associations, some of which are subjective and autobiographical, and some of which appear to be objective and universal. Opal patterns evoke similar associations. This explains why opal appreciation is so complex. An opal is an inkblot test combined with an abstract work of art.

This, perhaps, explains why, in my experience, opal aficionados are a distinct class. People either love opal or they hate opal. And those who love opal tend to give their full allegiance to opal to the exclusion of all other gemstones.

299 Damien Cody, personal communication, 2002.

300 Downing, *Opal Identification and Value*, p. 116.

Cuprian tourmaline diggings, Alta Lighona, Mozambique. Photo: Farouk Hashimi

TOURMALINE

*"One of the most magnificent known green [Mt. Mica]
tourmalines is . . . one inch long, three quarter inch broad and
one inch thick, and finer than any of the Hope gems."*

–G.F. Kunz, 1885

Tourmaline is possibly the most maligned and misunderstood of all gemstones. Until perhaps twenty years ago tourmaline was valued almost exclusively for its resemblance to other gems. Even its name, derived from the Singhalese word *turmali*, designates mixed parcels of unknown gems of dubious value. Despised in the East, tourmaline remained largely unknown to both European and indigenous New World cultures until the 16th century when Portuguese adventurers discovered it deep in the Brazilian hinterlands.[301]

When the Portuguese explorer Francisco Spinoza first stumbled upon green tourmaline in the mountains of Minas Gerais in 1554-55 he believed he had found emerald. The green stones were subsequently shipped to Portugal where they were set in the crown of *Nossa Senhora da Penha*. It took three hundred years before the mistake was corrected and "Brazilian emerald" was finally identified as tourmaline.[302]

With the advent of World War II, large quantities of mica, feldspar, and lithium minerals were needed for the war effort. Tourmaline was found as a byproduct of the mining. The Germans, who had begun mining in the Brazilian state of Minas Gerais

---

301 The *lyngourion* stone mentioned by Theophrastus (372-287 BC) was almost surely tourmaline. Theophrastus described *lyngourion* as having the pyroelectric "power of attraction just as amber does," a property unique to only tourmaline and amber. He further describes it as "cold and very transparent," stating that "seals are cut from this and it is very hard." The oldest tourmaline gemstone known is a green transparent seal stone carved with a likeness of Alexander the Great, dated between the 3rd and 2nd centuries BC. (See Ogden, *Jewellery of the Ancient World*, p. 170.) Other known uses in jewelry include a red cabochon set in a gold finger ring of early Nordic origin (1000 AD) and two other red tourmalines set in rings of European manufacture and dated to the 13th and 14th centuries. The earliest use of the name *turmale* or *turmalin* can be found in a 1707 manuscript by Johann Georg Schmidt. F. Benesch and B. Wohrmann, "Toramalli: A Short History of the Tourmaline Group," *The Mineralogical Record*, vol. 16, September-October 1985, pp. 331-338.

302 Keith Proctor, "Gem Pegmatites of Minas Gerais," *Gems & Gemology*, Summer 1984, pp. 78-81.

Fine examples of red, blue and green tourmaline from Minas Gerais, Brazil. The green gem in the upper left is similar to the best of the stones from Mount Mica, Maine. Photo: Harold & Erica Van Pelt, courtesy of Kalil Elawar.

around the turn of the century, were the first to curry an interest in tourmaline as a gemstone. Once outed, this gem species was largely ignored until the latter part of the 20th century.

Tourmaline Sources Around the World

Tourmaline occurs in almost every color, and the sources of tourmaline are more numerous and varied as its palette. Although Brazil remains the most important source, commercially viable tourmaline deposits have been found in Afghanistan, Pakistan, Mozambique, Nigeria, Namibia, Tanzania, Afghanistan, and the United States. The two principal sources in the United States are Maine and California.

Tourmaline deposits were first discovered at Mount Mica, outside Paris, Maine, in 1820. This deposit produced steadily until the mid-twentieth century.

In 1972 a huge pocket of tourmaline was found thirty miles away in Newry, Maine, at Plumbago Mountain. These two mines are part of a single gem bearing geological formation called a pegmatite which runs in a straight line from Brunswick, Maine, on the Atlantic coast through the towns of Paris and Newry to the New Hampshire border.

A selection of faceted tourmaline from Maine's Havey Quarry. Photo: Jeff Morrison.

Approximately seventy-five percent of the gems mined at these two sites were the lovely pastel pink and red gems for which Maine is justly famous. The remainder occurs in various shades of pastel from apple green to dark forest green. Mount Mica, reopened in the summer of 1990, has produced sporadically.[303]

Beginning in 2009, another rediscovered mine located twenty-eight miles north of Portland, formerly the Havey Quarry, now called the SparHawk Mine, has been producing some lovely medium-dark hued, slightly bluish "minty" green material.

For a short period, in the late nineteenth century, California was an important producer of gem tourmaline. The Tourmaline Queen, King, and Pala Chief mines in San Diego County were major producers of fine pink and red stones. Much of the gem material produced from this area was exported to China by the ton. It was coveted as a carving material. Three mines, the Oceanview, Mountain Lily and Carmelita mines, all located in the Pala District, are currently being worked and are producing small amounts of gem tourmaline.

Brazil is still a major world source for gem tourmaline though surface deposits are mostly depleted. Stones from Brazil come in many hues with green, pink, and red predominant. Brazil is the primary source for the rare blue variety known as indicolite and, of course, is the source of the legendary super saturated gems from Paraíba.

The highest concentration of Brazilian tourmaline is found in two roughly adjacent areas in the northwestern part of the state of Minas Gerais. One group of mines is clustered around the town of Govenador Valadares, the other north of Teofilo Otoni in the Aracuai district.

North of Govenador Valadares, on the east slope of the *Serra Safira* (Sapphire Mountains), is one of the oldest known and most prolific of Brazil's tourmaline deposits, the Cruziero Mine. Cruziero, a complex of several mines, produces tourmaline in a range of hues, including a red that rivals the best of Oro Fino as well as the unusual "watermelon" crystals that are pink in the center with a green outer rind. Cruziero is also the source of the dark, slightly bluish, emerald green gems first described by Portuguese adventurers.

Site of the famous Plumbago tourmaline strike. Newry Maine. Photo: Alan Plante.

---

303 Richard W. Wise, "Oldest Mine in the U. S. Reopens," *Colored Stone Magazine*, July/August 1992, cover, pp. 8-9. Only three tourmaline nodules of note have been found since the reopening in 1990. These **include** a 17.56-carat, a 9.40-carat, and a matched pair of 2.5-carat stones. The last two, which I purchased, were of a medium tone (sixty to seventy percent) minty green hue.

Tourmaline Grading

Traditionally, the three main hues of tourmaline – green, blue, and red – have been evaluated by what might be called the look-alike standard; that is, by how closely the green variety resembled emerald, the red ruby, and the blue sapphire. Until recently, this approach to connoisseurship continually reinforced tourmaline's status as a second-rate gem. Tourmaline, however, is a unique gem species with a unique chemical composition and crystal structure that only superficially resembles these more famous gem varieties. Tourmaline is found in the broadest range of hues of any gem species except diamond. It occurs in every color of the rainbow and in most intermediate hues as well. Perhaps the only hue that has not been found is a completely achroic or colorless stone. Mineralogists classify tourmaline as a group, like garnet. Tourmaline has a complex chemical formula, the alteration of any part of which can produce a stone with markedly different characteristics.[304] Until the late 20th century, tourmaline hues other than ruby, sapphire, and emerald look-alikes have been ignored.

The Perennial Daystone

Incandescent lighting is tourmaline's nemesis! The hue may or may not bleed but whatever mask is present will come out in the harsh yellow glow of a light bulb. Green through blue hues will usually exhibit a gray mask that is sometimes so dark as to appear black. Pink through red hues may show either gray or brown. Incandescent lighting will muddy the stone and decrease the transparency (crystal). Nightstones, gems that do not show this tendency, are the super novae in the tourmaline universe. The outlawing of the incandescent bulb may eventually render this distinction moot, but it is always advisable to view any gem in all available lighting environments.

---

304 The chemical formula for tourmaline is $(Na,Ca)(Mg,Li,Al,Fe^{2+},Fe^{3+})_3(Al,Mg)_6(BO_3)_3Si_6O_{18}(OH,O,F)_4$. Cf. Richard Dietrich, *The Tourmaline Group* (New York: Van Nostrand Reinhold, 1985), pp. 67.

Chapter 36

Cuprian Paraíba Tourmaline

"'Asking $1000 per carat for tourmaline is outrageous----no matter how beautiful it is' groused New York dealer Ary Reith, just as prices for Paraíba goods hit $2,000 per carat in Brazil. Reith refused to stock the gem until its prices 'come down to earth.' They never have."

–David Federman, 1992

Winner of an AGTA Cutting Edge award this 14.70 carat matched pair of extraordinary cuprian gems from Alta Ligonha, Mozambique. The pair exhibits the finest hue (rarely found from this source) along with flawless clarity and limpid crystal which characterizes the best of cuprian tourmaline. Photo courtesy A. Kleiman & Co.

In early 1989 a new variety of copper based or cuprian tourmaline was discovered in the Brazilian state of Rio Grande do Norte outside the village of São José da Batalha, near the border of the State of Paraíba in northwestern Brazil. In early 1990 this material began to dribble into the Brazilian market. This new variety was destined to permanently raise tourmaline's status in the world of gemstones.

Cuprian (copper) colored tourmaline from Paraíba hit the gemworld like a neutron bomb, slowed to a trickle and then it was gone. The material derives its color from minute amounts of copper and gold. The combination of these two trace elements yielded a light- to medium-toned blue to green gem of unrivaled saturation. So vivid in hue, the material immediately acquired the sobriquets of "neon" and "electric" blue. In 1991 similar material was discovered at two other locations in the Paraiba State within ten miles of the town of Parelhas at *Mulungu* and *Alto dos Quintos*, but São José de Batalha produced the major part of the finest material.

In 2005 just as Brazilian Paraiba was passing into legend, a new source of cuprian tourmaline was discovered in Mozambique. Previously, small quantities of cuprian material had been found in West Africa in 2000 near Ilorin, Ibadan State, southwestern Nigeria.[305] The Mozambique strike, however, hit the market in substantial quantities, together with a well thought out marketing plan. Found in an area called Alta Ligonha about one hundred kilometers southwest of the town of Nampula, this new source yielded material that was lighter in tone and generally less saturated, but larger and cleaner and more crystalline than gems from the original Brazilian source.

The hottest colors of cuprian tourmaline are not sapphire, ruby or emerald-like. They are vivid medium toned pastel hues in the blue to green range with tonal values between forty-five and sixty percent. Terms like "neon blue," "Caribbean blue," "electric green" and "Windex blue" aptly describe these gems.

The effect of these discoveries was twofold: first, it established a tourmaline aristocracy, giving tourmaline something it had previously lacked: snob appeal. Secondly, attention was shifted from the ruby-sapphire-emerald look-alike grading standard to a new appreciation of so-called Paraíba look-alikes. That is, non-copper bearing light to medium-toned stones in the blue to blue-green hues became highly sought after. Tourmaline had finally emerged from the shadows. The gem is

0.87 carat Brazilian Paraiba tourmalne exhibiting the Carribean Blue hue that is most desired by collectors. Photo: Jeff Scovil, courtesy R. W. Wise, Goldsmiths

now appreciated as important, beautiful and valuable in its own right.

Hue

The most sought after hues of cuprian tourmaline stones are the Windex or Caribbean blue, a hue reminiscent of the vivid hues of the shallow waters of the Caribbean Sea. This is a visually pure blue of between forty-five and seventy percent tone.

The secondary hue in cuprian blues is green, and gems without some hint of green are rare. A bit of green (5-10%) will have little impact on price. Add a bit more and the price will suffer. Green stones are found in a similar tonal range from visually pure chrome green to a slightly yellowish green, a hue similar to iceberg lettuce, with the slightest hint (five percent) of yellow secondary hue. A few reddish purple stones

305 The CIBJO definition would also preclude the use of the term *Paraiba* to describe cuprian gems of colors other than blue and green, specifically the lovely purple gems from Mozambique. *Cf.* Ahmadjian Abduriyim, et al., "Paraíba"-Type Copper Bearing Tourmaline from Brazil, Nigeria and Mozambique, Chemical Fingerprinting by LA-ICP-MS." *Gems & Gemology*, The Gemological Institute of America, Spring 2006, p.6.

were also produced by the Brazilian source. Alta Ligonha added a vivid purple, a color heretofore unknown in tourmaline.

Saturation

Vivid saturation is the defining characteristic of Paraíba type tourmaline. Gray is the normal saturation modifier or mask found in Paraíba tourmaline. A large

A magnificent 6.69-carat oval Paraíba tourmaline from Brazil. The stone is a greenish blue (a bit greener than the finest hue) with exceptional saturation and uncharacteristically fine crystal. Photo: Allen Kleiman, courtesy A. Kleiman & Co.

percentage of stones from Brazil and some stones from Mozambique exhibit a distinct grayish mask. This is particularly true of stones with tonal values above seventy percent. The asking price for such stones is often high simply because of the Paraíba pedigree. As in all such cases, beauty, not pedigree, should be the aficionado's guide. Non-cuprian tourmalines of similar well-saturated pastel hues from locations such

as Afghanistan can be both more attractive and far less expensive and hence a better option for the collector seeking a beautiful pastel blue green to green gem.

The best of Paraíba tourmaline shows no mask in either natural or incandescent lighting, resulting in an extremely vivid saturation. This fact explains why Paraíba tourmaline was catapulted to the heights of the gemstone pantheon in a relatively short space of time. The vivid saturation is found only in one other stone, Burmese ruby. What is even more amazing is that Paraíba stones, unlike the famous rubies of Burma, have achieved this level of saturation without the aid of ultraviolet fluorescence.[306] Paraíba tourmaline is normally inert to both short- and long-wave ultraviolet light.

Tone

As stated above, the normal tonal range for fine cuprian tourmaline is forty-five to sixty five percent. Darker-toned stones are frequently heat enhanced to produce a lighter-toned hue.

Clarity

Paraíba tourmaline from the Brazilian mines is normally visually included. Eye-clean stones from this source will command a substantial premium. Gems from Mozambique, however, are normally eye clean. This variety of tourmaline shows excellent crystal, which tends to lend any

306 Ruby, specifically Burma type formed in low-iron environments will exhibit strong pink/red fluorescence under ultraviolet light. This distinct fluorescence is in part responsible for the characteristically vivid saturation of this type of ruby (see Chapter 15).

inclusions even greater prominence.

Crystal

An exceptionally high degree of diaphaneity — transparency or crystal — is characteristic of the finest gems from Mozambique. Gems from the original source will often suffer by comparison.

A Potentially Costly Confusion of Terms

The point has been repeatedly made throughout this volume that beauty is the ultimate criterion of connoisseurship and that geographic origin should not be part of the equation. However, geographic sources do influence market pricing. Cuprian gems from the original Brazilian source sell at dramatically high premiums, and gem labs are able to separate gems from different geographic sources. It is important that the aficionado exercise care when contemplating acquiring a cuprian or Paraiba type gem.

Shortly after the Mozambique strike, a consortium of dealers intent on bathing in the reflected luster of the gems of São José de Batalha, successfully lobbied he *World Jewelry Federation* (CIBJO), the international organization charged with naming new gems, to mandate the use of the term "Paraíba" for all tourmaline with "a green to blue color caused by copper," regardless of source.[307] This was a mistake, a major break with tradition and the first occasion that CIBJO has used a geographic appellation as a naming convention.

Gem labs have largely followed CIBJO's

lead. In all cases where a major purchase is being contemplated, the collector should insist on a lab report and read it carefully. The major gem laboratories do have the ability to differentiate cuprian tourmaline from various countries using quantitative chemical analysis,[308] but geographic origin is normally optional on grading reports. The lack of a statement of geographic origin should be seen as a red flag which probably means that the stone is not from the Brazilian source and should be priced accordingly. In all cases, the aficionado should insist upon a grading report (certificate) from a major lab with geographic origin clearly stipulated.

Mozambique has produced some really exceptional gems other than blue to blue-green. The CIBJO ruling has also served to keep a tight lid on prices of other hues of Mozambique cuprian gems resulting in potential buying opportunities for the astute aficionado.

Heat Treatment Legitimized

Like tanzanite, cuprian tourmaline is almost universally heat-treated.[309] Approximately eighty percent of gems from Brazilian sources are treated as are most from Mozambique, and current market prices assume heat treatment. The original Brazilian mines produced darker-toned (eighty to eighty-five percent) stones

307 Ahmadjian Abduriyim, et al., ibid, 2006, p.19.

308 Ahmadjian Abduriyim, et al., ibid, 2006. pp. 19-20. Chemical analysis is not sufficiently sophisticated to differentiate between gems from the three Brazilian locations.

309 Both the tanzanite and cuprian tourmaline mines produce tiny amounts of natural unheated gems of fine quality. Untreated gems are certainly rarer, but the price structure assumes heat-treatment.

with a seventy-five percent primary hue of blue and a twenty-percent secondary hue of green resembling sapphires from Australia and Thailand. These stones were immediately heat treated. Those that turned a lighter-toned neon blue promptly quadrupled in value.[310]

The Rarity Factor

Paraíba tourmaline is rare in any size. Gems from the original Brazilian mine are rarely found above two carats. Mozambique stones can be quite large, up to twenty carats and even larger. Therefore, larger gems from this source do not command the premiums associated with Brazilian gems. Generally speaking, rarity and price tend to increase at one, three, and ten-carat and twenty carat sizes.

---

310 I held a small parcel of dark blue Paraíba stones for several years before having them heat-treated. The heat enhancement effectively quadrupled the value of the stones. This treatment is carried out under very low heat. In Brazil, I witnessed a Paraíba tourmaline lighten in color in a test tube under the heat of an alcohol lamp. Low temperature heat treatment is normally undetectable.

Creative cutting in fine green troumaline, medium tone exhibiting excellent crystal. Photo: courtesy Nomads.

Chapter 37
Green Tourmaline

"Green tourmaline is the most widely distributed of the precious varieties of this mineral and consequently is lower in price. It is rarely emerald-green but when this is the case its colour lacks none of the depth of that of the true emerald."

–Max Bauer, 1904

A fine traditional green tourmaline shown with several rough crystals. The hue shows just a touch (5%) blue. Photo courtesy Nomads.

Green is tourmaline's most common hue. There are two varieties of green tourmaline: elbaite, the most common, is found throughout the world; chrome tourmaline (dravite) is found mainly in East Africa. Elbaite is discussed in this chapter. The chrome variety will be discussed in detail in the next chapter.

Although Brazil remains the most important source, fine green tourmaline is found in Nigeria, Pakistan, Afghanistan, Namibia, and the states of Maine and California.

Historically there were essentially two standards used in the grading of green tourmaline. The traditional standard grades the stone on the purity of its primary green hue; in short, on how closely the stone resembles emerald (see the introduction and overview to this section). The second standard might best be termed the Paraíba standard and is discussed in detail in the previous chapter.

Hue

Tourmaline with a pure green key color is extremely rare and, despite the quotation at the beginning of this chapter, doesn't resemble emerald very much at all. Its refraction is comparatively feeble when compared to emerald and, with the exception of the chromium green variety,

A particularly saturated and exceptionally well-cut green Afghan tourmaline. Non-cuprian green to blue-green gems from this source rival in saturation the finest cuprian Paraiba from Mozambique at significantly lower prices. Photo courtesy, Precision Gem.

lacks the pure verdant hue. The green hue achieves its maximum saturation at about seventy-five percent tone. So, as with tsavorite garnet and emerald, a medium dark-toned, visually pure green is optimal. However, vividly saturated, medium toned green gems which resemble cuprian Paraiba-like stones will command yet a higher price. So, we are really dealing with a tonal range, medium to medium dark (40-75%), vivid visually pure green.

The secondary hues normally encountered in green tourmaline are yellow and blue. Although a bit of yellow is attractive, more than five percent of a secondary yellow is less so, particularly in medium dark -gems. Add ten percent yellow to a bit of gray mask, and you have

the unattractive olive or what is often called camouflage or army green. Most commercial grade tourmaline is exactly this hue. A ninety percent green with no more than ten percent blue runs a close second to the pure verdant gem. This is the look-alike standard, and there is no denying that a visually pure green stone is beautiful and desirable. A distinctly bluish stone is definitely more desirable than an overly yellowish gem.

In medium-toned gems (40-60%), the impact of yellow is more positive, but in medium toned greens with more than fifteen percent yellow, the hue becomes a bit too lime-like. As with all green gems, blue trumps yellow. A bit of the blue is always a plus — it warms things up.

Saturation and Tone

Stones with tonal values above eighty percent are visually blackish and overcolor. As mentioned above, lighter-toned, pastel to mint green tourmalines can also be quite beautiful. Many of the stones found at Mount Mica and Havey, Maine, and more recently in Afghanistan, fit this description and sell at higher prices than even medium dark-toned greens.

The mask (saturation modifier) in the green variety is normally gray, which grades to black in darker-toned gems. This mask is greatly enhanced in incandescent lighting, adding a sooty quality to darker toned gems.

In the lighter-hued Paraíba-like stones, what the collector should look for is a good balance between hue and tone. Some of the lighter-toned gems have a bright liquid quality that makes them very desirable.

Multicolor Effect

Tourmaline is the most dichroic of all gem species. Multicolor effect, sometimes called dichroic effect, is to be expected and should be embraced. The effect is less prominent in darker-toned (eighty percent) green stones. In lighter-toned gems it can be quite prominent, so much so that occasionally some medium-toned bluish green tourmaline will actually break up the primary and secondary hues and exhibit blue and green scintillation on adjacent facets. This can be a breathtakingly beautiful effect and is in no way a fault.

Crystal

Dark-toned green tourmaline never exhibits the glow, the limpid transparency of emerald. With all due respect to Max Bauer, it is precisely this quality that is often found in emerald but rarely in green tourmaline. Tourmaline is a daystone. Darker-toned green tourmaline tends to be quite turbid; incandescent lighting greatly increases this proclivity. Perhaps smoky or even sooty is a more visually accurate description — it's as if the stone had been blackened over the flame of a kerosene lamp. Transparency is greatly decreased. Gems which exhibit good crystal in incandescent light may be said to be the true gems of this gem variety.

Clarity

Green tourmaline is normally eye-clean. Visually included stones, excepting cabochons, are virtually unsalable and are of no interest to the connoisseur.

The Rarity Factor

Though rare in any size, fine green tourmaline is readily available in fairly large sizes. Prices tend to decrease on a per carat basis over twenty carats.

1.85 carat chrome tourmaline custom cut by Gene Flanagsm.
Photo: courtesy Precision Gem.

Chapter 38
Chrome Green Tourmaline

"Dreaming of tourmaline is supposed to insure success through superior knowledge but there seems to be no evidence that the residents of tourmaline mining areas in Maine or California, where such dreams are particularly common, are any better off as a result."

–John Sinkanas 1971

Chrome tourmaline from Tanzania with a vivid, slightly bluish green of about 80% tone. Photo: Tino Hammid.

Chrome is a special variety of green tourmaline that owes its vivid hue to trace amounts of chromium and vanadium. These are the same elements that also impart to emerald and tsavorite garnet their distinctive pure hues. It should not be surprising, therefore, that the finest chrome tourmaline tends to resemble the finer examples of emerald and tsavorite. Chrome tourmaline is a distinct tourmaline variety called chrome dravite, found in East Africa. Chrome tourmaline is often associated with tsavorite garnet; miners will usually concentrate their efforts on tsavorite, as it fetches higher prices.

Hue

Like emerald and tsavorite garnet, fine chrome tourmaline is a visually pure forest green with slightly yellowish to bluish secondary hues. The blue will normally show itself in incandescent light; the yellow will be more visible in daylight. The same criterion applied to tsavorite garnet and emerald is applicable to chrome tourmaline. A blue secondary hue is preferred to yellow. Chrome tourmaline, unlike emerald, can never be said to be too blue. A visibly pure to slightly (five to fifteen percent) bluish green gem between seventy and seventy-five percent tone is the most desirable.

Saturation and Tone

Gray grading to black is the normal saturation modifier or mask found in chrome tourmaline. Due to its chemistry, chrome tourmaline is normally highly

saturated. An overabundance of chromium/vanadium appears to be the culprit. Larger stones tend to be overcolor; that is, so dark in tone as to be virtually opaque. Even though the ideal tone or the gamut limit of green is seventy-five percent, chrome of eighty percent tone can be quite beautiful, due to its vivid hue. Stones with tones of eighty-five percent and above are definitely overcolor. Gems of this description will appear to have a virtually opaque black body color punctuated by the occasional flash of vivid green key color.

Multicolor Effect

As tourmaline is the most dichroic of gemstones, a strong multicolor effect is to be expected. However, perhaps due to its unique chemical composition and its distinctively dark tone, multicolor effect is rarely observed in the chrome green gemstone.

Crystal

A high degree of transparency is very rare in chrome tourmaline because of its normally dark tone and its behavior in incandescent lighting. Chrome tourmaline, like other members of the species, is a daystone. The stone's crystal has a tendency to close up — to turn sooty like the chimney of an oil lamp — under incandescent light.

It seems particularly sensitive to certain kinds of halogen lighting. If the stone tends toward a gray mask, the yellowish light of the light bulb will also exacerbate that tendency.

Generally speaking, chrome tourmaline is simply more opaque than emerald or tsavorite garnet. Its hue is vivid but it is also dense. A chrome tourmaline which meets the criteria discussed and that exhibits good crystal — that is, one which exhibits a limpid transparency and retains it in incandescent light — is the crème de la crème of this species.

Clarity

Chrome tourmalines are normally sold eye-flawless. Stones with visible inclusions sell at very substantial discounts and are of little interest to the connoisseur.

The Rarity Factor

Chrome tourmaline is quite rare generally, and particularly rare in sizes over one carat. Therefore, the collector should expect a large percentage increase in the price of stones in carat-plus sizes. The next jump in rarity occurs at five carats. Stones of fine quality above ten carats are extremely rare, so rare in fact that I have never seen one.

Chapter 39

Blue Tourmaline

"Seeing a true blue tourmaline is practically a once-in-a-lifetime experience for most jewelers and probably most dealers."

–David Federman, 1992

Sapphire blue? Not really, but an exceptional example of blue tourmaline with a rich, medium blue hue and fine crystal from a secret source located somewhere in Africa. Photo courtesy Nomads.

Sapphire blue tourmaline! If they exist at all, such stones are very rare. The search for such a stone is a bit like the pursuit of the Holy Grail—it remains somehow just out of reach.

Historically, two Brazilian mines, the Manoel Mutuca in Aracuai and the Golconda Mine northwest of the city of Govenador Valadares, are the sources of the fabled sapphire blue indicolite. The best of these stones have an intense blue primary hue and a turquoise secondary hue, and are known as *Mutuca Blue*. Stones of this quality are extremely rare. Indicolite is found sporadically throughout the entire region of Minas Gerais. In recent years, African mines in Namibia and elsewhere have produced some stunning blues.

Hue

If cut on the C axis of the crystal, tourmaline with a primary blue hue will face-up blue, but not without just a slight hint of a secondary hue of green (ten to twenty percent). Stones cut in this manner will always be dark in tone because the C axis of the tourmaline crystal is itself very dark. In more conventionally cut stones, the primary hue is indeed a rich medium to deep hue reminiscent of sapphire, but the gem will always show a distinct green secondary hue. Gems with an eighty-five percent blue primary hue with no more than a fifteen percent green secondary hue should be considered fine.

As with all tourmaline, incandescent lighting will bring out the secondary hue. Stones which appear almost pure blue in sunlight or incandescent lighting will pick up a distinct greenish secondary hue under the light bulb. All other factors being equal,

A range of hues from green to blue tourmaline from Neu Schwaben, Namibia. The oval opposed bar cuts at the top left and bottom right are exceptionally rare hues. Photo: Tino Hammid, courtesy Christopher Johnston.

the bluer the stone the better the stone.

Saturation and Tone

Indicolite can be found in all tonal ranges. The stone is most attractive as the tone approaches seventy-five to eighty percent, the ideal (gamut limit) tone for blue. Gray is the normal saturation modifier found in indicolite. As with most tourmaline, the gray is punched up in incandescent lighting.

Clarity

Tourmaline in the green-blue range of hues is usually visually flawless. However, an exceptional indicolite, given its extreme rarity, may be forgiven a few minor flaws. The presence of any visual inclusions should, however, dramatically lower the price.

Crystal

Indicolite tends to turn grayish, lose transparency, and appear muddy in incandescent lighting. Again, tourmaline is a daystone and puts its best foot forward in natural daylight and daylight-equivalent fluorescent lighting. Despite the difficulty in finding one, the collector should always seek to view blue tourmaline under the incandescent light bulb before making a purchase. Although the gem is often compared to sapphire, blue tourmaline rarely exhibits a transparency comparable to the finest sapphire.

As with all transparent gemstones, all other factors being equal, diaphaneity – transparency or crystal – will separate the fine from the merely good, the beautiful from the merely pretty. Finer examples of blue tourmaline which hold their hue and which do not close up or turn sooty in incandescent lighting are exceedingly rare and are the most desirable examples of this gem variety.

The Rarity Factor

Blue tourmaline is very rare and desirable in any size. Prices will tend to level off above twenty carats.

Chapter 40

Red/Pink Tourmaline

"Rubellite may be of various shades of colour, from pale rose to dark carmine red, sometimes tinged with violet. The colour may be so like that of certain rubies that it is difficult, even for an expert, to discriminate between these stones on mere inspection"

–Max Bauer, 1904

A fine medium dark toned oval pinkish-purplish red tourmaline together with a lovely rough crystal. The gem is about 70-75% tone Photo: Jeff Scovil.

The Ouro Fino Mine in the Brazilian state of Minas Gerais, about ten miles east of the Manoel Mutuca mine, was the legendary source of some of the world's finest red tourmaline, called rubellite. Stones from this source have a pure red primary hue and a purplish secondary hue, yielding a lovely purplish red which has been compared to the color of the Bing cherry. Although Ouro Fino has been closed for decades, the term *Oro Fino Red* has come to designate the very finest red tourmaline, regardless of source.

Tourmaline occurs in every conceivable tonal value from pink through red. Despite the misleading appellation rubellite, red tourmaline looks like red tourmaline, not like ruby at all. The chief difference in appearance is the multicolor effect. Tourmaline is the most dichroic of gemstones and always exhibits a pronounced multicolor effect. In the finest gems, the colors will be red/purple.

Hue: Ruby-Like Tourmaline

Stones with a pure red hue plus the absolute minimum of secondary hue (normally pink to purple) are, without question, the most valued in the marketplace. However, red tourmaline will always have some mixture of secondary hue, particularly when the viewing environment is shifted from natural or fluorescent to

Matched pair of Nigerian red tourmaline of about 80% tone. Photo: Jeff Scovil, courtesy R. W. Wise, Goldsmiths.

Tone

Red reaches its most vivid saturation, or gamut limit, at about eighty percent tone, which is, not surprisingly, also the ideal tone in ruby. However, in red tourmaline, slightly lighter tones of red (70-75%), when combined with a higher percentage of secondary hue, will often result in a marvelously beautiful gemstone. Gems with tonal values between forty and fifty percent will appear rosy; pinkish stones in the fifty to sixty percent range are the color of maraschino cherries. Darker-toned red stones with a purple secondary hue are best described as magenta. This wonderful range of hues provides much latitude for the tourmaline aficionado.

incandescent lighting. Under the light bulb, this tourmaline will almost always appear distinctly pinkish or violetish to purplish red in darker toned gems. Secondary hues of twenty percent or more are characteristic and lend to tourmaline a distinctive appearance which can be quite beautiful although, with apologies to the late great gemologist Max Bauer, not particularly ruby-like at all.

Saturation

Though brown is possible, gray is the normal saturation modifier or mask found in red tourmaline. Red, like all tourmaline, is a daystone. The tendency to gray is normally exacerbated by viewing the stone in incandescent lighting. The hue will often appear to close up, losing both saturation and transparency (crystal) under the light of a flame. Some stones will gray only slightly and appear violetish under the light bulb.

Crystalline Nightstones: the Crème de la Crème

Diaphaneity — transparency or more commonly "crystal" — is the true fourth C of colored gemstone evaluation. As mentioned, tourmaline is normally a daystone. That is, it looks its best in natural lighting. Incandescent lighting produces a negative effect. A grayish or sometimes a brownish mask shows up under the light bulb and, if it is strong, causes the stone to lose transparency, to close up under this type of light.

Finer red-pink tourmaline will turn violetish, purplish or slightly grayish violet. These stones tend to hold their transparency, and the violet secondary hue will enhance the overall appearance of the stone. The collector should check any pink-red tourmaline for this property before deciding on an acquisition. It is important when considering a purchase to observe the stone carefully in incandescent light. This is the crucial test. A stone that is a limpid pinkish red in daylight but becomes a dull grayish or muddy brownish red in incandescent is less than desirable.

Tourmaline is one gem species where a contrarian approach to collecting can yield big dividends. It is recommended that the aficionado forget comparisons to ruby and pink sapphire and consider the stone. A strongly violetish to purplish red tourmaline with little or no brownish mask is much more desirable than a muddy ruby-like stone and should be available at the same or perhaps even a lower price.

Clarity

Pink-red tourmaline is often visually included. Stones with a few small visible inclusions that affect neither the durability nor the beauty of the gemstone and are acceptable. Eye-visible inclusions have been the norm, particularly in Brazilian reds. However, a strike in Nigeria brought a large number of eye-flawless stones into the market in the late 1990s, and we have seen no major strikes since.

Cut and Crystal

Tourmaline, as gemologist Max Bauer pointed out in 1904, has a "somewhat feeble" refraction.[311] This is particularly apparent when the gem is compared to ruby. That said, red tourmaline is one of the few examples of this gem variety that can be advantageously cut into round, oval, and pear-shaped mixed brilliants. This is because the C axis of the red crystal is normally not as dark and dense as it is in the green and blue varieties. Red tourmaline potentially may exhibit much better crystal than either the blue or green material. To punch up the refraction, the pavilion of red tourmaline is often cut with multiple tiny facets called a Portuguese cut. The scintillation produced by this faceting style shows the gem's multicolor effect to great advantage.

The Rarity Factor

Pink-red tourmaline can be found in large sizes. Although clean red stones of any size are rare, gems over twenty carats tend to decrease in price on a per carat basis.

311 Max Bauer, *Precious Stones*, pp. 366-368. Despite his propensity to compare blue, green, and red tourmaline to sapphire, emerald, and ruby, he also takes notice of the tendency of some tourmaline to appear "quite dark and imperfectly transparent."

8.14 carat unenhanced purple Mozambique cuprian tourmaline.
Photo: Jeff Scovil, courtesy R. W. Wise, Goldsmiths.

Chapter 41

The Other Colors of Tourmaline

"So to the present day, though tourmaline is considerably used in jewelry, it is rarely ever called by that name. The green varieties are often known as Brazilian Emerald, chrysolite, or peridot, some varieties of blue as Brazilian Sapphire, others as indocolite, and the red as rubellite and even as ruby."

–Oliver Cummings Farrington, 1903

A lovely "cognac," natural brown tourmaline. Here brown plays the part of a hue. Note the distinct tonal gradations (multicolor effect) in the face-up mosaic of the gem. Photo: Jeff Scovil, courtesy R. W. Wise, Goldsmiths.

It is fair to say that tourmaline comes in many other hues — perhaps every other hue. Yellow orange and purple tourmaline stones are exceedingly rare. Prior to the first edition of this book (2003) strike at Alta Ligonha, Mozambique I had never seen a tourmaline with a primary purple hue. That color was found in Mozambique in 2005.

Lemon yellow tourmaline is quite beautiful; however, yellow is usually tinged with green. I haven't seen pure yellow stones above two carats. Most orange is tonally dark and appears brown. Tourmaline of this hue is beautiful in flavors darkening from mocha to chocolate to cognac. Brown stones are usually available at very reasonable prices.

The tourmaline crystal will often show distinct zones of different hues. Stones purposely cut to include two or more hues are called bicolor. The most common bicolors are brown/green. The color alteration results from the depletion of certain chemicals during the crystal's formation. This depletion changes the environment and destabilizes the crystal, so that multiple fractures are created within the rough and, therefore, are in the cut stone. Clean bicolored gems are rare. An eye-flawless gem that includes a visually pure red and green in approximately equal amounts is the most sought after, and will command the highest price.

A magnificent 120 carat flawless Brazilian bicolor tourmaline. Photo: R. W. Wise, courtesy R. W. Wise, Goldsmiths.

fault only in sapphire blue, ruby red, and emerald green tourmalines. In lighter-hued tourmalines and non-look-alike hues, the only issue is whether this particular multicolor effect detracts from the appearance of a specific gemstone.

It is safe to say that tourmaline which approaches a pure spectral hue — whether red, orange, yellow, green, blue, violet, or purple — will be more desirable in the market than stones which fall between these pure hues. This is generally true of all species and varieties: a pure hue is more desirable than a mixed hue. A pure yellow stone is more sought after than a greenish yellow gem. The characteristically strong multicolor effect in tourmaline dictates that stones of this description are truly scarce. The mixture of hues, which distinguishes this gem, contributes to its distinctive beauty. Tourmalines must be evaluated stone by stone.

Multicolor Effect

Tourmaline is the most dichroic of gemstones, usually showing a pronounced multicolor effect. The multicolor effect is considered a fault in sapphire, ruby, and emerald, gemstones that are chiefly valued for the purity of their primary hue. Thus, multicolor effect is considered a

appearance of a specific gemstone.

Connoisseurship in Tourmaline: a Contrarian Approach

Tourmaline does have a traditional grading structure. However, that structure is limited to just a few varieties, or color ranges, of the gem. Historically, stones which fall outside these narrow parameters have been largely ignored. Given tourmaline's propensity to occur in almost any hue, this leaves broad areas open to inexpensive contemplation by the connoisseur. The key issues in evaluating tourmaline are discussed above. I recommend, if the primary/secondary hue is pleasing, that the collector focus on the presence and strength of the gray/brown mask, multicolor effect, and crystal when evaluating a specific stone. If a stone is well cut and earns high marks in both day and incandescent light, it is a fine rare gem and worthy of consideration.

Chapter 42
Tanzanite

"Tanzanite looks like sapphire wishes it could look."

–Barry Hodgin, 1988

A 15.42 carat gem tanzanite photographed in daylight equivalent fluorescent lighting. Note the gem's rich just slightly purplish blue hue and the extraordinary saturation. Photo: Wimon Manorotkul, courtesy Lotus Gemology..

Tanzanite was discovered twenty-five miles south of the dusty frontier town of Arusha, Tanzania, in 1967. Ndugu Jumanne Ngome, a Masai warrior brought a ten-thousand-carat chunk of transparent purplish blue material to Navrottoni Pattni at his office in Nairobi. Pattni, Kenya's first gem cutter, thinking that anything that big had to be glass, dismissed the Masai without purchasing the material. As subsequent events would prove, Pattni had turned away an entirely new gem material. It was eventually named tanzanite by Henry Platt, then vice president of Tiffany's, in honor of the East African country where it was first discovered. Pattni speculates that the piece he was offered could have been cut into a gem in excess of the 220-carat beauty currently in the collection of the Smithsonian. Pattni was offered the stone for fifty dollars. Such are the wages of hesitation.

Tanzanite is the gem variety of the mineral zoisite. Tanzanite is a trichroic gemstone, meaning that it exhibits one of three colors depending on which way the crystal is viewed: blue, amethystine to red, and green-yellow to brown. Although some of the stones found on the surface are purple blue because of the heat of the equatorial sun, almost all tanzanite comes out of the ground a gray to root beer brown. Relatively gentle heating will drive off the gray brown, leaving the lovely violet to purple blue behind.[312] Tanzanite may also occur in green and yellow hues, but these

---

312 Peter C. Keller, *Gemstones of East Africa*, Geoscience Press, Tucson, 1992, pp. 68-69.

colors are so rare that they are not part of this discussion.

Because tanzanite possessed a passing resemblance to sapphire, Platt decreed that the finest color of tanzanite would be the one that came closest to looking like the finest quality of sapphire. This is another example of the dated look-alike standard. Tanzanite does look a bit like sapphire or, as one of my first mentors in the gem business once put it, "Tanzanite looks like sapphire wishes it could look."

Hue

A fine 5.55-carat oval tanzanite showing its bedtime color in incandescent light. The blue hue (left) represents an exceptional hue in daylight. To the right shows multicolor effect in incandescent lighting. Photo: Jeff Scovil, courtesy R. W. Wise, Goldsmiths.

In daylight, tanzanite exhibits a range of hues from a light-toned violet to purple to a rich dark-toned blue (eighty to eighty-five percent) with a moderate (fifteen to twenty percent) purple secondary hue. The slightly purplish blue color, which Platt defined

as the finest, will exhibit a pure blue in daylight, an environment that suppresses the purple. The secondary purple becomes visible as soon as the stone is placed in an incandescent lighting environment. Incandescent lighting will always bring out the secondary hue, be it violet or purple.

The finest stones will show a blue primary hue with just a hint (ten percent) of purple secondary hue by day. By night, under a light bulb, the purple secondary increases to fifteen to twenty percent. Some connoisseurs prefer to see a bit of purple even in daylight. The purple adds a velvety quality to the hue. Some stones will appear almost a visually pure blue — only slightly purplish — in daylight, and change dramatically to a visually pure purple in incandescent. Some connoisseurs, in defiance of the sapphire look-alike standard, find this range of hues nearly as desirable.

The quality of the secondary hue depends to a large extent on the purity of the red trichroic hue and on the orientation of the crystal prior to cutting. Tanzanite can be violet to purple or violetish to purplish blue depending upon the amount of red. Purple lies precisely halfway between blue and red on the color wheel. Violet lies halfway between purple and blue.

Some dealers, along with organizations such as the Tanzanite Foundation, often have trouble seeing the color purple. The problem is so acute that the Tanzanite Foundation doesn't seem to recognize the

color at all. In their buying guide they describe the finest hue as violetish blue. This confusion of terms goes back a long time. Prior to the movement to standardize nomenclature in the 18th century, color names, like spelling, were a matter of choice. Jean Baptiste Tavernier, for example described the great blue diamond he sold to Louis XIV as " beau violet," because in 17th century France, violet was a synonym for blue.

Traditionally, sapphire is downgraded if it loses or bleeds color when the lighting environment is changed from daylight to incandescent. How important this is in a world where incandescent lighting seems to be going the way of the Dodo is an open question. Tanzanite will always shift color when the viewing environment is switched from daylight or daylight-fluorescent to incandescent lighting. Tanzanite does not bleed color; it changes color, losing nothing in the way of saturation. Tanzanite is not sapphire. This hue-shift is characteristic and breathtaking, and partly defines the beauty of the stone.

Saturation

Gray is the usual saturation modifier or mask in tanzanite. Stones without the gray modifier will exhibit an extremely vivid, velvety hue. To the untutored eye, the gray mask is quite difficult to see, particularly in daylight. Incandescent lighting will bring out the gray. A gray mask visibly dulls the hue, imparting a cool aspect. Gray is usually more prominent in stones of less than sixty percent tone.

Tone

Like sapphire, tanzanite achieves its optimum hue at between seventy-five and eighty-five percent tone. Although color science holds that eighty-five percent tone is optimum for blue, stones with tonal values of seventy-five to eighty percent have greater transparency (crystal). Stones of eighty-five percent tone appear slightly inky. The difference between blue at seventy-five percent tone and blue at eighty-five percent tone is what separates the finest color from second best. As with sapphire, gemstones of eighty-five percent tone cross the line; they become too dark and are, in fact, overcolor.

Tanzanite Versus Sapphire

Tanzanite is often compared to Kashmir sapphire.[313] This is not a particularly helpful comparison. Comparisons between two gemstone varieties are usually to the detriment of one of the victims. Each species, each variety of precious gemstone, has its singular virtues. Although the hue of a fine tanzanite, like the finest Kashmir sapphire, can be described as velvety, tanzanite will always have a more distinct purple secondary hue.

Purple sapphire may more closely resemble tanzanite. Most, though not all, purple sapphire behaves exactly like tanzanite in changing lighting environments, appearing almost gem blue in daylight and picking up a distinct purple secondary hue in incandescent light.

313. Keller, *Ibid.*, p. 72.

Crystal

Tanzanite has a wonderful crystalline character that is barely affected by the heat-enhancement process. Blue sapphire of a similar tone (seventy-five to eighty-five percent) will often appear murky by comparison. Unheated sapphire from Burma will often show excellent crystal, but in fine tanzanite good crystal should always be present.

Most tanzanite is purple or violetish to purplish blue. Stones with a purple primary hue normally occur in lighter tones, usually no darker than seventy percent. Purple sapphire is normally darker, between seventy-five and eighty-five percent tone. Purple sapphire with tonal values less than seventy-five percent do resemble tanzanite of a similar tone. Such stones may be difficult to separate visually from tanzanite.

Durability

If tanzanite has a downside, it is durability. Tanzanite is relatively soft, ranking at six and one half on the Mohs scale. From a toughness perspective, the stone is somewhat brittle. Gems with a hardness of less than seven are subject to abrasion if wiped with a dusty cloth. Dust is mainly a silicate and its hardness is seven on the Mohs scale. Still, six and a half is approximately the hardness of steel and is the cutoff for membership among the new precious gemstones. Tanzanite is problematic as a ring stone, particularly if worn every day, and should be cleaned carefully using solutions manufactured for the purpose.

The Rarity Factor

Tanzanite is the single exception to the usual relationship between size and rarity. Most gem varieties become rarer and hence more valuable as size increases. Tanzanite appears to require greater mass to achieve its finest color. For this reason a one-carat gem-quality tanzanite is rarer than a five-, ten-, or twenty-carat stone of the same quality; a sub-carat fine gem is rarer still. One would expect that larger stones would command a higher per carat price. For years, however, this was not the case. Smaller fine quality gems sold for lower per carat prices than larger ones. Only recently have dealers begun to recognize this fact and to adjust prices accordingly.

Tanzanite prices have seesawed wildly over the past decade due to the fluctuating political climate in Tanzania. Prices seemed to bottom out in 1997, with stones selling for as much as sixty percent off of prices current in the early eighties. In 1998 tragedy struck the mines, all of which are in Merelani. Cave-ins took the lives of a number of miners and choked off rough supplies. A false rumor that tanzanite sales were used to finance terrorism caused another precipitous fall in prices after the destruction of the World Trade Center on 9-11 2001. As of this writing, prices for tanzanite have stabilized. Still, a fine tanzanite can be purchased for pennies on the dollar when compared to sapphire. Considering the gem's beauty, tanzanite, today, must be considered one of the true bargains in the gem world.

FANCY COLOR DIAMONDS

"The colouring of diamonds is seldom intense, pale colours being much more usual than deeper shades. Diamonds that combine great depth and beauty of colour with perfect transparency are objects of unsurpassable beauty"

–Max Bauer, 1904

Fancy color diamonds pose a real challenge for the connoisseur. Unlike their colorless brethren, fancy color diamonds are genuine rarities.

The "Colour Variety Collection" assembled for a private collector by colored diamond expert Stephen Hofer. Each of the thirteen colors are represented. Photo: Stephen C. Hofer.

Color in diamond is rare because diamond is composed of a single element, carbon. Diamond is, in fact, the only gemstone composed of a single element. Carbon has an extremely tight atomic structure, which explains the stone's legendary hardness. It is very difficult for impurities, the atoms that cause color, to make their way into the diamond's crystal lattice.[314] Agents such as vanadium, chromium, and iron, the causes of color in many other varieties of gemstones, are composed of atoms simply too large to replace carbon atoms inside the crystal lattice. In fact, of the one hundred or so known elements, the atoms of only three, nitrogen, boron, and hydrogen, are small enough to work their way into the compact structure of the diamond crystal lattice. Nitrogen, the most common impurity atom, is the cause of yellow, violet and some pink diamonds. Boron is responsible for the gray blue colors. Hydrogen is the coloring agent for some red, olive, violet, and blue colors.

Color in diamond has two additional sources: atomic radiation, which is responsible for the color of green diamonds, and the physical deformation of the atomic structure of the crystal, known as plastic deformation. Plastic deformation is responsible for the color in red and pink diamonds.

In his seminal book on fancy color diamonds, Stephen Hofer maintains that fancy color diamonds occur in twelve basic hues: red, orange, yellow, green, olive, blue, violet, purple, white, brown, gray, and black.[315] Also, in fancy color diamonds, white, gray, and black may function as primary hues.[316] Olive, too, has a distinguished lineage.[317]

Chapter 5 includes a detailed discussion of the criteria used to grade fancy color diamonds. It is recommended that the reader consider this section in conjunction with the essays that follow.

---

314 The crystal lattice is the three-dimensional arrangement of carbon atoms which is the basic unit of structure in the diamond crystal. The diamond lattice is composed of a tetrahedral arrangement of carbon atoms with one atom at each corner and one in the center.

315 Stephen C. Hofer, *Collecting and Classifying Fancy Coloured Diamonds* (New York: Ashland Press, 1998). This book is essential reading for anyone interested in collecting fancy color diamonds. Recently at least one diamond with a violet primary hue has been found.

316 The reader will recall that throughout this volume the color of gemstones has been described using just eight hues: red, orange, yellow, green, blue, violet, purple, and pink. Gray and brown are considered masks that dull the brightness of the hue (see Chapter 3). The terminology may be confusing, since GIA classifies as hues all of the above, *plus* brown and gray, but excludes olive, which is termed gray-yellow-green. Given the importance of laboratory grading in the fancy color diamond market, it seems appropriate to stay as close to GIA-GTL terminology as possible in this section.

317 Stephen Hofer, personal communication, 1999.

Chapter 43
Red Diamonds

"The true red diamond is valuable according to the glorious beauty of its perfection . . . it feeds the eyes with much pleasure in beholding, and hence discovered to us the Excellencies of super-celestial things."

–Henry Nichols, 1651

Red diamond is arguably one of the rarest substances on earth. Certainly it is the most expensive. The concept of a pure red diamond, like the pure red ruby, probably exists somewhere in the Platonic paradise of pure idea and not in this world at all. The Gemological Institute of America (GIA-GTL) has documented the existence of fifteen red diamonds. Of these, only four have received the grade of *fancy red*, which classifies them as pure red diamonds with no secondary modifying hue.[318] Australia's Argyle Mine, famous for its production of pink diamonds, produced one thousand and sixty five pink diamonds between 1995-2014. Of this group, twenty-seven were graded purplish red, thirteen were graded pure red.

Hue, Saturation, Tone

Red is a deep dark pink and inversely pink is light-toned pale red. In the

Pictured at the top of the photograph and weighing in at just 0.95 carats carries a GIA-GTL grade of Fancy Purplish Red. The middle stone weighing 0.59 carats is a Fancy Purplish Pink. The stone pictured at bottom weighing 0.54 carats was graded Fancy Reddish Purple. This photograph illustrates how subtle the nuances of hue can be and the importance of the wording of a GIA-GTL grading report. In this case, purplish pink is particularly difficult to separate from purplish red. Photo courtesy GIA.

---

318 King et al., "Natural-Color Pink Diamonds," p. 134. These fifteen stones are described by GIA-GTL as being "in the public domain," meaning that other stones graded red could not be included due to confidentiality agreements with its clients.

 315

connoisseurship of red diamonds this is an important fact to keep in mind. Any dividing line between pink and red must be in some sense arbitrary. In the marketplace the distinction is extraordinarily important.

Given the tens of thousands of dollars, sometimes hundreds of thousands of dollars in price, the terminology used in the lab report becomes crucial.

As with ruby, pink and purple and orange are the most attractive secondary hues in red diamonds. Of those reds that exist, pinkish stones are necessarily lighter in tone than purplish stones.[319] The most expensive diamond ever sold, the 0.95-carat Hancock Diamond, is a fancy purplish red.[320] As with colored gems generally, chromatic secondary hues are more desirable than neutrals (masks) such as gray and brown. Brown is the normal saturation modifier or mask found in red

The 0.88 carat Fancy Purplish Metzger Red. The stone was sourced in Brazil. Photo: Jeff Scovil, The Edward Arthur Metzger Gem Collection, Bassett, W. A. and Skalwold, E. A.

diamonds. Orangy reds tend to include an element of brown.[321] Brownish reds tend to be darker in tone and, of course, duller of hue.

GIA-GTL revised its diamond color grading nomenclature in 1994. The new terminology retains the original designations *faint, very light, light, fancy,* and *fancy intense,* adding two additional classifications, *fancy vivid* and *fancy deep.* As of this writing, GIA-GTL has not found the need to use any term beyond *fancy* to describe red diamonds graded in its laboratory.[322]

Multicolor Effect

Multicolor effect is a characteristic found in red diamonds. Often this shows up as tonal variations of purple and pink visible in portions of the face-up mosaic of the gem and can be very pronounced indeed. As discussed in Chapter five, in the section on grading fancy color diamonds, pinkish red is not a term used by GIA-GTL

319 The term *pinkish red* is not part of GIA-GTL's fancy diamond grading nomenclature. A pinkish red would simply be graded fancy red.

320· Robert E. Kane, "Three Notable Fancy-Color Diamonds: Purplish Red, Purple-Pink, and Reddish- Purple," *Gems & Gemology,* Summer 1987, p. 91..

321 Stephen Hofer, personal communication, 2016.

322. King et al., "Natural-Color Pink Diamonds," p. 135.

on its fancy color diamond grading reports. Since pink is a lighter-toned paler red, a stone of this description would be graded fancy red. This is another example of how the certificate-driven world of fancy color diamonds can leave the budding aficionado scratching his or her head. I examined a 0.52-carat radiant cut diamond with a GIA-GTL report that described the stone as fancy purplish red. The stone was certainly red, but exhibited strong pink multicolor effect along the edge of the crown.

Clarity

Color in red diamonds is a result of plastic deformation of the crystal lattice. As a result, the color in red and pink diamonds may occur in thin closely packed zones called grain lines. This may result in the stone showing some texture in the face-up position. As in sapphire, texture, or lack of uniformity, in the face-up color is a negative.

It is difficult to talk about negatives with a stone of such breathtaking rarity. As with any colored gem, a brownish mask muddies the hue and should be considered a negative. Texture is a negative. However, it is difficult to conceive of any collector turning down a red diamond, assuming he could afford to purchase it, regardless of its clarity grade. Clarity grade has little or no impact on price in the rarer hues of fancy color diamonds. The fancy color diamond market is driven by rarity; beauty is, at best, a secondary consideration.

8.72 carat (type IIa) Historic Pink diamond. Formerly owned by Princess Mathilde Bonaparte. Sold at Sotheby's, Geneva in 2015 for $15.9 million in 2015. Photo courtesy Sotheby's.

Chapter 44

Pink Diamonds

"The first was a stone I came across in the reign of Pope Clement, a diamond literally flesh-coloured, most tender, most limpid, it scintillated like a star, and so delightful it was to behold that all other diamonds beside it, however pure & colourless, seemed no longer to give any pleasure and lose their gratefulness."

–Benvenuto Cellini, 1568

The 23.88 carat Graff Pink Diamond. The diamond was purchased in 2010 for 46.5 million dollars, the highest price ever paid for a gemstone. This stone is graded Fancy Vivid Pink. The stone most probably hails from the ancient Indian mines near Golconda. Photo courtesy Sotheby's.

Prior to 1985, when pink diamonds from Australia's Argyle Mine began to filter into the market, pink diamonds were an extraordinary rarity. The few pink stones known came from the famous alluvial deposits of southern India, southern Africa, and the Brazilian state of Minas Gerais. A good proportion of those stones were very light in tone and pale in saturation, and were likely to garner designations of faint to light on the GIA-GTL fancy diamond grading scale. The Argyle pinks changed all that. Between 1985-2014 Argyle has sold one thousand and sixty five pink diamonds. Of these, ninety five percent were graded Fancy Deep, Intense or Vivid.[323] To put the numbers in perspective, the total number of pinks produced at Argyle represent less than one tenth of a percent of the mine's total production. Dealers and connoisseurs await with much anticipation Argyle's once-a-year tender of pink diamonds.

Unfortunately Argyle pinks are relatively small. Only two cut stones over three carats have been produced. Famous large pinks are from India, Brazil, or southern Africa.

323 John M. King, et al, *ibid*, 2014. P. 277

The color in pink diamonds occurs in thin, needlelike color zones. Darker zones, closely packed, produce a deeper-toned hue. These zones, commonly called pink graining, can be visible to the naked eye. Graining is caused by deformation of the diamond crystal while it is in a semisolid state, a process known as plastic deformation. This deformation of the crystal lattice is the cause of color in pink diamonds. Profuse zoning can lend the pink diamond a whitish fuzzy or hazy appearance, known as *texture*. An even face-up color is the ideal, as in all gemstones. Visible zoning is not desirable.

Hue/Saturation/Tone

Pink is not a distinct hue. It is a pale or less saturated light tone of red. Thus the visual division between red and pink is in some sense arbitrary. Pure pinks are particularly rare. The secondary hues found in pink diamonds are red, purple, and orange. Argyle stones are known for their purplish secondary hues. Light purplish secondary hues are sometimes erroneously described as violetish. In the marketplace, pure pinks and reddish pinks are the most desired, followed by purplish and orangy pink.

Brown is the normal saturation modifier or mask found in fancy pink diamond though gray is sometimes present. In fact, brownish purplish pink is considered the signature color in pink diamonds from the Argyle mine. In the marketplace, brownish stones are generally preferable to grayish ones. Brownish stones generally appear warmer, and the brown is often prominent enough to rival the pink hue, adding to the overall depth (tone) of the color. Grayish stones read simply as a dull-hued pink.

Pink stones are, by definition, pale (light) in tone and this creates difficulties for the connoisseur judging color. Deep-toned pinks approach red. Dull pinks shade into brown. Warm orangy pinks may be confused with brownish pinks. Light-toned pinks are easily confused with light-toned purples. Lighting is critical. To evaluate color in fancy color diamonds, GIA-GTL uses a special neutral gray box called the Judge II and a 6,500-kelvin daylight fluorescent light source manufactured by the MacBeth Corporation. This kelvin temperature is equivalent to daylight lighting at noon. However, unlike natural lighting, the light box is not subject to the modifying effects of changing latitudes and air quality.

Multicolor Effect

Isn't a reddish-pink red? In the certificate-driven world of fancy color diamonds, such questions are answered by laboratory analysis, with the most important document by far the one issued by GIA's Gem Trade Lab (GIA-GTL). In an attempt to avoid confusion, GIA-GTL does not use the term reddish pink in its grading reports. This is not to say that reddish pink diamonds do not exist — they most definitely do. As discussed in Chapter 5, pink stones often have portions of the face-up mosaic of the gem which exhibit a definite red primary hue. The certificate grade, however, will name the dominant hue that which encompasses the largest percentage of the gem's face, ignoring the multicolor effect.

In 1994 GIA modified its colored diamond grading nomenclature — *faint, very light, light, fancy, fancy intense, fancy vivid,* and *fancy deep.* A majority of pink diamonds will fall into the first four categories of saturation/tone. According to fancy color diamond expert Stephen Hofer, of the three hundred forty-five pink diamonds sold at the major houses between 1959 and 1999, only three merited the designation *fancy intense.* Hofer further states, "In the last couple of years, a growing number of Argyle stones are being awarded the designations *vivid* and *deep* pink to satisfy the market's fascination with exemplary pink colors.[324]" Hofer's statement is a circumspect way of saying that these designations may have more to do with marketing than beauty.

Origin and Price

Most gemstones are formed in the earth's crust and differing local geology puts a stamp in the form of inclusions specific to the local environment. Diamond is the sole exception. Diamond is formed in the earth's mantle and is, therefore, almost impossible to determine country of origin. Pink diamonds are something of an exception. Most pink diamonds come from Australia, specifically the Argyle mine located in the rugged northwestern part of the country. Recent studies (2015) suggest that Argyle pinks have characteristics and inclusions specific to the Australian location.[325] Pink diamonds from older sources - India, Brazil and South Africa

– are mainly type IIa stones with little or no measurable nitrogen. Argyle pinks, by contrast, are type Ia diamonds and do contain nitrogen. Type Ia diamonds come from a number of sources including Russia, Venezuela and Canada. Type Ia pink diamonds may sell at something of a discount to type IIa gems, particularly those with provenance traceable to India's legendary Golconda Mines.

Size and Rarity

A majority of pink diamonds are available in melee sizes (0.01-0.19 carats). Most Argyle pinks are available in these sizes. A pink diamond weighing over a fifth of a carat (0.20) should be considered rare. A pink diamond above one carat is exceptionally rare and a gem above two carats extraordinarily rare. As size increases, prices will increase geometrically.

324 Stephen Hofer, personal communication, 1999.

325 Branko Deljanin, et al., "The World of Pink Diamonds and Identifying Them,"*Incolor Magazine,* The International Colored Gemstone Association, (Spring 2015). pps. 32-40.

Three purple diamonds: Right to left, Fancy Vivid Purple, Fancy Intense Pinkish Purple Pink and Fancy Pinkish Purple. Photo courtesy: Leibish & Co.

Chapter 45

Purple Diamonds

"Pure purple diamond with no secondary color {modifier} is very rare . . . when such diamonds come on the market they are invariably small — almost never more than 1 carat and usually under 50 points . . . there has yet to be a large, historically important pure purple diamond."

–Harvey Harris, 1994

A 0.85 carat Fancy Pinkish Purple (FpkP) diamond. Photo: Jeff Scovil, The Edward Arthur Metzger Gem Collection, Bassett, W. A. and Skalwold, E. A.

urple is a modified spectral hue existing halfway between red and blue on the color wheel. Purple is often confused with violet, which is a pure spectral color between purple and blue on the color wheel. Thus, purple teeters on the spectral seesaw, balanced between red and blue.

Purple diamonds are extreme rarities. A number of these stones, lilac in color and mostly less than one carat, appeared in Antwerp in the spring of 1989. Rumor has it that they were Russian in origin. Before the appearance of these stones, fancy purple diamonds were virtually unheard of.[326]

Hue, Saturation, Tone

Color science tells us that the purple hue reaches its optimum saturation, its gamut limit, at about sixty percent tone. For this reason, medium dark-toned purples are more visually satisfying, more beautiful. Unfortunately most purple diamonds occur in tones less than fifty percent and, therefore, are lacking in saturation, appearing quite pale.

Color in purple diamonds is often uneven. The color occurs in purple zones alternating with colorless zones. At times the color may be limited to grain lines, resulting in uneven or blotchy color, or

326 David Federman, "All in the Family: Inside the Diamond Spectrum," *Modern Jeweler*, (October, 1990), p. 51.

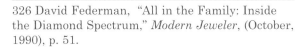

texture, when the gem is viewed face up. Uneven color is much less desirable than even color in this and in all gemstone species.

Pink and red are the normal secondary hues found in purple diamonds. Red is obviously the more desirable, as it is rare and by definition a darker tone than pink. Reddish purple is also prized more in the marketplace, even more than a pure purple hue. A touch of red lends the primary hue both vividness and warmth; also, due to its rarity, any time the word red is mentioned, bells start ringing in the world of fancy color diamonds. Pure purple hues are rarer, but this is one of the more unusual cases in colored gemstone connoisseurship where a pure spectral hue is not the most desired. Pink is somewhat less desirable as a secondary hue, and it occurs in lighter-hued stones.

At lighter tones, it is extremely difficult for even the trained eye to differentiate between a pink and purple. This problem is exacerbated when the two hues are mixed, as in pinkish purple or purplish pink diamonds. The use of the correct lighting environment is especially critical when evaluating color in diamonds in the pink to purple color range. North daylight at noon is the standard. GIA-GTL uses a 6,500-kelvin fluorescent light manufactured by the MacBeth division of the Kollmorgen Corporation. Given the importance of the GIA-GTL color "cert" in the marketplace, the serious connoisseur is well advised to invest in one of these lamps.

Either brown or gray can occur as a saturation modifier or mask in purple diamonds. Brown is more common. Experts use the term smoky to refer to purple diamonds that exhibit a mask. Many, perhaps most, light-toned purple stones do not show a mask. A grayish mask in lighter-toned purple diamonds will give the stone a cool aspect; brown will appear warmer. If the diamond is pale, dull, and cool, a grayish mask can be inferred. Incandescent lighting will bring out the brown. Daylight and daylight-equivalent fluorescent lighting will bring out the gray. Although noon daylight is the theoretical standard, the aficionado is advised to evaluate the gem in both daylight and incandescent light.

Chapter 46

Green Diamonds

"Diamonds of a green color are distinctly rare, only a few examples being known. The most beautiful green diamond known is a transparent brilliant weighing 48 ½ carats preserved in the 'Green Vaults' at Dresden."

–Max Bauer, 1904

The green diamond follows red in overall rarity. According to the records of the major auction houses, just sixty-four natural green diamonds have been offered at auction between 1959 and 1997. Of this total, only twenty-nine were judged to be pure green.[327] During this same period, the major auction houses sold one hundred seventy-five of the next rarest fancy color diamond, natural color blues.

Most known green diamonds in historical collections are type Ia and were mined in Brazil during the 19th century. The most famous green diamond of them all resides in the famous Green Vaults of Dresden. Known as the Dresden Green, this forty-one-carat, slightly grayish medium-toned fancy green is the largest green diamond in existence.[328] It has lain in the Green Vaults since its purchase in 1741 by Friedrich Augustus II, Elector of Saxony.

There are two types of green diamond: those with a green skin and those with an

The legendary Dresden Green diamond; set in its original mounting, a hat ornament made by the Prague jeweler Diessbach, circa 1768. Modern measuring techniques indicate that the Dresden Green weighs approximately forty-two carats. Though visually grayish, the diamond would carry a GIA-GTL grade of fancy green. Photo: Shane F. McClure © Gemological Institute of America.

327 Hofer, *Fancy Coloure Diamonds*, p. 7.

328 Robert E. Kane et al., "The Legendary Dresden Green Diamond," *Gems & Gemology*, (Winter, 1990), pp. 248-265.

inherent green body color. Natural green diamonds derive their color from the association with natural atomic radiation. The length of exposure and the type of radiation — alpha, beta or gamma rays — determine the depth of green color. Those exposed to gamma radiation for extremely long periods of time will exhibit a green body color. Green diamonds with inherent green body color are considered much more desirable than greens whose color is only, so to speak, skin deep.

1.08 carat Fancy Vivid Green-Blue, SI1 diamond. This gem was acquired in 2003. A bit of a tweak and you've got a blue-green. Photo: Jeff Scovil, The Edward Arthur Metzger Gem Collection, Bassett, W. A. and Skalwold, E. A.

produces a deeper warmer hue. Fancy green diamonds may also fluoresce green in ultraviolet light. This fluorescence sometimes laps over into the visible spectrum. This is one of the rare instances where visible luminescence may be considered in calculating the color grade on a diamond certificate.

Gray is the normal saturation modifier or mask found in green diamonds though brown is also possible. As with blue diamonds, the gray secondary is sometimes added to the green to pump up the grade (see Chapter 4). For example, GIA-GTL specialists describe the Dresden Green as "visually grayish," yet its certificate grade in 1990 would have been fancy green.[329]

Aside from the Dresden Green, there are no large historically important green diamonds. The largest notable fancy green is a 6.13-carat rectangle Fancy Intense Green which sold at Christie's Hong Kong in 2014 for $594,510 per carat.[330] Only

Hue/Saturation/Tone

Most green diamonds are pale of hue and light of tone. Due to the gem's rarity, only a small quantity of green (saturation) is required for the stone to make the grade. The normal secondary hues found in green diamonds are yellow and blue. It is difficult, with this degree of rarity, to speak of preferred secondary hues. A yellow secondary produces a brighter, more saturated hue overall. Two former record setting green diamonds sold at auction (1985 and 1995) were stones with a yellowish secondary hue. A blue secondary

329 Ibid.

330 Federman, *All in the Family*, p. 49.

two fancy vivid greens have ever been sold at auction. In 2009 a 2.52 carat sold for $3.07 million and in 2016 The *Aurora Green* a 5.03 carat Fancy Vivid Green sold for $16.2 million dollars.

Color Treatments; Buyer Beware

Though most artificially colored fancy color diamond may be separated by gemological labs using advanced testing equipment, as of this writing, the reliability of testing methods used to separate natural green from artificially irradiated green diamonds is a matter of some dispute.[331]

The Gemological Institute of America's gem trade lab (GIA-GTL) has issued natural grading reports. However, other prominent experts have questioned GIA's methodology. Color treatments began in 1905 using Radium. Diamonds treated by this ancient method may still be dangerously radioactive. Modern treatments normally involve linear accelerators. According to colored diamond expert Stephen Hofer, modern methods of treatment are so precise that it is impossible to distinguish between naturally irradiated and artificially irradiated gems. Another noted expert makes the following observation. "The separation of natural versus treated greens is more complex than most people assume, and even with known

green samples, it is difficult, as we do not have any known green stones in museums that have the kind of "super" saturated green colours which we are seeing in the market today."[332] Collectors seeking a green diamond for their collection may be on extremely unstable ground.

---

332 Stephen Hofer, personal communication, 2014.

---

331 Thomas Hainschwang, personal communication, 2014. Hainswang, who was in the final phase of his Ph.D. under Emmanuel Fritch at The University of Nantes, maintains that only diamonds in historical museum collections can be considered natural. See also: Frank Notari and Thomas Hainschwang "Issues With Treated and Natural Green to Greenish Blue Diamonds," *Gem Market News,* (May/June 2014), pp. 1-4.

12.03 carat Blue Moon of Josephine Diamond sold at Sotheby's, Hong Kong for $48.4 million. Photo: courtesy Sotheby's. .

Chapter 47

Blue Diamonds

"Although writers describe these stones (blue diamonds) as possessing . . . the beauty of fine sapphires, no comparison can really be instituted, their blue color being peculiar to themselves, dark, verging on indigo possessing a characteristic intensity which differs materially from the mild, soft hue of the Sapphire."

—E.W. Streeter, 1879

Blue diamonds are one of nature's great rarities. Unquestionably the most famous blue is the 45.52-carat Hope Diamond on permanent display at the Smithsonian Institution in Washington, DC. The Hope has an interesting history. The celebrated seventeenth-century gem merchant Jean Baptiste Tavernier originally purchased it in India sometime before 1667. Tavernier sold it to Louis XIV in 1668. The stone originally weighed 112.25 carats and was known as the Tavernier Blue. It was subsequently recut to 67.12 carats and rechristened *le diamant bleu de la couronne de France* or more simply: The French Blue. Stolen in 1792 during the French Revolution, the stone did not resurface again until 1830, recut in its present form. According to the story, the stone was then sold to the English financier Thomas Philip Hope.[333]

Although this famous gem has been described in the literature as "sapphire

FThe Fancy Deep Grayish Blue (FDpB) Hope Diamond under intense lighting. Photo: Tino Hammid.

blue" and "superfine deep blue," these descriptions are a bit wide of the mark. The Hope is of a "fancy deep grayish blue" hue.[334] Visually, the gray modifier is so dark it appears black, leading to the term some experts use to describe the Hope: "inky blue." Fancy deep–hued stones of this type comprise less than ten percent of all blue diamonds.

---

333 There is a very good chance that Henry Phillip Hope purchased The French Blue himself and then had it recut. C.f. Wise, Richard W., *From The Sun King To The Smithsonian, The Epic Story of The French Blue.* www.thefrenchblue.com.

334. King et al., "Characterizing Natural-Color Type-IIB Blue Diamonds," *Gems and Gemology* (Fall, 2015), p. 262.

The 31.06 carat, Fancy Deep Blue (FDpB) Wittelsbach-Graff Diamond. Photo: Gary Roskin.

The question is often raised as to which GIA-GTL designation — Fancy Vivid or Fancy Deep — confers the greatest value on a blue diamond.[335] Were it not for the prevalence of a gray mask in blue diamonds this question would be fairly easily answered, as blue normally achieves its optimum saturation between eighty and eighty-five percent tone. Fancy Deep is the designation describing stones that are moderate to dark in tone. As it stands, it is the percentage of gray mask that provides the answer. The gem with the lowest percentage will be the more beautiful and hence the more valuable.

A good example of this is the 35.56 carat Wittelsbach Blue Diamond. First

mentioned in 1667 as part of the collection of the Empress Margarita Teresa, it was purchased in 2008 by diamond dealer Lawrence Graff for a record price of $24.3 million. Originally graded Fancy Deep Grayish blue, the diamond was recut by Graff to 31.06 carats and resubmitted to GIA-GTL, which regraded it as Fancy Deep Blue. Visually when viewed side by side with the Hope, it doesn't resemble the Hope at all. The newly minted Wittelsbach-Graff appears to be a medium-toned bluish gray in contrast to the Hope's inky dark grayish blue hue. Despite the gem's true appearance, the gray modifier/mask is ignored or simply added to the overall tone.[336]

Next to red and green, blue diamonds command the highest prices of any substance on earth. Of the ten highest per carat prices paid for fancy color diamonds, six have been for blue diamonds.

Hue/Saturation/Tone

The great majority of blue diamonds are very light in tone and pale in hue. Blue diamonds are extremely rare in nature, so rare that almost any quantity (saturation) of blue hue will be sufficient for a laboratory report to qualify the stone as a blue diamond. As stated, most blue

335 John M. King, "Gem Trade Lab Notes," *Gems & Gemology*, Spring 2002, p. 80. A similar situation exists with green diamonds. See the section on grading fancy color diamonds in Chapter 5.

336 The author personally viewed he Hope and Wittelsbach diamonds at The Smithsonian Institution displayed side by side and unset. According to the late Chip Clark, the Smithsonian's official photographer: "The color itself, the base color is the same as the Hope." By base color I assume he meant the same hue because he goes on to say; "The Hope is a lot darker." So the Hope is Fancy Deep, but subtract the gray and the Wittelsbach would appear a very light toned blue indeed. Vid. http://roskingemnewsreport.com/an-evening-with-the-blues-part-v/.

The 10.10 carat De Beers Millennium Jewel. Graded Fancy Vivid Blue and Internally Flawless (IF) by GIA-GTL. Sold at Sotheby's Hong Kong on April 5, 2016 for 22.5 million, the highest priced oval blue diamond ever sold at auction. The Blue Moon of Josephine sold at Sotheby's Geneva in November 2015 and retains the record for the highest priced blue. Photo: Courtesy Sotheby's.

diamonds are noticeably gray or grayish; so much so, in fact, many so-called fancy blues are actually slightly bluish gray. At very low tonal levels, grading reports issued by the Gemological Institute of America treat this gray modifier as a maximizer, adding the gray to the blue, thus pumping up the grade. Most blue diamonds do exhibit a gray mask, leading to such descriptions as steel blue, another term often used to describe the Hope. The Hope, however, is more than steely, it is so dark in tone it is inky.

This is not to say that visually pure blue stones do not exist; they do, but most are slightly steely, and a gem with fifteen percent or less of a gray mask is a stone of great beauty and will command

an astonishing price. Green is the normal secondary hue found in blue diamonds and occasionally one with a violetish secondary hue will be seen.[337] As with most gemstone varieties, a visually pure primary hue is the ideal. After a visually pure blue, greenish blue to green blue are the most desirable combinations of primary and secondary hues.[338] As with other colored gemstones discussed in this book, a gray mask is a distinct negative. Thus, the lower the percentage of gray the more beautiful and more desirable the ston

337 The green secondary hue is caused by carbon vacancies in the atomic structure. The violet is the result of hydrogen impurities. A Fancy Light Greenish Blue (FLgB) will likely be called Fancy Light Blue (FLB) on the GIA report. At very low levels of saturation/tone, GIA-GTL ignores the secondary hue. Christopher Smith, Personal Communication, 2015.

338 As with gray, the description on the GIA-GTL report will not call chromatic secondary hues in stones graded Fancy Light Blue (FLB).

Fancy Vivid orange diamond viewed under a 10x loup. Photo: cour-
tesy Liebish & Co.

Chapter 48

Orange Diamonds

"Of the many pure (fancy) orange diamonds reported in the literature most actually have a tinge of yellow. . . . Since yellow and orange colours mix so intimately with each other, this often makes it difficult to decide visually if a subtle yellow modifier is present. "

–Stephen Hofer, 1998

In October 1997 The *Pumpkin*, a 5.52-carat Fancy Vivid orange diamond sold at Sotheby's for $238,718 per carat or a whopping $1,322,497 for the stone. [339] In November 2013 a 14.82-carat Fancy Vivid Orange, reputed to be the largest orange diamond on earth, sold at Christie's for $36 million or $2.4 million per carat. Prices like this demonstrate just how costly fancy color diamonds have become.

Hue/Saturation/Tone

Orange is a spectral hue lying halfway between the hues yellow and red on the color wheel. GIA-GTL does not use the terms Faint, Very Light or Light to describe saturation/tonal levels in orange diamonds. Pure orange diamonds are exceedingly rare. The grading of orange diamonds is further complicated by the fact that the color we call brown is not a distinct spectral hue; it is in fact a deep, dull orange. When grading darker-toned stones, it is difficult to determine where orange ends and brown begins. Is it a fancy orange brown or a brown orange? The

The world's largest known 14.82 carat Fancy Vivid orange diamond of about 35-40% tone. Note the yellow in The Orange. Bridgeman Images.

difference can mean thousands, even tens of thousands of dollars. There is no such thing as a dark-toned, dull orange diamond: diamonds of this description are brown. So, given orange's low gamut limit, 20-35% tone should be considered ideal[340]

339 "Fancy Vivid" is GIA-GTL's highest designation. (See Chapter 5, Introduction and overview to fancy color diamonds.)

340 Hofer, Fancy Coloured Diamonds, p. 325.

Red, pink, yellow, and brown are the normal secondary hues found in orange diamonds. Green is sometimes found as well. As is true in most gemstone varieties, a visually pure orange is the most desirable hue, followed by stones that are reddish, pinkish, yellowish, greenish, or brownish, in that order.

Brown, as might be guessed, is the most common saturation modifier or mask found in orange diamonds. As noted, brown adds measurably to the confusion, particularly in fancy color diamonds, because in the GIA-GTL system of grading fancy color diamonds, brown is classified as a hue rather than as a saturation modifier or mask as defined in this book. Thus a dull dark-toned orange diamond is a brown diamond and, as will be noted in the following chapter, is graded on a different scale.

Multicolor Effect

Orange diamonds often exhibit multicolor effect, usually in the form of alternating flashes of orange and brown. Due to the varying lengths of the light path in both brown and orange diamonds, some light rays will absorb more color than others, resulting in a distinct multicolor effect in the face-up mosaic of the gem. This is further complicated by the addition of yellow, which is the most common secondary hue. It can be hard to decide if the gem is orangy brown or brownish orange. A diamond that has a primary orange hue will, of course, bring a much higher price in the marketplace.

Day Stone

Orange diamond is normally a daystone. If there is a brownish modifier, incandescent lighting will bring it out. As always the gem should be examined in both natural and incandescent lighting.

The Rarity Factor

Orange diamonds are so rare that any size orange stone will be pricey. One carat, two carats, five carats and ten carats would be the expected price breaks, but there are so few larger stones known that these size breaks are, for the most part, theoretical.

Chapter 49

Brown Diamonds

*"A diamond when brown unless of a deep and pleasing colour,
is very undesirable, as it absorbs much light and appears dirty
by daylight and dark and sleepy by artificial light."*

–Frank Wade, 1918

*The 7.69 carat Autumn Diamond. Graded Fancy
Deep Orangy Brown by GIA-GTL and measured
as Brownish Orange by colorimeter, the lovely
orangy/red tones are not included in the description on the grading report. Photo: Jeff Scovil.
The Cora N. Miller Collection, Peabody Museum,
Yale, courtesy R. W. Wise, Goldsmiths.*

rown is not a unique spectral color; it is a dark-toned dull hue of orange. This, as noted in the foregoing chapter, complicates the grading equation. Light-toned brown diamonds create an additional complication. Brown-"ish" diamonds, like yellowish diamonds, are considered off-color and are graded on the Gemological Institute's colorless diamond scale. This scale could also be called the colorless to light yellow, brown, and gray scale, as it evaluates brown in almost exactly the same way as it does yellow. This is based on a tradition which historically has considered brown stones to be of little value as gems.

Still, there are a number of famous brown diamonds, most of them from South Africa. The Earth Star, a 111.59-carat pear-shaped orangy brown, was cut in 1967 from a 248.90-carat rough found at the Jagersfontein Mine. This stone was sold to a collector in 1983 for $900,000, or $8,065 per carat.[341] Other famous brownies include the 104.15-carat pear-shaped bronze called the Great Chrysanthemum and the 55.09-carat emerald-cut Kimberley.

Market resistance to brown-hued diamonds began to break down in the 1980s as Fancy color diamond prices began

---

341 Harris, *Fancy-Color Diamonds*, p. 142.

A finely cut Fancy Deep Orangy Brown diamond. Photo: Jeff Scovil, courtesy R. W. Wise, Goldsmiths.

When Australia's huge Argyle Mine came on line in the early 1990s, brown diamonds became more available in the market. In fact, fully one-third of Argyle's production is brown stones. The mine's owners initiated a brilliant marketing program to sell its brown diamonds. Rechristened "champagne" and "cognac," these formerly despised colors became the new darlings of the diamond market.

A Brown is a Brown is a...

Diamonds with a faint brownish body color to a light brownish body color are graded L through Z on the colorless diamond scale. Brown diamonds graded K through M are termed Faint Brown; those graded between N and R are called Very Light Brown, and stones graded S through Z are termed Light Brown. Brown stones considered fancy color diamonds begin with

to escalate, and changed forever in the late 1980s when a 4.02-carat Fancy Brown brought $4,925 per carat at auction in 1986 and a 8.91-carat fancy orangish brown

C–1   C–2   C–3   C–4   C–5   C–6   C–7

The Argyle Mine's "champagne" diamond grading scale, with tonal values from five to ten percent called light champagne (C-1 and 2), twenty to forty percent termed medium champagne (C-3 and 4), fifty to seventy percent dark champagne (C-5 and 6), and fancy cognac (C-7). Note the tonal jump from forty to sixty percent between C-4 and C-5. C-7 is approximately eighty percent tone. Photo courtesy Stephen Hofer.

diamond sold at Christie's for $82,497 or $9,259 per carat in 1987.[342]

the grade Fancy Brown at about thirty to thirty-five percent tone. Diamonds carrying letter grades on the colorless scale are not

342 *ibid.*

considered fancy color diamonds; they are off-color colorless diamonds and will carry a negative premium.

It is important to remember that the letter grades on the GIA-GTL colorless scale reflect the body colors, not the key color, of the stone. A diamond with a Z grade should face up brownish rather than brown. GIA-GTL uses master stones to grade diamonds on this scale. For a yellow, gray, or brown diamond to achieve fancy status, its key color must be more saturated than the key color of the GIA-GTL Z-grade master stone viewed face up. The aficionado is advised to examine diamonds graded X through Z on the GIA-GTL colorless scale closely. They may face up with more color than the actual grade would suggest.

Hue/Saturation/Tone

Any hue except green is possible as a secondary hue in brown diamonds. In fact, red followed by orange, pink, purple, and yellow are the usual modifying hues and, in descending order, are the most desirable in the marketplace. Reddish brown is distinctly beautiful. Diamonds of this hue have the quality of fine old wine aged in oak casks. Orangy, along with yellowish brown, is the most common. A visually pure chocolate brown is also quite desirable and difficult to find. These are followed by the non-spectral, or mask, hues of gray, white, and black. As is the case with other gems, gray tends to mask or dull the color, reducing saturation.

Night Stone

Tonally, brown can occur in light to dark tones. Brown diamonds will often appear both darker of tone and deeper of hue in incandescent lighting. As with all gems, the aficionado is advised to view brown diamonds in both daylight and incandescent lighting.

Multicolor Effect

Brown diamonds often exhibit multicolor effect, usually in the form of alternating flashes of orange and brown. Due to the varying lengths of the light path in both brown and orange diamonds, some light rays will absorb more color than others, resulting in a distinct multicolor effect in the face-up mosaic of the gem. This is further complicated by the addition of yellow, which is the most common secondary hue. It can be hard to decide if the gem is orangy brown or brownish orange or, for that matter, a yellowish orangy brown. A diamond that has a primary orange hue will, of course, bring a much higher price than a brown hue in the marketplace.

Rarity

In the world of diamonds, brown diamonds are not particularly rare. Although rarer than colorless diamonds, they are, along with yellows, the most common of the fancy color diamonds.

0.70 carat Fancy Vivid Orange Yellow diamond. Photo: Jeff Scovil, courtesy R. W. Wise, Goldsmiths.

Chapter 50

Yellow Diamonds

"Equally unfortunate for the cape yellows is that they have often been described in the literature as "off-colour" or "byewater" diamonds, especially by gemologists who fail to appreciate these delicate pastel colour tones in favor of the more saturated "fancy" cape yellows promoted in the marketplace. "

–Stephen Hofer, 1998

ellow diamonds are among the most beautiful of the yellow gems. They are unquestionably the most expensive. In April 1997, a 13.83-carat, Fancy Vivid yellow diamond sold at Sotheby's for $3.3 million or $238,792 per carat. In 2013 a 74.53-carat gem graded a mere Fancy yellow sold for over $40,000 per carat. Pure yellows can be found in a number of other gem species, including sapphire, topaz, and citrine. Yet nowhere else can be found the unique combination of pure hue and exceptional brilliance possible in a yellow diamond. Yellow diamonds are relatively common in nature but due to strong demand they are rare in the marketplace.

Most yellows are type 1a diamonds, but the much rarer type 1b stone produces the purer hued more saturated canary yellow hues. Though both contain nitrogen, the element responsible for the yellow hue, the nitrogen content of type 1b is more evenly distributed throughout the visible spectrum. Type 1b diamonds will often exhibit a weak yellow or orange fluorescence under short wave ultraviolet. Type 1a diamonds will normally fluoresce blue.[343]

Almost all diamonds have a hint of yellow in the body color. When diamonds were

7.05 carat Fancy Vivid Yellow (FVY) diamond. Cora N. Miller Collection, Peabody Museum, Yale. Courtesy R. W. Wise, Goldsmiths.

343 King, John, M., et al., "Characterization and Grading of Natural-Color Yellow Diamonds," *Gems & Gemology, Colored Diamonds in Review,* The Gemological Institute of America, (2006), P.193.

first found in southern Africa in the late nineteenth century, it was noted that most of these stones, when compared to Indian and Brazilian diamonds, were yellowish. Quality grading in colorless diamonds is all about the elimination of yellow. Beginning with the letter H and proceeding to Z, the body color of the stones becomes progressively more yellowish. This is not always reflected in the face-up view. Dealers may be heard to say that a stone "faces up white," meaning that it has a slightly yellowish body color not visible in the key color. This is the result of cutting. The inverse is also possible! As discussed in the chapter on blue-white diamonds, a stone with a light yellow body color may show more yellow face up. This is another good reason to examine the stone carefully and avoid buying the cert. Speaking generally, diamonds graded H through Z are not yellow, but yellowish. Dealers use the term off-color or byewater to describe a yellow hue which lacks the saturation and tone for the yellow to be truly present.

Hue/Saturation/Tone

Yellow is a primary spectral hue lying between green and orange on the color wheel. Green and orange are the normal secondary hues found in yellow diamonds. Of the two, green is the more prevalent, orange the more desirable. Greenish stones may appear murky. A visually pure yellow hue is more desirable still. The writer prefers stones with an orange secondary. The more orange the better! Because orange reaches its gamut limit at a slightly darker tone than does yellow, the addition of orange adds richness and warmth. Pure

yellow diamonds by contrast often exhibit a ghostly greenish secondary hue.[344] Yellow diamonds are unusual in that they may show either a brown or gray mask. Brown is the more common and the more desirable. Grayish yellow stones appear cool, dull, and listless, compared to the warmth of brownish stones.

Color science tells us that the yellow hue achieves its maximum saturation in transparent media at about twenty percent tone. The vividness of the yellow hue tends to drop off dramatically over forty percent tone. This is the lightest optimum tone of any gem hue. Translated into terms useful to the aficionado, this means that in yellow-hued gemstones, a lighter (twenty to thirty percent) tone is a brighter stone, and a brighter stone is a better stone. Stones of this description fall into the coveted GIA-GTL categories of Fancy Intense and occasionally Fancy Vivid, the highest accolade.

Pure dark-toned yellows do not exist. As tonal values increase, yellow becomes either brownish or greenish, usually the former.

Night Stone

Yellow diamond will often appear more saturated under the light of the light bulb. This makes yellow diamond something of a nightstone. It is recommended that the aficionado view yellow diamonds in both daylight and incandescent light before considering an acquisition.

---

344 Fancy yellow diamonds exhibiting a very pure yellow hue, typically type 1b, will often have a green luminescence, which though invisible to the naked eye will produce a visible ghostly, slightly greenish hue.

Given the low optimum combination of saturation and tone, stones in the last few color grades on the colorless scale, i.e. X through Z, are often quite attractive face up. These are called cape colors, a reference to the fact that diamonds found in southern Africa (Cape of Good Hope) often have a yellowish tinge. Pure yellows in these grades display a delicate hue which goes beyond the designation off-color; that is, they are a lovely light yellow.

GIA-GTL grades yellow diamonds as Fancy if the key color is darker in tone and deeper in saturation than the GTL-GTL Z-grade master stone (when examined face up). As previously mentioned, stones graded Faint, and Very Light fall within the colorless diamond scale. Stones in these grades are very low in price because they are considered off-color. But, as previously noted, diamonds can face up either darker or lighter than their body color would indicate. The aficionado interested in a bargain is advised to examine with care the key color of yellow diamonds in these last three grades.

The 2.83 carat Argyle Violet diamond. Photo: courtesy Rio Tinto.

Chapter 51

Violet Diamonds

"Because HGBV diamonds typically have very low color saturation and dark tone, no single hue or color description succinctly captures the color grades exhibited by this group."

–Carolyn H. van der Bogert, et al, 2009

The Violet diamond is one of the extreme rarities of the gem world. Prior to 2003 only a very few violet diamonds were known. Since 1990 the Argyle Mine in Northwestern Australia has produced most of those found. Argyle offers its fancy colors at an annual tender. One of the earliest, a 2.34-carat *dark violet gray* (GIA) or *dark grayish violet* (Hofer) or *fancy grayish blue* (HRD) emerald cut was offered in the 1998 tender. The 1998 tender also included a 0.59 carat violet stone. The 2000 tender included a 0.63-carat. All of these stones had a strong element of gray. Between 1993 and 2008 the Argyle Mine has offered twenty violet diamonds at its annual tenders[345].

Colored diamond expert Stephen Hofer measured his first visually pure violet diamond in early 2000. Up to that point he had measured a few fancy gray diamonds with violet secondary hues but none with

0.73 Fancy Deep Violet (FDpV) Metzger Violet Diamond. Note the distinct grayish mask. Sold in 2002 at Sotheby's Geneva, then Hong Kong, it was, at that time, the largest violet diamond known. Photo: Jeff Scovil. Courtesy: The Edward Arthur Metzger Gem Collection, Bassett, W. A. and Skalwold, E.A.

a primary violet. He discovered his first true violet diamond, a 0.55-carat marquise, amongst a mixed parcel of fancy color diamonds sent to him by an Indian dealer. Hofer used a colorimeter to scientifically measure the color. I was privileged to examine this gem in 2002. GIA-GTL subsequently graded this gem Fancy Deep

---

345 Van der Bogert, Carolyn, et al, "Gray-To-blue-To-Violet Hydrogen Rich Diamonds From The Argyle Mine, Australia", *Gems & Gemology*, The Gemological Institute of America, Spring 2009, p.20

Violet (FDpV).[346]

Origin

Violet is the result of hydrogen atoms substituting for carbon in the diamond crystal lattice. The higher the concentration of hydrogen, the more saturated the violet hue. These stones are a rare type of diamond known as type IaB. Thus far all of the known violet diamonds were sourced at the Argyle Mine. This mine, located in Western Australia, was once (1990) the world's most productive diamond mine and is the source of a majority of the fancy color pink and brown diamonds available in the market.

Hue/Saturation/Tone

Ninety percent of all violet diamonds are dark of tone and low in saturation. Pure violet diamonds comprise about two percent of all known violet diamonds. Blue has been documented as a secondary hue in violet diamonds. These hydrogen rich gems are of low saturation and dark tone. As of 2009, none have been graded faint to light by GIA-GTL. Some few violet gems achieved GIA-GTL grades of Fancy Dark (FDkV)) or at best Fancy Deep (FDpV). None have as yet been graded either Fancy Intense or Fancy Vivid[347]. A grayish violet diamond may, on rare occasions, exhibit a blue secondary hue and on even rarer occasions a blue-violet has been seen. Gray is the saturation modifier or mask found in violet diamonds. Ninety-eight percent of known violet diamonds will exhibit a distinct grayish saturation modifier or mask occasionally modified by a bit o' the blue.[348]

Day Stone/Night Stone

Fluorescent lighting tends to bring out the blue secondary hue and incandescent will bring out the violet. So, technically speaking, violet diamond is a bit of a night stone. GIA-GTL grades colored diamonds using a 6500-kelvin fluorescent bulb.

Size and Rarity

As of this writing (2016), there are, perhaps, two hundred true violet diamonds known. Hofer, a critic of GIA-GTL, suggests that an objective measurement of these gems would reduce that number by approximately fifty percent. All violet diamonds are relatively small, usually measuring less than 0.60 carats. A few three to five carat gems are known.

0.55 carat Fancy Deep Violet Diamond measured by Stephen Hofer in 2000. Photo courtesy Stephen Hofer.

346 Stephen Hofer, personal communications, 2002, 2016.

347 Van der Bogert, *ibid.* p.21.

348 Hofer, *ibid.*

Chapter 52

Lapis Lazuli

"The workmen enumerate three descriptions of ladjword (lapis lazuli). These are Neeli, or indigo colour; the Asmani, or light blue; and the Suvsi, or green. Their relative value is the order in which I have mentioned them."

— John Wood, 1841

Sumerian lapis lazuli bull amulet from the early dynastic period (2650-2350 BC). Photo, © 2001 Christie's Images.

Lapis lazuli: the name evokes a sense of mystery and has a Biblical ring. This is simply because the gem has an ancient lineage. Lapis lazuli, like carnelian, was one of the precious gems of antiquity and was highly esteemed through the Middle Ages, though it has fallen into relative disfavor in modern times. Lapis was almost certainly the sapphire mentioned in the Bible. The Roman historian Pliny described what he called sapphire thusly: "Sapphiros contains spots like gold. It is also sometimes rarely blue tinged with purple. It is never transparent." Obviously this is not an accurate description of sapphire; it is, however, a dead-on description of lapis lazuli.

Lapis lazuli — the name means blue stone — was highly regarded by the great civilizations of ancient Egypt, Babylon and Sumer.

The main source of both ancient and modern lapis is in the far northeastern reaches of Afghanistan, north of the Hindu Kush, in the remote province of Badakhshan. Archeological evidence suggests that these mines were worked as early as 8000 BC. The English traveler, John Wood described the road to the mines in 1841: "The summit of the mountains is rugged, and their sides destitute of soil or vegetation. The path by which the mines are approached is steep and dangerous."[349]

Lapis was traded throughout the Mediterranean during the late Bronze Age (1100-1600 BC) In those times Badakhshan was one of the most reliable sources of tin, a strategic Bronze Age metal. Imagine, long caravans, hundreds of Bactrian camels transporting flattened tin ingots, carnelian and lapis lazuli from "Meluhha, the black land", to the ancient cities of Ur and Babylon.[350] Lapis was considered sacred to Enlil, the father god of the Sumerian pantheon. His house (Ekur) was built of lapis lazuli.[351]. The earliest known worked lapis can be traced to the site of the ancient city of Mehrgarh in the Indus Valley, but the finest collection of ancient lapis was unearthed at the site of the royal tombs of Ur. Jewelry in the form of elaborate beaded necklaces and carvings of the finest lapis, the property of a Sumerian

queen, has been dated back approximately forty-six hundred years.

Russia and Chile are the two other major sources of lapis. Russian lapis has greater amounts of pyrite, and Chilean lapis is rich in calcite. The very finest Chilean and Russian material is said to rival the Afghani. However, I have never seen material from either source that even comes close.

Hue/Saturation/Tone

This polished slab shows the rich midnight blue hue characterizes the finest color in lapis lazuli. Photo: Tino Hammid.

349 Wood, John., *A Personal Narrative of a Journey to The Source of the River Oxus...in the years 1836, 1837, 1838*. John Murray, (London, 1841), 1st Edition, pps. 263-4.

350 One of the ancient names (2150 BC) of the Afghan province of Badakhshan (Bactria), vid. Cleuziou, S., & Burthoud, T., "Early Tin in The Near East, A Reassessment Based on New Evidence From Western Afghanistan," *Expedition Magazine*, (Fall 1982), pp. 14-15.

351 Samuel Noah Kramer, *The Sumerians, Their History, Culture and Character*. Loc. 1547 (Kindle eBook). "The Ekur, the lapis lazuli house, the lofty dwelling place, awe inspiring, its awe and dread next to heaven." Sumerian hymn.

Lapis is an opaque mineral, a rock with a varying composition of lazurite, pyrite, calcite, and diopside. Except that the stone is opaque rather than translucent, the color of lapis lazuli is directly analogous to blue sapphire in hue, saturation, and

tone. Crystal is, obviously, not part of the equation in the evaluation of lapis. From a connoisseur's perspective it might as well be called opaque sapphire. Lapis was considered by the ancients in the Near East to be the stone of heaven and in its finest qualities resembles the night sky. The finest lapis, like sapphire, is a visually pure (ninety percent) royal blue hue, sometimes modified, as Pliny noted, with a secondary hue of about ten percent purple.

Tone is a critical part of the quality equation, as lapis is opaque and, unlike sapphire, lacks transparency (crystal) to mitigate its deep tone. Again like sapphire, the finest lapis teeters on the edge of overcolor. The optimum tone in lapis is between seventy-five and eighty-five percent. At ninety-percent tone it loses saturation and becomes decidedly blackish and dull. Often a blackish appearing stone (ninety percent tone) will be a bit too purple; pure blue stones tend to be lighter (seventy-five percent) in tone. Close your eyes and picture the night sky at midnight under the light of a full moon. That is the finest color lapis lazuli.

Inclusions

Calcite and iron pyrites are the normal inclusions found in lapis lazuli. Calcite occurs as either whitish veins or as tiny dust-like particles. In either case it is considered a flaw. The Afghan government in 1973-1979 described the finest grade lapis as "royal blue, no calcite or pyrite inclusions."[352] Lapis from Chile is known

for its distinctly dull and dusty appearance caused by concentrations of small calcite particles.

Pyrite, though considered a flaw by some, may sometimes materially contribute to the beauty of the gem. The slightly yellowish pyrite, commonly known as fool's gold, when sprinkled throughout the stone gives the appearance of golden stars scattered across the evening sky. An artful dusting of golden flecks of pyrite can be very desirable. Lapis with pyrite inclusions is highly sought after, particularly by collectors in the United States.

Cut

Lapis lazuli is almost always cut *en cabochon*, although lapis beads are sometimes faceted. This sort of faceting has nothing to do with creating brilliance; it is simply a decorative effect. Since the gem is opaque it is never cut into traditional faceted shapes. The shape of the cabochon has little impact on value, except that a domed stone is usually preferred to one that is flat.

Treatments

The aficionado should beware of color-treated lapis lazuli. Lapis simulants date back to the ancient Egyptians, who made imitations using faience or frit, a type of fused glass (see Chapter 1). Today the stone is often dyed. Another simulant is reconstituted lapis. In this case lapis dust is mixed with an adhesive, allowed to harden, then polished to mimic the solid stone. Given the relatively low cost of even the finest lapis lazuli, these sorts of imitations

352 Bowersox, Gary W., & Chamberlin, Bonita E., *Gemstones of Afghanistan*, Geoscience Press, 1995, p.60.

are rarely seen in the United States, and any
of them can be readily separated from the
genuine article by a seasoned gemologist.

The Rarity Factor

Fine lapis is available in large sizes.
Stones over twenty carats tend to decrease
in price on a per carat basis.

SPINEL

"Spinel has endured a mostly downhill slide on the rocky road to popularity. Spinel might be called "the dealer's stone" because so many gem traders have fallen in love and tried to promote it to the public with little success."

–Secrets of the Gem Trade, 1st Edition, 2003

Town gem market, Luc Yên District, Yên Bái Province, Vietnam. Photo: R. W. Wise.

"Fifty shades of green," gem dealer Joe Belmont says wiping the sweat from his eyes. The rice fields stretch out long verdant fingers toward a horizon flanked by jagged, jungle-covered hills with foundations of pure white marble. Next month wood-wheeled ox carts will creak along paths raised between patties exhibiting a vivid hopscotch pattern of gold patches across a green quilt signaling the time to begin the harvest.

The sun is a blistering hot orb as May ushers in the summer monsoon in northeastern Vietnam. The roads are by turns, good, passable and not at all. Besides rice, crops like sweet corn adorn hillsides cloaked with tall thick clumps of bamboo undulating like the long-skirted figure of a young girl in the hot breath of an afternoon breeze.

Many of the farmers when not planting rice are engaged in part-time gem mining. Peasants shovel the gluey lateritic soil from shallow holes in their fields, usually bordered by a stream which will wash away the mud to reveal smooth pebbles

FFarmer mining in the rice patties, Luc Yên District, Yên Bái, Vietnam. Photo: R. W. Wise.

of pink, lavender and blue glowing amongst the gravel.

The heady days of big strikes and major finds are long past in Luc Yên. Only the most well-connected and tenacious dealers with clients who will accept a broad range of size and quality can do business. The big glamor stones are a distant memory and what is available is poorer colors in smaller sizes. Diminishing supply has been helped somewhat by prices that have skyrocketed over the past fifteen years.

What a difference a decade makes. The quote at the head of the chapter accurately described the status of spinel for most of the last century. But things change and the changes in spinel's status in the gemworld have been dramatic.

From its first discovery in the 9th century, red spinel was known as balas ruby and was among the most highly valued and sought after gems.

The Twin Stars: A magnificent 4.70 carat cobalt blue spinel from Luc Yên, Vietnam and a 2.11 carat pinkish red spinel from Namya, Burma. Note the hazy scintillation of the blue gem. Photo: Wimon Manorotkul. Courtesy K. V. Gems (private collection).

In 1572, the Spanish goldsmith Juan de Arfe y Villafañe published a series of tables comparing the values per carat of the five most precious gems; diamond, ruby, oriental emerald, Colombian emerald and red spinel. Between one and ten carats, red spinel is consistently valued at approximately half the price of a diamond of the same size.[353] Writing of the crown jewels of France, Bernard Morel notes, "Up to the end of the 16th century (it was not until the 17th century onwards that spinels became less

---

353 Juan de Arfe de Villafañe, *Quilatador de la Plata, Oro y Pedras*, Alonso & Diego Fernandez, Cordoba, 1572. pp. 72-74.

valuable) the 'balas rubies' were the glory of royal treasuries."[354] However, Harry Emmanuel writing in 1865 states that spinel's value is "extremely uncertain and variable...and it is impossible to say what the intrinsic worth of this gem may be."[355]

Spinel was first synthesized in the first decade of the 20th century, and colorless synthetic spinel was promoted as the diamond simulant of choice between 1920-1947. Unfortunately, this association with synthetics and simulants sullied the name spinel, completing the gem's downward spiral. Spinel sunk into deep obscurity until two discoveries just on the cusp of the 21st century combined to change the world.

The New Twin Stars of the Firmament

In 2007 an artisanal miner digging in a farmer's field outside the Tanzanian town of Mahenge hit pay dirt in the form of a massive, fifty-two kilogram pink/red spinel crystal.[356]

Gems cut from this crystal ranged from a hot pink to a pinkish red of such astonishing saturation that they set a new standard in the appreciation of spinel and catapulted the gem from the virtual obscurity into the dizzying heights of international stardom. Spinel, specifically pink Tanzanian spinel, became an overnight sensation.

Across the Indian Ocean, at about the same time, small quantities of a new spinel variety colored blue by cobalt from a source located in the hilly jungles of Yên Bái Provence, North Vietnam, began trickling into the Bangkok market.[357] Unlike the traditional dark grayish blue spinel, these cobalt flavored honeys occur in hues from a medium-light toned sky to a rich-medium dark royal blue.[358] They also exhibit an exceptionally vivid saturation. The best material comes from mines located in Yên Bái outside the town of An Phú and in Luc Yên District, surrounding the tiny village of Bai Son.

Sparked by these two discoveries, spinel's increase in popularity and breathtaking acceleration of price has awakened interest among prospectors, and there have been significant new discoveries in Burma, one of spinel's most ancient sources. Highly

354 Bernard Morel, *The French Crown Jewels*, Fonds Mercator, (Antwerp, 1988), p. 93. The 1774 inventory of the French Crown Jewels includes fifty-eight balas rubies (red spinel), the finest, weighing 58,18 carats valued at 50,000 livres, approximately half the value of a diamond of a similar size.

355 Harry Emmanuel, *Diamond and Precious Stones, History, Value and Distinguishing Characteristics.* John Cameron Hotten, (London, 1865), p.123. Emmanuel gives an example of a 40-carat spinel of good quality that sold in 1856 for £400, in 1862 at public auction for £80 and "lately"(1864) for £240.

356 Michael C. Krzemnicki, *Spinel, A Gemstone on the Rise*, 2010 Swiss Gemmological Institute SSEFSwitzerland:http://www.ssef.ch/fileadmin/Documents/PDF/650_Presentations/HK2010March_Spinel.pdf.

357 Edward Gubelin and Franz Xavier Erni, *Gemstones: Symbols of Beauty and Power*, Geoscience Press, Tucson, 1998. According to Professor Gubelin, "The cobalt spinel can be expected to develop into a cult stone for collectors."

358 Jean Baptiste Senoble. "Beauty and Rarity – A Quest for Vietnamese Blue Spinels", *Incolor Magazine*, The International Colored Gemstone Association, (Summer 2010), p.3.

saturated hot pink to red spinel, rivaling the best of Tanzania, has been found northeast of Burma's famous Mogok Valley in Kachin State at Namya and Man Sin.[359]

---

359 Vincent Pardieu, "Hunting Jedi Spinels in Mogok," *Gems & Gemology*, The Gemological Institute of America, Spring, 2014.

"Most of the mining in each area is done by small groups of local farmers using mini-generators and water hoses. Activity fluctuates according to agricultural cycles. Excavators and high-pressure water hoses are used in some places, but most of the miners use basic tools, washing with rattan buckets and picking the gemstones by hand."

–Pham Van Long, 2013

Red Spinel Versus Ruby

H istorically, red spinel has been compared and confused with ruby since it was first discovered in the 9th century.[360] Spinel crystals are often found in gem gravels mixed with ruby and for millennia the balas ruby, as red spinel was first known, was consistently mistaken for true ruby. Chemically, spinel is $MgAl_2O_4$:Cr and ruby is Al_2O_3:Cr. Both ruby and spinel owe their blazing red hue to tiny amounts of chromium. The difference is spinel is singly refractive, crystallizes in the cubic system and contains magnesium. Aside from sharing the same coloring agent, their specific gravity (density) often overlaps and the two gems have a similar hardness; 8.5 for spinel and 9.0 for ruby. Ruby will, therefore, take on a slightly superior polish, the sole objective clue to their separation in ancient times.

The 17th century traveler Jean Baptiste

An exceptional 12 carat balas ruby from the old mines at Kuh-i-Lal. The best red spinels from this source show little orange secondary hue and were misidentified as rubies for many centuries. Photo: Jeff Scovil. Cora N. Miller Collection, Yale Peabody Museum. Courtesy R. W. Wise, Goldsmiths.

360 This purported inability to discriminate between ruby and spinel is a bit overblown. Al Beruni writing in the 10th century understood the difference, as did al Tifaschi (1253). In Europe, Arfe Y. Villafañe, writing in 1572, makes a clear distinction between ruby and balas.

Tavernier tells an old and venerable tale. A large red stone represented to be ruby was sold by a Hindu merchant to Ja'far Khan, uncle of Aurangzeb, the Great Mogul of India, for the princely sum of ninety-five

thousand rupees.[361] The stone was then presented to the monarch in the form of an annual tribute. The emperor regarded the gift with suspicion and wished to know if the red gem was a true ruby or a balas ruby (spinel). He sent the stone to Shah Jahan, his father, the former emperor, whom he had overthrown and imprisoned in the Red Fort at Agra. The deposed emperor and builder of the Taj Mahal identified the stone as spinel, stating that it was worth no more than five hundred rupees. At which point Aurangzeb demanded a refund.[362] Shah Jahan, alas, remained a captive and subsequently died at the fort with a view of his beloved Taj just down river.

Discovered sometime in the 9th century, the ruby red spinels of medieval times were found in a mine along the Panj River at Kuh-i-Lal in what is now Tajikistan, an area known in ancient times

as Badakshan.[363] Some of these gems made their way to Europe where they were readily embraced as ruby.

With the birth, in the mid-19th century, of modern scientific gemology, many of

Three Mahenge spinels. Left to right: 4.62, 7.30, 5.64 carats. Note the slight orangy secondary hue in the center gem. Photo Courtesy Vladyslav Y. Yavorskyy.

these famous Badakshan "rubies" have been identified as spinel. Examples include the celebrated 140-carat Black Prince's Ruby and the 352.50-carat Timur Ruby.[364] Some of these gems, for example the famous Three Brothers – the 212.44 carat, *Côte de Bretagne*, the 247.62 carat, *Oeuf de Naples* and the 124.32 carat *A Romain* ruby-spinels

---

361 The canny Aurangzeb would often sell jewels from his collection to Agra merchants just before the annual tribute, knowing that court officials would purchase and present these same jewels to him as their yearly gift. In so doing, he received the merchant's money and also got his jewels back again.

362 Tavernier, *Travels*, Vol. II, p.270

363 Al Beruni, *The Book Most Comprehensive on Precious Stones*, (Adam Publishers and Distributors, New Delhi, 2007), pp. 67-68. Al Beruni, writing in the 10th century, is the first to mention this gemstone.

364 Richard W. Hughes, *Ruby & Sapphire*, p. 247. New evidence confirms that the Arabs, at least, were aware of the gemological distinction between ruby and spinel as early as the 13th century. Ahmad ibn Yusuf al Tifaschi, in his *Best Thoughts on the Best of Stones*, comments that spinel is a distinct gem, can be scratched by ruby, and is worth only about half the value of ruby. See Huda, *Arab Roots of Gemology*, p. 112.

– had graced royal collections since the mid-15th century.[365]

It is fair to say that prior to the Mahenge discovery, spinel was compared to ruby using the look-alike standard described in the first half of this book. In the decade since a double standard has emerged.

As noted, spinel and ruby form in tandem an identical geological environment; spinel crystallizes first and ceases only when the magnesium present in the immediate environment is exhausted. With a refractive index of 1.71, spinel mirrors ruby (1.76) in its refractive qualities. Spinel, however, is singly refractive and rarely exhibits the dramatic multicolor effect or the color bleeding normally seen in ruby (see Chapter 4).

The Egyptian scholar, al Tifaschi, writing in the 12th century, recognized four categories of red spinel: mu'aqrabi or scorpion (deep red), atash (of lesser red), inari (the color of the pomegranate), and niaziki (of lesser red, possibly pink).[366]

365 Morel, *ibid*, pp.86-95. Large numbers of spinels were imported into Europe beginning in the 10th century. Because they were large, they were the red gems of choice amongst royalty. The orangy-red Côte de Bretagne spinel, the oldest gem amongst the French Crown Jewels,was originally owned by Marguerite de Foix (1449-1486). It was eventually carved into the shape of a dragon (107.88 carats) on the orders of Louis XV and set along with the French Blue diamond into the jewel of The Order of The Golden Fleece. It is presently on display in the Louvre.

366. Huda, *ibid*. St. Edward's Crown worn by Henry III of England at his second coronation in 1220 is described as "set with rich stones of great balases, rubies and emeralds." See Blair, C., et al., *The Crown Jewels, The History of The Coronation Regalia In the Jewel House of The Tower of London,* 2 Volumes, The Stationary Office, London. Volume I, p.291.

Hue/Saturation: Red Spinel

In the case of the reds from the old source at Kuh-i-Lal, comparison with ruby is unavoidable. Historically, red spinels were indistiguisable from rubies. The spinels from Kuh-i-Lal, do bear an uncanny resemblance to ruby. Red reaches its ultimate saturation, or its gamut limit, at about eighty percent tone. As with ruby, a fine red spinel tends to be medium dark (80%) in tone. Reds over eighty percent tone rapidly lose saturation and are generally considered overcolor.

Orange is the usual and ubiquitous secondary hue found in red spinel from Kuh-i-Lal and most other sources, though purple is also sometimes present. In stones from the traditional source, gems lacking a strong (20% or more) secondary orange are difficult to obtain. Stones with fifteen percent or less orange or purple secondary hue (or in combination) should be judged very fine.[367]

Pink Spinel

The hue we call pink is simply a lighter-toned, paler red; therefore the demarcation between a pink and a red spinel is in some sense arbitrary. The boundary tends to move up and down, depending on whether it is the buyer or the seller who is making the call.

Without question, the newer sources of spinel from Tanzania, Vietnam and Burma have complicated the connoisseurship equation. These supercharged pinkish reds are beautiful but not at all ruby-like

367 Joseph Belmont, personal communication, 2014.

in the traditional sense. Gems from these sources are lighter in tone, more transparent and will normally exhibit an orangy/pink secondary hue. This mixture is beautiful, quite exotic and, as stated, the saturation of hue is of another order entirely. These colors are hot, hot hot! According to one spinel expert, a red stone with no more than twenty percent pink as a secondary hue is the new top color.[368]

"Jedi Pink" spinel. So hot they seem to vibrate. Note the two perfectly formed, highly transparent, crystals to the left and right. Photo courtesy Nomads.

Saturation, i.e. vividness of hue, becomes a relatively more important factor in the evaluation of red to pink spinel from these sources. It is fair to say that a highly saturated red, pinkish-red and pink are the most desired colors in that order. A true red is very rare. Caveat emptor; it is sometimes very difficult, particularly for the untrained eye, to separate medium to medium-dark toned purplish pink from true red gems.

Saturation and Tone

Gray is the normal saturation modifier or mask found in spinel although brown is more prevalent in darker-toned red gems from Kuh-i-Lal. In fact, most of the other hues of spinel from older sources are undesirable simply because they are distinctly grayish and quite dull.

With pink stones, the darker the tone, the more saturated the hue will become. Thus, a darker-toned pink hue is a more desirable pink. That said, the lighter-toned,

highly transparent pinks possess a delicacy of hue that is compelling. In light toned gems, the gray mask is often difficult to see. A hot pink is a purer pink, a cool pink is a dull pink and the result of a hidden grayish mask.

According to gemologist Vincent Pardieu: "The best of these pinks are more highly saturated than those from Mahenge. " Pardieu further notes that the Namya material is redder and that the Tanzanian stones "look greyish" when compared to the Namya hot pinks, which Pardieu considers so otherworldly he has labeled them "Jedi pink."[369] Saturation in gems from this source are so vivid that even rough crystals appear to vibrate and dance before the eye and are difficult to capture in a photograph. On a recent trip to Bangkok and Vietnam (2015), this writer has seen astonishingly

368 *Ibid.*

369 Vincent Pardieu, personal communication, 2014. Pardieu further states that the Namya gems are relatively low in iron when compared to the pinks of Mahenge. Iron quenches the natural fluorescence in spinel and ruby.

beautiful examples of fine stones from each of these areas which would be difficult to tell apart.

Crystal

At its finest, spinel is dramatically more diaphanous than corundum. To borrow a phrase from the great 17th century dealer Jean Baptiste Tavernier, at its finest, its *water* is very pure. A true red spinel, due to its exceptional crystal, will compare in beauty to even the finest ruby. Both blue and pink spinel will sometimes exhibit a blue haze which adds a soft misty quality to the gem's scintillation. The effect, while positive in blue gems, is a bit distracting in pinks and will negatively affect value.

Ruby/spinel dealer, Luc Yen, Vietnam. Photo: R. W. Wise

Chapter 54
Cobalt Blue Spinel

"The best spinels have a wonderful transparency, an attribute that is difficult to find with doubly refractive stones, like ruby, which are typically included and silky. It is this combination of good cut and polish, and high transparency and attractive color that make large, clean spinels beautiful."

–Vincent Pardieu, 2009

𝓑lue spinel had been known for centuries, but most blue spinel is colored by iron and is either distinctly greenish or decidedly grayish or both and has little value as a gemstone. The recent discovery of cobalt blue spinel, however, has dramatically altered the connoisseurship equation. The finest cobalt blues exhibit a particularly vivid, pure blue hue. cobalt blues are decidedly low in Iron and the cause of color has been shown to be measurable amounts of cobalt (Co^2) substituting for the normal Manganese ($Mg^{2)}$) in the crystal structure of this variety of spinel.[370]

A cobalt blue spinel. This gem would be described as medium-toned (70%) blue. A high degree of transparency (crystal) visually separates spinel from sapphires of similar tone. Photo courtesy: Vladyslav Y. Yavorskyy.

Hue/Saturation/Tone:

Cobalt blue spinel is comparable to sapphire in hue and tone. Tonally, cobalt blues occur in the entire range of blue from a faint near colorless to a very light 30-45% toned to a 50-75% medium sky blue to the darker 75-80% toned royal blue hues often with a secondary hue of violet to purple. As

with sapphire, the negative secondary hue is green. A hint of green will dramatically reduce the asking price.

Though comparable in hue, the best of the cobalt blues are capable of a saturation that goes well beyond most sapphire. The presence of iron dampens that saturation in many gem varieties and less iron combined with the presence of trace amounts of cobalt is responsible for an exceptionally vivid saturation in the finest examples of this gemstone. Gray is the normal

---

370 Boris Chauviré, et al., "Blue Spinel from The Luc Yen District of Vietnam," *Gems & Gemology*, The Gemological Institute of America, (Spring, 2015), p.15.

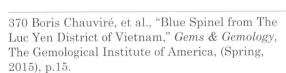

Tiny cobalt blue rough from Yen Bai. The top color is known among dealers as "Bic-pen" blue. Photo courtesy Nomads.

saturation modifier or mask found in blue spinel. Grayish blues will exhibit a distinctly cool appearance.

Color Shift/Daystone

Some cobalt blues will exhibit a phenomenon known as color shift. That is, they will shift from a rich blue in daylight to a distinct violet or purplish to purple in incandescent lighting. Gems that exhibit this phenomenon contain measurably less cobalt than those of a more vivid hue.[371] This phenomenon is quite striking, but cobalt blue spinel is a daystone. The nighttime violet is normally grayish and less attractive than the day color. The nightime color also lacks the purity and intensity of hue. Gems with a color shift will carry with them a discount over gems that hold their blue in all lighting environments and those that have a purer hued bedtime color. At

times, the aficionado may observe a red flash which seems to emanate from the heart of the stone. This phenomenon can be quite striking and should not be confused with color shift; it is a result of *dispersion* and is similar to the rainbow effect in diamond.

Crystal

Compared to sapphire, spinel exhibits superior crystal. Exceptional diaphaneity is one of the defining qualities of spinel. Gemologist Richard W. Hughes describes the color of cobalt blue spinel as "unique in the Gemworld." This characterization is, however, a tad wide of the mark. Both spinel and sapphire come in the complete range of blue. It is not the hue, but rather the water. A particularly vivid hue combined with a high degree of transparency (Chapter 4) is responsible for the cobalt gem's unique appearance. Spinel will sometimes exhibit a blue haze that adds a soft misty quality to the gem's scintillation. The effect is not unlike the velvety appearance of some Kashmir sapphire and can add measurably to the gem's beauty. French gem dealer Jean Baptiste Senoble says it well: "The Vietnamese blue is particularly sought after for its extremely soft and pure color combined with fantastic crystal or "water."[372]

371 SSEF Research, "Blue Cobalt Spinel from Vietnam," *Facette, Institut Suisse de Gemmologie*, No. 21, (February, 2014), P.14.

372 *Ibid.*, p.3-4

As with blues and reds, regardless of hue mix, gems which exhibit a rich vivid color and are of eye clean clarity and highly crystalline with no visible gray are highly desirable and often sell for a substantially than red and blue gems.

A 10.37 carat slightly orangy pink "padparadscha" spinel of particularly fine water. Photo: Jeff Scovil, courtesy R. W. Wise, Goldsmiths.

Spinel: A Gem of Many Colors

The discussion has focused on red and blue spinel. However, like sapphire, fine spinel occurs in the full range of spectral and mixed hues, save perhaps yellow. From a delicate orangy pink Padparadscha to green; from a plummy pink to a pastel lavender blue to lavender pink, and from violet to pure vivid purple. With their exceptional diaphaneity, gems in these color ranges are difficult to characterize and categorize but can be true beauties. Lavender is in high demand,[373] particularly lavender blue, but generally these mixed hues will sell at something of a discount from the pure hued reds and blues. Mixed hues are, relatively speaking, a buying opportunity. The aficionado is advised to allow the eye to browse through this wonderful cornucopia of hues, to appreciate and enjoy.

373 Joseph Belmont, personal communication, 2014.

achroic Without color; a colorless gemstone.

achromatic As above

adularescence A phenomenal effect; e.g., the billowing "moonglow" effect in moonstone.

akoya pearl A pearl from the saltwater akoya-gai oyster (Pinctada martensii); "Japanese pearl."

baroque Any pearl that is not symmetrical, round, or teardrop, oval, or button shaped.

bead nucleation The use of a shell bead, usually spherical, implanted in the oyster to stimulate the growth of a cultured pearl, and forming the center of the pearl.

bellied Describing a stone purposely cut with extra weight around the girdle, yielding a bulbous outline.

bicolor A stone with two distinct, separate hues, common in tourmaline.

bleed color The loss of saturation and tone when the viewing environment is shifted between natural and incandescent lighting.

body color The color of light transmitted through a gem, as distinguished from key color, the color of refracted light.

brilliance The total quantity of light refracted and reflected (from a gemstone) back to the eye of the viewer.

brilliant cut Usually refers to a full-cut brilliant of fifty-eight facets, with thirty-two facets and table above the girdle, twenty-four facets and a culet below; used almost universally in cutting larger round diamonds.

byewater Off-color; poor color and transparency; see also water.

"buying the cert" A purchase based not on an analysis of the beauty of the stone, but on the language of the grading report (certificate).

cabochon, en cabochon Gem with a rounded top, without facets. French, "little head."

carat Unit of weight in gemstones, one-fifth of a metric gram.

cat's-eye Phenomenal effect in cabochon cut gemstones resembling the iris of a cat's eye.

clarity One of the "four Cs" of quality grading, referring to the presence or absence of inclusions or flaws.

color See body color, key color.

crown The top half of a faceted gemstone; the portion above the girdle.

crystal One of the "four Cs" of gem connoisseurship, coined by author, referring to the transparency and diaphaneity of the gem. See also water and transparency.

culet The point at the very bottom of the pavilion of a gemstone.

cut The style in which a gem has been fashioned; e.g., emerald cut, brilliant cut. Also refers to a gem's proportions; e.g., well cut.

daystone A term coined by the author to describe a gem species or variety or single stone that looks its best in natural daylight.

diaphaneity The property of being transparent or translucent.

dichroic Of two colors; the characteristic of a transparent substance to divide refracted white light into two distinct rays.

diffraction The modification of white light as it breaks up into the color spectrum.

dispersion The division of white light into its constituent components as in light through a prism; the rainbow effect.

dog A poor quality gemstone in a parcel.

drusy Tiny quartz crystals growing on the surface of a gemstone.

en cameo Cut in relief. Opposite of cutting intaglio.

enhancement Any process applied to a gemstone to improve its color or clarity; also heat enhancement, burning, cooking; see also treatment.

extinction The dark gray to black portion of a face-up gemstone that does not refract light; usually caused by off-axis refraction.

eye The finest gemstone in a parcel.

eye-clean, eye-flawless Describing a gem with no inclusions when viewed with the naked eye (assumes 20/20 vision).

eye-visible Inclusions visible to the eye without magnification.

face up The view of a gem from the top or crown.

fancy color In diamond, any color other than colorless viewed face up; in diamond any color is a fancy color.

fish-eye A dark gray to black (achroic) spot at the center of the gem caused by improper proportions from poor cutting; see also extinction.

flawless Describing a gemstone with no visible inclusions under 10X

magnification.

flour Tiny inclusions that cause a "sleepy" effect in the key color of a gemstone.

fluorescence A glow or color visible under ultraviolet light.

fortification The technical term for stripes of color in chalcedony.

"four Cs" The four factors used to analyze and discuss the beauty of a gemstone: color, clarity, cut, and crystal.   Altered by author.   Traditionally, color, cut, clarity and carat (weight)

gamut limit The point on the tonal (light to dark) scale at which a given hue produces the most vivid saturation.

The outer edge of a faceted stone, the area of greatest diameter, usually the part where the prongs are placed when the gem is set.

"holding the carat" A cutting strategy designed to produce a finished gem above a certain whole carat weight.

hue The descriptive technical term for color; e.g., red, pinkish orange, and chartreuse are hues.

ideal cut A specific set of proportions discovered by Marcel Tolkowsky in 1919 thought to produce the greatest brilliance and maximum dispersion in a round brilliant cut diamond.

imperial Archaic term used to describe certain hues of topaz.

incandescent Light produced by a flame, candle, campfire, or light bulb.

inclusion Anything visible to the naked eye or under magnification in a gemstone, such as a foreign body or crack. A stone with an inclusion is described as included. See also eye clean, loupe clean

intaglio Designs cut in a gemstone that appear in relief when the stone is pressed into a soft substance like clay. Opposite of cutting en cameo.

iridescence The exhibition of prismatic (rainbow) colors on the surface of a gem. See also orient, overtone.

kelvin A unit used to measure light temperature.

key color The color of the light refracted out of a faceted gemstone; the color of the gem's brilliance or sparkle.

loupe clean Describing a flawless gem; no visible inclusions under 10X magnification.

luster Reflections off the surface of a gem or pearl.

mask A modifying color, usually brown or gray, that diminishes the saturation (brightness or vividness) of a gem's hue.

master stone Stone of a known color and quality used for comparing other gems of the same species or variety; also a diamond of known color grade used to determine the grade of other diamonds.

modifier A color that changes the appearance (hue) of another, e.g., a primary hue such as red can be modified by a secondary hue such as orange, yielding an orangy red hue. Saturation modifer, see mask.

Mohs scale A relative scale of gem hardness; talc is 1, diamond is 10.*mosaic* The complex visual scene in the face-up gem. See also multicolor effect.

multicolor effect The display of divergent colors or tonal variations of the same color on a gem viewed in the face-up position.

nacre The mother-of-pearl secretions of the mollusk; pearl essence.

nailhead Dark center in a gem; see fish-eye.

native cut A poorly proportioned gemstone supposedly fashioned by primitive means; considered pejorative.

nightstone A term adapted by the author to describe a gem species or variety that looks its best in incandescent light, showing its most saturated hue. Traditionally, an opal that retains a strong play-of-color in low light environments.

nonchromatic Without color; a colorless stone, without chroma; see also achromatic, achroic.

off-color Insufficient color, or saturation, in a diamond to be considered good color. Used to describe colorless diamond with a strong tint of yellow.

opaque Impenetrable by light; neither transparent nor translucent. Opacity is the quality of being opaque.

orient The iridescent effect visible in finer quality pearls; also called overtone.

overcolor A stone with a hue that is overly dark in tone, usually above eighty-five percent.

overtone See orient.

padparadscha Color of the lotus; a light to medium-toned pinkish orange to orangy pink sapphire.

pair Two gems matching in color, cut, clarity, crystal, and diameter.

parcel Gems sold as a group.

pavilion The portion of a faceted stone beneath the girdle.

Peruzzi cut An early brilliant style cut, the first brilliant.

phenomenal stone A stone that exhibits a phenomenal effect, such as a star, cat's-eye, or adularescence.

pick A selection from a parcel of gemstones.

pink To heat topaz at a low temperature to turn the hue to pink.

play of color The iridescent effect in opal.

precious An archaic term used to describe certain hues in topaz.

primary When referring to hue, the dominant hue in a gemstone; a pinkish red gem has a red primary hue.

reflection Light reflected from a surface.

refraction The bending or deflection of light passing from one transparent media to another; e.g., from air to water, from a gem into the air.

refractive index A measure of the angle of the deflection of light as it passes from one substance to another.

reinforcement Alternating bands of color in chalcedony (agate). Coined by author to describe a memory aid used to recall the color of a specific gem, i.e. using taste to reinforce sight.

rutile A crystal (inclusion) that forms in golden hairlike masses.

saturation The quantity of color in a gem which translates into the color's vividness or dullness.

scintillation The breaking up of light into tiny constituents, a function of the facets of a gem.

seal stone A gem carved in intaglio producing a design in relief, when pressed into clay, usually the owner's signature or emblem.

secondary When referring to hue, the second or modifying color; an orangy red gem has an orange secondary hue.

simpatico An affinity between a certain color pearl and the skin of the potential wearer.

skylight Light diffused around the body of the viewer with the back turned to the sun.

sleepy Having a fuzzy or misty quality in the brilliance of a gem, a lack ofcrispness. Defining quality in Kashmir sapphire

smoky Dirty or sooty looking. Also refers specifically to grayish fancy color diamonds.

sooty Blackish and fuzzy or smoky, refers to the crystal or transparency of the gem.

star A phenomenal effect in some gemstones, appearing as a six rayed figure.

suite Three or more matched gems.

super-d A misnomer used to describe an ultra-transparent diamond. A noted characteristic of Golconda diamonds.

texture In transparent gems, a description of color zoning visible face up. In pearls, a reference to any indentations or imperfections in the pearl's surface.

tint A hint of color, not sufficiently saturated to be a hue.

tissue nucleation The use of mantle tissue from a donor mollusk to stimulate the growth of a pearl. A tissue-nucleated pearl is a non-nucleated cultured pearl.

tone The third constituent of color, defined as lightness to darkness.

translucent Allowing light to pass through, but preventing the clear viewing of images.

transmission luminescence Emission of light effect by a substance caused by the excitation of light rays.

transparent Allowing light to pass through so that objects may be clearly seen.

treatment, heat treatment The heating of a gemstone to improve its color or clarity.

trichroic The breaking up of light into three constituent rays, each containing a portion of the visible spectrum (rainbow).

water An archaic term that refers to the combination of color and transparency in gemstones; used hierarchically: first water (gem of the finest water), second water, third water, byewater.

window The center of a gem cut too shallow. Produces a read through effect that lacks brilliance; also called len s effect.

zone, zoning Alternating sections of color inside a gemstone.

BIBLIOGRAPHY

Books:

Al Beruni. 2007. *The Book Most Comprehensive in Knowledge on Precious Stones*. New Delhi: Adam Publishers and Distributors.

Ashby, C.R., trans. 1967. *The Treatises of Benvenuto Cellini on Goldsmithing and Sculpture* (1568). New York: Dover Publications.

Bauer, Max. 1904. *Precious Stones*. Translated by L.J. Spenser. 2 vols. Reprint. 1968. New York: Dover Publications

Basset, W. A. & Skalwald, E. W. 2012. The Edward Arthur Metzger Gem Collection. Herbert F. Johnson Museum of Art, Cornell University, Ithaca..

Beesley, C. R. 1985. *Color Scan Training Manual, Part 1, The Systematic Analysis of Color Quality in Gemstones*. New York: American Gemological Laboratories, Inc.

Berge, Victor, and Henry W. Lanier. 1930. *Pearl Diver: Adventuring Over and Under Southern Seas*. New York: Garden City Publishing Company.

Berlin, Brent, and Paul Kay. 1969. *Basic Color Terms; Their Universality and Evolution*. Berkeley: University of California Press.

Boodt, Anselmus. De. 1647 *Gemmarum et Lapidum Histori...*Tertia edition longe purgatissima...with Laet, Johannes de. *De Gemmis et Lapidibus libri duo. Quibus praemittitur Theophrasti liber de lapidibus Graece & Latine cum brevebus annotationibus.* Leiden, Johannes Maire,

Cline, Eric H. 2014. *1177 BC. The Year Civilization Collapsed*. Princeton: Princeton University Press.

Conklin, Lawrence H. 1986. *Notes and Commentaries on Letters to George F. Kunz*, The Tiffany Edition. New Canaan: Matrix Publishing Co.

Cox, Christopher R. 1996. *Chasing the Dragon: Into the Heart of the Golden Triangle*. New York: Henry Holt, 1996.

Demakopoulou, Katie, ed. *1996. The Aidonia Treasure; Seals and Jewellery of the Aegean Late Bronze Age.* Athens: National Archaeological Museum.

Dietrich, R. V. 1995. *The Tourmaline Group*. New York: Van Nostrand Reinhold Co. Ltd.

Donkin, R. A. 1972. *Beyond Price: Pearls and Pearl Fishing Origins to the Age of Discovery*. Philadelphia: American Philosophical Society.

Downing, Paul. 1972. *Opal Identification and Value*. Tallahassee, Florida: Majestic Press.

Dubin, Lois Sherr. 1987. *The History of Beads*. New York: Harry N. Abrams.

Emmanuel, Harry. 1865. *Diamond and Precious Stones, History, Value and Distinguishing Characteristics*. London: John Cameron Hotten.

Epstein, David S. 1995. *The Gem Merchant*. Self-published.

Epstein, Edward J. 1982. *The Rise and Fall of Diamonds: The Shattering of a Brilliant Illusion*. New York: Simon & Schuster.

Federman, David and Tino Hammid. 1990. *A Consumer's Guide to Colored Gemstones*. Lincolnshire, Illinois: Vance Publishing Corporation.

Gilbertson, A. 2007. *American Cut: The First 100 Years, the Evolution of the American Cut Diamond*. The Gemological Institute of America.

Greenbaum, Toni. 1996. *Messengers of Modernism: American Studio Jewelry 1940-1960*. New York: Flammarion.

Gubelin, Eduard, and Franz-Xavier Erni. 1999. *Gemstones; Symbols of Beauty and Power*. Tucson, Arizona: Geoscience Press.

The Guide. 2002-2003. Northbrook, Illinois: Gemstone International, Inc.

Halford-Watkins, J. F., 2012. *The Book of Ruby & Sapphire, Being a Description of the Minerals of the Corundum Group Used for Gem Purposes.*

Bangkok: RWH Publishing.

Harris, Harvey. 1994. *Fancy-Color Diamonds.* Liechtenstein: Fancoldi Registered Trust.

Hofer, Stephen C. 1998. *Collecting and Classifying Coloured Diamonds.* New York: Ashland Press.

Huda, Samar Najm Abul. 1998. *The Arab Roots of Gemology: Ahmad ibn Yusuf al Tifaschi's Best Thoughts on the Best of Stones.* London: The Scarecrow Press.

Hughes, Richard W. 1998. *Ruby & Sapphire.* Boulder, Colorado: RWH Publishing.

Hughes, Richard W. 2014. *Ruby & Sapphire, A Collector's Guide.* Gem Institute of Thailand.

Kanfer, Stefan. 1993. *The Last Empire: De Beers, Diamonds, and the World.* New York: Noonday Press.

Kant, Immanuel. *The Critique of Pure Reason.* Kemp-Smith, N., ed. & trans. 1966. London: St. Martins Press.

Kautilya. *The Arthashastra.* Rangarajan, L. N., ed. & trans. 1987. India: Penguin Books.

Keller, Peter. 1992. *Gemstones of East Africa.* Tucson, Arizona: Geoscience Press.

Kessel, Joseph. 1960. *Mogok: The Valley of Rubies.* London: Macgibbon & Kee.

King, John M., ed. 2006. *Gems & Gemology in Review, Colored Diamonds.* The Gemological Institute of America.

Kornitzer, Louis. 1933. *Trade Winds: The Adventures of a Dealer in Pearls.* London: Geoffrey Bles.

Kunz, George F. 1913. *The Curious Lore of Precious Stones.* Reprint. 1971. New York: Dover Publications.

Kunz, George F. 1916. *Shakespeare and Precious Stones.* Philadelphia: J.B. Lippincott Company.

Kunz, George F., and Charles H. Stevenson. 1908. *The Book of the Pearl: The History, Art, Science, and Industry of the Queen of Gems.* Reprint. 1998. New York: Dover Publications.

Landman, N. H. et al. 2001. *Pearls: A Natural History.* American Museum of Natural History, New York: Harry N. Abrams.

Lane, Kris. 2010. *The Colour of Paradise; Colombian Emeralds in the Age of Gunpowder Empires. New Haven:* Yale University Press.

Lenzen, G. 1983. *Diamonds and Diamond Grading.* London: Butterworths.

Moholy-Nagy, L. 1947. *Vision in Motion.* Chicago: Paul Theobold.

Nassau, Kurt. 1984. *Gemstone Enhancement: History, Science and State of the Art.* London: Butterworths.

Ogden, Jack. 1982. *Jewellery of the Ancient World.* New York: Rizzoli.

Raabe, H.E. 1927. *Cannibal Nights: The Reminiscences of a Free-lance Trader.* New York: Payson & Clarke Ltd.

Ringsrud, Ronald. 2009. *Emeralds: A Passionate Guide, the Emeralds, the People, Their Secrets.* Oxnard, California: Green View Press.

Rose, Gustav.1984. *Humboldt's Travels in Siberia, 1837-1842: The Gemstones.* Translated by John Sinkankas. Edited by George Sinkankas. 1994. Phoenix, Arizona: Geoscience Press.

Roskin, Gary. 1994. *Photo Masters for Diamond Grading.* Northbrook: Gemworld International.

Semrád, Peter. 2011. The *Story of European Precious Opal from Dubnik.* Czech Republic: Granit Ltd.

Schmetzer, Karl. 2010. *Russian Alexandrites.* Stuttgart: Schweizerbart Science Publishers.

Shipley, Robert M. 1974. *Dictionary of Gems and Gemology.* Sixth edition, Santa Monica:

Gemological Institute of America.

Sinkankas, John. 1981. *Emerald and Other Beryls*. Radnor, Pennsylvania: Chilton Book Company.

Sinkankas, John. 1997. *Gemstones of North America Volume III*. Tucson: GeoScience Press.

Smith, G. F. Herbert. *Gemstones*. 9th edition. 1940. London: Methuen & Co.

Sperisen, Francis J. 1961. *The Art of the Lapidary*. New York: The Bruce Publishing Co.

Strack, Elisabeth, 2006. *Pearls*. Stuttgart: Ruhle-Diebener-Verlag.

Streeter, Edwin. 1879. *Precious Stones and Gems*. London: Chapman & Hall.

Tagore, Souindro M. *Mani Mala; A Treatise On Gems*. 2 vols. Reprint, 1996. Nairobi: Dr. N. R. Barot publisher.

Tavernier, Jean. Baptiste. 1676. *The Six Voyages to India*. Translated by V. Ball. Edited by V. Ball. 1977. New Delhi: Oriental Books Reprint Corporation.

Themelis, Ted. 2000. *Moguk, Valley of Rubies and Sapphires*. Los Angeles: A&T Publishing.

Tillander, Herbert. 1995. *Diamond Cuts in Historic Jewellery 1381-1910*. London: Art Books International.

Villafañe, Juan Arfe Y. 1572. *Quilatador, de la Plata, Oro Y Piedras*. A & D Fernandez de Cordoba.

Villiers, Alan. 1940. *Sons of Sinbad: Sailing with the Arabs in their Dhows*. London: Hodder & Stoughton.

Wade, Frank B., *Diamonds: A Study of the Factors that Govern their Value*. New York: G. P. Putnam & Sons.

Ward, Fred. 1994. *Pearls*. Bethesda, Maryland: Gem Book Publishers.

Wise, Richard W. 2010. *The French Blue: A Novel of the 17th Century*. Lenox, Massachusetts: Brunswick House Press.

Wood, John, (Lieut) 1841. *A Personal Narrative of a Journey to the Source of the River Oxus...in the years 1836, 1837, 1838*. London: John Murray.

Yavorskyy, Vladyslav Y., with Hughes, Richard W. 2010. *Terra Spinel, Terra Firma*. Self-Published.

Yavorskyy, Vladyslav Y., with Hughes, Richard W. 2014. *Terra Garnet*. Self-published.

Ying-yai Sheng-lan. *Ma Huan, The Overall Survey of the Ocean's Shores* [1433]. 1970. Bangkok: White Lotus Press.

Zucker, Benjamin. 1987. *Gems and Jewels: A Connoisseur's Guide*. New York: Thames and Hudson.

Articles:

Ahmadjian Abduriyim, et al. Spring 2006. "Paraíba-Type Copper Bearing Tourmaline from Brazil, Nigeria and Mozambique, Chemical Fingerprinting by LA-ICP-MS." *Gems & Gemology*, The Gemological Institute of America.

Benesch, F., Wohrmann, B. September-October 1985. "A Short History of the Tourmaline Group." *The Mineralogical Record*, vol. 16: 331-338.

Bhat, Dhananjaya, July 22, 2012. "The Baroda Pearl Necklace-Costliest in History." *The Kashmir Times*.

Van der Bogert, Carolyn, et al. 2009. "Gray-To-Blue-To-Violet Hydrogen Rich Diamonds From The Argyle Mine, Australia." *Gems & Gemology*, The Gemological Institute of America.

Chauviré, Boris; Benjamin Rondeau, Emanuel Fritch, Devidal Jean-Luc Ressigeac. Spring 2015. "Blue Spinel from The Luc Yen District of Vietnam," *Gems*

& *Gemology*, The Gemological Institute of America.

Christie's Hong Kong, Jadeite Jewelry Auction Catalog. October 30, 1995. "A Chinese Approach to Jadeite." 18-23.

Cleuziou, S., and T. Burthoud. Fall 1982. "Early Tin in The Near East, A Reassessment Based on New Evidence from Western Afghanistan." *Expedition Magazine.*

Cowing, Michael D. 2010. "The Overgrading of Blue-Fluorescent Diamonds: the Problem, the Proof and the Solutions." *The Journal of Gemmology*, Volume 30.

Crowningshield, Robert. Spring 1983. "Padparadscha: What's In a Name?" *Gems & Gemology* 31.

Damour, A. Alexis. 1846. "Analayse du jade oriental réunion de cette substance à la Tremolite." *Annales de Chimie et de Physique*, 3rd ser., Vol. 17.

Federman, David. October 1990. "All in the Family, Inside the Diamond Spectrum." *Modern Jeweler.*

Frazier, Si, and Ann Frazier. 2012. "Amethyst, Wine, Women and Stone." *Amethyst, Uncommon Vintage.* Lithographie, Ltd., Boulder, Colo. 4-7.

Gilg, H. Albert. 2012. "In The Beginning: The Origins of Amethyst." *Amethyst, Uncommon Vintage*, Lithographie Ltd, Boulder, Colo. 10-13.

Hammid, Mary Murphy. October 23, 1990. "Golconda Diamonds." *Christie's Magnificent Jewels*, October 23, 1990, catalog: 301-302.

Hemphill et al. Winter 1998. "Modeling the Appearance of the Round Brilliant Cut Diamond: An Analysis of Brilliance." *Gems & Gemology* 158-183.

Hughes, Richard, et al. Spring 2000. "Burmese Jade The Inscrutable Gem., Part II, Jadeite Trading, Grading and

Identification." *Gems & Gemology.*

Kane, Robert E. Summer 1987. "Three Notable Fancy-Color Diamonds: Purplish Red, Purple-Pink, and Reddish-Purple." *Gems & Gemology* 90-95.

Kane, Robert E., et al. Winter 1990. "The Legendary Dresden Green Diamond." *Gems & Gemology* 248-265.

King, John M., et al. Summer 2002. "Characterization and Grading of Natural-Color Pink Diamonds." *Gems & Gemology* 134-147.

King, John M., et al. Winter 1998. "Characterizing Natural-Color Type II-B Blue Diamonds." *Gems & Gemology* 246-268.

King, John M., et al. Winter 1994. "Color Grading of Fancy Color Diamonds in the GIA Gem Trade Laboratory." *Gems & Gemology* 222-242.

King, John M., et al. Winter 2008. "Color Grading "D-to-Z Diamonds at the GIA Laboratory." *Gems & Gemology* 296-321.

King, John M., Shigley, James E., Claudia Iannucci. Winter 2014. "Exceptional Pink to Red Diamonds: A Celebration of the 30th Argyle Diamond Tender," *Gems & Gemology* 268-279.

Kjarsgaard, B.A., and A.A. Levinson. Fall 2002. "Diamonds in Canada." *Gems & Gemology* 208-237.

Levinson, A.A., et al. Winter 1992. "Diamond Sources and Production: Past, Present and Future." *Gems & Gemology* 234-253.

Liu,Yan; Shigley, J.E., and K.N. Hurwit. December 1999. "What Causes Nacre Iridescence?" *Pearl World* 1, 4.

Mok, Dominic. Pamphlet: *Fei Cui Jadeite Jade Smart Grading System*, Asian Gemological Institute and Laboratory Ltd.

Moses, T. F. Winter 1997. "A Contribution

to Understanding the Effect of Blue Fluorescence." *Gems & Gemology* 244-259.

Peretti, Adolf, et al. Spring 1995. "Rubies from Mong Hsu." *Gems & Gemology* 2-25.

Procter, Keith. Spring 1988. "Chrysoberyl and Alexandrite from the Pegmatite Districts of Minas Gerais." *Gems & Gemology* 26-28.

Procter, Keith. Summer 1984. "Gem Pegmatites of Minas Gerais." *Gems & Gemology* 78-81.

Reinitz, Ilene M. et al. Fall 2001. "Modeling the Appearance of the Round Brilliant Cut Diamond: An Analysis of Fire, and More about Brilliance." *Gems & Gemology* 174-1977.

Stockton, Carol, and Vincent Manson. Winter 1985. "A Proposed New Classification for Gem-Quality Garnets." *Gems & Gemology* 205.

Tashey, Thomas E. Spring-Summer 2000. "The Effect of Fluorescence of the Color Grading and Appearance of White and Off-White Diamonds." *The Professional Gemologist* 5.

Ward, Fred. July/August/September, 2002. "The Wisdom of Pearls." *Pearl World, The International Pearling Journal*, vol. 11, no. 2: 11.

Ward, Fred. "World Jade Resources." *Arts of Asia*, Volume 29, No. 1: 68-71.

Wise, Richard W. December 1993. "Burma Ruby Making a Market Comeback." *National Jeweler Magazine.* 34, 36.

Wise, Richard W. 1989. "The Colors of Africa." *Jeweler's Quarterly Magazine* Designer Color Pages, 7-8.

Wise, Richard W. March/April 1998. "Diamond Cutting; New Concepts, New Millennium." *Gem Market News* 3-4.

Wise, Richard W. May/June 2000. "Gariempiero Dreams." *Gemkey Magazine* 42-44.

Wise, Richard W. 1998. "Light Up Your Life." *Asia Precious Magazine, Guide to Industry Services*: iii-iv.

Wise, Richard W. July/September 2001. "A Meditation on Pearls." *Pearl World* 10-11.

Wise, Richard W. November/December 1992. "The New Face of Chinese Freshwater Pearls." *Colored Stone Magazine* 30-31.

Wise, Richard W. July/August 1992. "Oldest Mine in the U.S. Reopens." *Colored Stone Magazine*: Cover-8.

Wise, Richard W. Spring 1993. "Queensland Boulder Opal." *Gems & Gemology* 4-15.

Wise, Richard W. 1989. "In Search of the Burma Stone." *Jeweler's Quarterly Magazine, Designer Color* 8-11.

Wise, Richard W. May/June 1991. "Tourmaline: A Modest Proposal." *Colored Stone Magazine* 6-7.

Wise, Richard W. 6/23/2007. "The Golconda Diamond." *GemWise Blog.* http://www.thefrenchblue.com/rww_blog/2007/06/23/the-golconda-diamond/

Wise, Richard W. 5/27/2009. "Golconda or type IIa Diamonds, Bring Big Type IIa Diamonds Bring Big Price at Auction." *GemWise Blog.* http://www.thefrenchblue.com/rww_blog/2010/12/09/type-iia-diamonds-big-prices-at-auction/

Wise, Richard W. 1/1/2010. "The Wittelsbach, All Tarted Up and Ready to Sell." *GemWise Blog.* http://www.thefrenchblue.com/rww_blog/2009/12/30/the-wittelsbach-all-tarted-up-and-ready-to-sell/

Wise, Richard W. 2/9/2010. "The Wittelsbach, Old Stone, New Myths." *GemWise Blog* http://www.thefrenchblue.com/rww_blog/2010/02/09/

maline, 304

for evaluating tourmaline, 288

for grading diamonds, 155–156, 155n175

impact on crystal, transparency, 53–54

and multicolor effect, 56n73

multiple environments, viewing stones in, 58n76

and photographs of gemstones, 64–65

type of light and color temperature, 47–48

loupe clean diamonds, 45, 157. *See also* flawless diamonds

luster

evaluating, 80

in jadeite, 266

and nacre, 78–79

natural nacreous pearls, 194

South Sea cultured pearls, 216

Tahitian black cultured pearls, 208

MacBeth lights (Kollmorgen), 49n58, 75, 324

Madagascar, gemstones from

blue sapphire, 234

malaya garnet, 178

new discoveries, 88

ruby, 245–247, 249

tsavorite garnet, 173n196, 174

magical/medical properties of gemstones, 7, 9

Maine, tourmaline from, 89, 286–287, 287n303

malachite, 133–134. *See also* gem chrysocolla

malaya garnet

clarity, 179

color shifts, 178

hue, 178–179

refractive index, 178, 181

saturation and brilliance, 178–179

size and rarity, 177–179

sources, 177–178

mandarin garnets, 183

market rarity, defined, 17

Martinez, Sergio (Martinez mine), 107–108

masks

alexandrite, 97

amethyst, 102

aquamarine, 110

blue agate, 142

blue diamonds, 331

blue sapphire, 236

chrome green tourmaline, 300

chrysoprase, 136

cobalt blue spinel, 360

cuprian Paraíba tourmaline, 291

emerald, 115

gem chrysocolla, 133

grading, 296

green diamonds, 325–326

grey vs. brown, 27

orange diamonds, 334

padparadscha sapphire, 240–241

peridot, 220

pink diamonds, 320

pink spinel, 356

purple diamonds, 324

red-pink tourmaline, 304

ruby, 249

spessartite garnet, 183

tanzanite, 311

violet diamonds, 344

matched pairs and suites, 199

McLaurin-Moreno, Douglas, 187n202

melon jade, 267–68

Menge, Johann, 273

Mexico, gem chrysocolla from, 133

microscopic standard, 45

Mikimoto, K., 204

Miller, Reggie, 95n110

Minas Gerais, Brazil, 105–108, 111, 285–287, 301, 303

modified spectral hues, 25

Mogok, Burma, sapphire/ruby from, 40, 107, 227–229, 243–244, 248–249

Mohave blue agate, 130, 141

Moholy-Nagy, Laszlo, 41

Montana, gemstones from

blue sapphire, 234–235

padparadscha sapphire, 242

moonstone. *See also* adularia moonstone

cuts, 122–123

sources and characteristics, 121

Morel, Bernard, 350

Morse, Henry Dutton, 150, 150nn164–165

Moses, Tom, 161

Mozambique, gemstones from

aquamarine, 109

cuprian Paraíba tourmaline, 289–292

new discoveries, 87–88

ruby, 246n249, 247n251, 249, 252

multicolor effect

amethyst, 101–102

brown diamonds, 337

causes, 250

chrysoprase, 300

dichroic effect vs., 56n73

faceted topaz, 222

face-up evaluations, 55–56

in GIA-GTL grading reports, 73–74

green tourmaline, 297

orange diamonds, 334

padparadscha sapphire, 241

pink diamonds, 320–321

pleochroism, 56n72

red diamonds, 316–317

red and pink spinel, 355

and refraction, 56–57

ruby, 250

tanzanite, 309, 311

tourmaline, 307–308

Munsteiner, Bernd, 40

nacre (mother of pearls)

and bead nucleated cultured pearls, 211

chemical composition/structure, 78–79

source, 202

in South Pacific pearls, 208

thickness of, evaluating, 80–81

Naftule, Roland, 177

Namibia, gemstones from

blue tourmaline, 301

carnelian, 140

demantoid garnet, 167–168

spessartite garnet, 178, 182–184

natural light, for evaluating gemstones, 38. *See also* lighting

natural nacreous pearls

asymmetry faults, 191n207

body color, 192